D1287172

Contributors

WALTER ANGST
GISELA EPPLE
CHARLES R. GEIST
STEVEN GREEN
DAVID AGEE HORR
WILLIAM K. REDICAN
PETER STEERE
DAVID A. STROBEL
ROBERT R. ZIMMERMANN

Primate Behavior

Developments in Field and Laboratory Research

Volume 4

Edited by

Leonard A. Rosenblum

Primate Behavior Laboratory
Department of Psychiatry
Downstate Medical Center
Brooklyn, New York

1975

ACADEMIC PRESS New York San Francisco London
A Subsidiary of Harcourt Brace Jovanovich, Publishers

ACADEMIC PRESS, INC.
111 Fifth Avenue, New York, New York 10003

United Kingdom Edition published by
ACADEMIC PRESS, INC. (LONDON) LTD.
24/28 Oval Road, London NW1

LIBRARY OF CONGRESS CATALOG CARD NUMBER: 79-127677

ISBN 0-12-534004-4

PRINTED IN THE UNITED STATES OF AMERICA

Contents

The Behavior of Marmoset Monkeys (Callithricidae)

Gisela Epple

Behavior and Malnutrition in the Rhesus Monkey

Robert R. Zimmermann, David A. Strobel, Peter Steere, and Charles R. Geist

The Borneo Orang-Utan: Population Structure and Dynamics in Relationship to Ecology and Reproductive Strategy

David Agee Horr

Basic Data and Concepts on the Social Organization of *Macaca fascicularis*

Walter Angst

List of Contributors

Numbers in parentheses indicate the pages on which the authors' contributions begin.

WALTER ANGST, Zoological Institute of Basel University, Basel, Switzerland (325)

GISELA EPPLE, Monell Chemical Senses Center, University of Pennsylvania, Philadelphia, Pennsylvania (195)

CHARLES R. GEIST,* University of Montana, Missoula, Montana (241)

STEVEN GREEN, The Rockefeller University, New York, New York (1)

DAVID AGEE HORR, Department of Anthropology, Brandeis University, Waltham, Massachusetts (307)

WILLIAM K. REDICAN,† Department of Behavioral Biology, and Department of Psychology, California Primate Research Center, University of California, Davis, California (103)

PETER STEERE,‡ University of Montana, Missoula, Montana (241)

DAVID A. STROBEL, University of Montana, Missoula, Montana (241)

ROBERT R. ZIMMERMANN,§ University of Montana, Missoula, Montana (241)

* Present address: University of Alaska, Fairbanks, Alaska.

† Present address: Department of Psychobiology and Physiology, Stanford Research Institute, Menlo Park, California.

‡ Present address: University of Georgia, Athens, Georgia.

§ Present address: Central Michigan University, Mt. Pleasant, Michigan.

Contents of Previous Volumes

Volume 1

Volume 2

Volume 3

Monkeys as Perceivers

Roger T. Davis, Douglas L. Medin, and Mary L. Borkhuis

Variation of Vocal Pattern with Social Situation in the Japanese Monkey (*Macaca fuscata*): A Field Study*

STEVEN GREEN

The Rockefeller University
New York, New York

* This research was supported by a behavioral sciences training grant (GM 1789) from the National Institute of General Medical Sciences, National Institutes of Health, U.S. Public Health Service, to the Rockefeller University, by the Rockefeller University Graduate Fellowship program, and by the New York Zoological Society. Partial support during manuscript preparation was provided by a National Science Foundation grant to the New York Zoological Society (GB 16606, Peter Marler, principal investigator).

I. INTRODUCTION

Communication is a social phenomenon. It is the most prominent feature of human speech and language. Man's complex societies are mediated by the ability of human beings to inform each other and are dependent on that ability.

Animal societies are equally dependent on the exchange of information. Any organism that lives in complex social groupings must rely on communicating some aspects of the status of each individual to others. Such an exchange of information, the process that defines a communication system, implies the existence of a common language or set of rules governing the encoding and decoding of signals in the communication system.

It is tempting to think of animal communication systems as being composed of simple invariant designators or external manifestations of some basic internal state such as hunger, pain, or reproductive readiness. For monkeys and apes, however, it is known that in addition to these states many other individual and societal factors, such as individual identities, kinship, roles, dominance relations, and coalitions play an important part in social organization and behavior (e.g., studies in DeVore, 1965). The complexity of many primate societies has kindled interest in the communication systems mediating social behavior (e.g., studies in Altmann, 1967a).

It is particularly in the vocal communication system of other primates that we can expect to find clues that illuminate the evolutionary background and biological heritage of human language. These kinds of clues, hints of the rules by which socially important information is encoded into and decoded from speech sounds, are especially relevant to hypotheses on the origins of human language since there is no fossil evidence available and comparative studies alone must be relied upon.

The uses of vocalizations and their relationship with social behavior may be investigated when both the audible and social parameters of behavior are available. In many primates, certain features of the social situations in which the sounds are given are accessible to the investigator. Carpenter (1934) heralded in his study of howlers the still current method of describing vocalizations and the situations in which they are used. In recent years descriptions of vocalizations have been augmented by applying sound spectrography to recordings (Borror and Reese, 1953). Rowell and Hinde (1962) were the first to characterize the vocal repertoire of a monkey by publishing sound spectrograms; in the same year, Andrew (1962) demonstrated their usefulness for interspecific comparisons. Winter, Ploog, and Latta (1966) added a quantitative dimension to the analysis by measuring acoustic features of the sounds recorded in their colony of squirrel monkeys. Struhsaker (1967) statistically analyzed the vocalizations recorded in his field study of vervets.

This study applies similar techniques of sound analysis and description of social situations to a primate in which the social parameters of vocal behavior are as readily accessible as the audible. In Japanese monkeys (*Macaca fuscata*) access can be gained to many details of social behavior with ease unparalleled in primate field studies. At a number of provisioned troops in Japan, knowledge of individual identities, kinship and matrilineal relationships, determinations of exact social status and dominance, and accounts of life histories have been pioneered by Hazama, Imanishi, Itani, Kawamura, Miyadi, Tokuda, and their associates (see accounts of early work in Frisch, 1959; Miyadi, 1965; Imanishi and Altmann, 1965; Sugiyama, 1965).

In addition to characterizing the vocal repertoire and noting the principal circumstances in which each sound is used, this report examines vocal pattern variants and deviations from typical usages. Rowell and Hinde (1962; Rowell, 1962) stated that in rhesus monkeys some of the variation in intergraded vocalizations reflects graded aspects of their use. Marler (1965) suggested that in general the variability in vocal patterns of primates may have communicatory significance. The differential use of related sound patterns reported here confirms this suggestion for Japanese macaques. Furthermore, this study revealed a regularity of the kind Marler called graded in the relationship between vocal patterns and social situations. Although there are specific uses of each kind of vocalization, the entire vocal repertoire may be arranged in a natural fashion reflecting gradations in the determinants of its use.

One kind of vocalization, the clear tonal *coo's*, selected here for detailed presentation, is discussed in Sections III, A and B. An analysis performed on these sounds relates their acoustic pattern to the demeanor of the vocalizing animal and to the functional and social aspects of the circumstances of utterance. A hypothesis is offered relating situations in which there is similarity in internal state of vocalizing animals to similarities in acoustic pattern of *coo* uttered in these situations.

A typological classification of the remainder of the repertoire is presented in Section III, C. Summary statements are presented for each class from the same kind of analysis as performed on the *coo* sounds. These statements indicate which of the acoustic patterns in each class differ in use from the others. They also describe the predominant social usage of each type.

Each circumstance in which sounds are given is listed in Section IV, A. Disregarding the differences among types of sounds in each class, the extent of use of sounds from each of the classes of the repertoire is indicated for each circumstance. These overall results are examined by arranging them synthetically so that the circumstances are grouped into different situations. Two such groupings are presented, each a reflection of a different hypothesis.

The first hypothesis reflects a traditional arrangement by the functional context alone. The second derives from the study of *coo* sounds and transitional utterances. By this latter hypothesis, the circumstances are grouped into situations according to similarities in the emitters' internal states as gauged by the monkeys' demeanor and motor patterns of behavior. This arrangement for deducing similarities also includes other features of the situations, such as the social relationship of the interactants and broad contextual function. Using the monkeys' observed preferences for sounds in each circumstance, concordance is measured between sounds used in the groupings dictated by each hypothesis; the concordance by the second hypothesis is twice as great.

Both the sounds and the situations are then each arranged by degree of relatedness, vocal or behavioral, according to the scheme discussed in Section IV, B. This arrangement appears on the axes of a contingency table with each cell given a preference score. The results display a trend of covariation; the strongest associations of sounds with situations tend to fall on a diagonal across the table. This trend indicates that sounds that are most closely related occur, as a rule, in situations where the physiological substrate underlying social behavior can be inferred to be most similar.

II. PROCEDURES AND METHODS

A. Field

Three study sites where Japanese monkeys are provisioned and individually known were visited repeatedly in the course of a 14-month field study from July 1968 through August 1969. These periods of study of the troops at Koshima, Miyajima, and Iwatayama (Arashiyama-A and Arashiyama-B troops) were timed to coincide with the seasonal peaks of birth and conception. Other troops were visited sporadically. Magnetic tape recordings were made using a Nagra III recorder operated at 7.5 in./second with a directional condenser microphone, the Sennheiser MKH 804.

Of the 6×10^4 vocalizations recorded in the 2×10^3 observation hours,

10^4 were separately cataloged as given by an identified animal in a situation where the other participants were known, the nature of the context and behavior was described, and sufficient ancillary data were available to characterize the situation. Sampling these utterances in the field was opportunistic, the recorder being activated whenever a situation was judged likely to yield a vocal event.

B. Laboratory

For an acoustic pattern analysis, the sounds were displayed on sound spectrograms produced by the Kay Electric Co.'s "Sonograph," model 6061B, using a 300-Hz filter to scan the 80–8000 Hz range. For fine discriminations of pitch the appropriate spectral region of the pitch scale was expanded with the Kay 6076C module, using a narrow 45-Hz filter; the 160–16,000 Hz scale was employed occasionally for sounds of very high pitch. The spectrograms were measured with the aid of a transparent acetate overlay calibrated to the machine and its filter.

Tonal sounds analyzed in this study varied in the minimum pitch of the fundamental from the base line of the spectrogram (indicating ca. 80 Hz or less) to 7800 Hz. For atonal sounds, the lowest concentrated band of energy, or formant, ranged from base line in some vocalizations to 10 kHz in others. The highest-frequency harmonic observed, at the scale limit of 16 kHz, was part of a rich series of overtones beginning at 1 kHz which may well have extended even higher. The highest pitch formant band was 14 kHz. The duration of units varied from the lower limit of detection, about 0.0025 second to 2.1 second.

Sounds associated with provisioning are excluded from this report; the remaining sample was pooled and the acoustic patterns of 2286 representative utterances were measured.

Sounds were inspected and measured for the characteristics listed in Table I. For sounds that are predominantly tonal, a number of additional measurements reflecting the fine details of structure and morphology were also made (Table II). These finer distinctions are used in describing all sounds with tonal energy but are used principally in separating and characterizing types within the classes containing predominantly tonal sounds.

Sounds were classified typologically on the basis of distinctive patterns in acoustic morphology. The repertoire was then divided into ten classes, each composed of similar sounds. The advantages of this kind of analysis have been pointed out by Winter *et al.* (1966).

The easiest patterns to recognize are those characterizing the *coo,* or simple tonal, sounds. All the energy in this class is part of harmonic

TABLE I

CHARACTERS USED TO CLASSIFY TYPOLOGICALLY THE VOCAL REPERTOIRE

I. From field observations
   Is the utterance part of a bout?
   Is any part produced during inspiration?
   Is there a concomitant articulatory process?
   What is the relative loudness and range of audibility?
II. From sound spectrography: structure of units
   A. General appearance
      1. Uniform: relatively constant from beginning to end
         a. Simple tonal: noise-free series of harmonically related bands
         b. Simple atonal: noisy hash with no tonal bands
         c. Complex, predominantly tonal: simultaneous mixture of tonal and atonal energy with tonality of greater intensity
         d. Complex, predominantly atonal: as c above, with atonality more intense
      2. Nonuniform: at least one marked change
         a. Modified tonal: tonal with transient deviations
         b. Predominantly tonal: simple tonal region compounded in sequence with regions of a different character whose total duration is less than the simple tonal one
         c. Predominantly atonal: simple atonal region compounded with regions of different character whose total duration is less than the simple atonal one
         d. Noisy, predominantly tonal compound: complex, predominantly tonal regions summing to more than half the duration in a compound
         e. Predominantly atonal compound with some tonal structuring: complex, predominantly atonal regions summing to more than half the duration in a compound
         f. Modified atonal: atonal with transient modifications
   B. Pitch factors
      1. Maximum frequency (detectable at −20 dB relative to average peak intensity)
      2. Minimum frequency (detectable at −20 dB relative to average peak intensity)
      3. Dominant frequency: frequency of zone or tonal band of greatest intensity
      4. Formant banding: relatively emphasized zones of the frequency spectrum are evident
         a. Bandwidth of formants
         b. Midband frequency of formants
      5. Measures of the fundamental of a harmonic series of tonal resonances (see Table II)
      6. Richness: relative intensity of harmonic multiples of computed fundamentals
   C. Temporal factors
      1. Total duration
      2. Time to point of change in character of nonuniform sounds
      3. Duration of each region of compound sounds
   D. Modulations
      1. Amplitude modulations
         a. Single or irregular intensity changes

TABLE I (Continued)

1. Plosive: sharp fronts of high intensity
2. Clicks: vertical lines
3. Dropout: momentary region −40 dB or less relative to surround

b. Regularly repeated intensity changes
1. Pulsations: waxing and waning with a relatively constant short period (ca. 0.04 sec)
2. Micropulsation: fine vertical striations (unless these are spectrographic artifacts such as beat notes)

2. Frequency modulations and slope changes of tonal bands
a. Inflection: local maximum or minimum
b. Warble: regularly periodic excursions
c. Pitch shift: tracing is displaced

III. From sound spectrography: temporal delivery
A. Further characterization of units: ordering within compounds, e.g., atonal followed by tonal portion or vice versa?
B. Are there units that occur in series separated by sounds of inspiration?
C. Sequences of units: are there units that characteristically appear in a series of similar kind or sequence of different kinds and should therefore be considered as a unitary behavioral pattern, i.e., one kind of utterance?

TABLE II

PITCH AND TIME MEASURES USED TO CHARACTERIZE TONAL SOUNDS

	Ten direct measurements Position on sonographic tracing at which measurement taken:					
Measurement of:	<u>B</u>eginning	<u>L</u>ow	<u>H</u>igh	<u>I</u>nterruption$_{kind}$[a]	<u>E</u>nd	<u>M</u>idpoint
Frequency	P(B)	P(L)	P(H)	P(I)	P(E)	P(½)
Duration from beginning to	(≡0)	D(L)	D(H)	D(I)	D(E)	$\left(\frac{\equiv D(E)}{2}\right)$

Additional measures derived from direct measurements

Pitch		Relative positions	
Range:	P(H) − P(L)	Low:	D(L)/D(E)
Mean:	$\frac{P(H) + P(L)}{2}$	High:	D(H)/D(E)
Change:	P(E) − P(B)	Dip:	D(I)/D(E)
Ratio:	P(H)/P(L)	Peak:	D(I)/D(E)

Rate of change	Rise and fall times
A "flatness" index; the upper bound of | average slope |: [P(H) − P(L)]/D(E)	Rise: least [D(H),D(I_{peak}), D(E)] Fall: D(E) − greater [D(H),D(I_{peak})]

[a] Including local minima (dip), maxima (peak), and other characteristics evidenced as a local change.

series of bands, the fundamentals of which are visible as dark tracings. The 226 measured utterances of this class showed few distinctive variations of acoustic patterns. For this reason they were considered to be one type with minor perturbations in pattern. A close examination of this variation led to delineating different types of acoustic pattern variants within this class. Analysis of this one class is presented here in detail to exemplify the approach used in relating acoustical variants of macaque sounds to the social context in which they are used.

Each utterance was scored from the field notes as having occurred in one of a number of defined social and functional situations. This scoring revealed that physically similar sounds often have overlapping uses. A statistical analysis was therefore performed on the distribution of occurrence of the types of sounds within each class.

The analysis is based on a contingency table that arrays each vocal event in a unique cell corresponding to its acoustic pattern and social circumstances. The relative degree of association of a particular type of sound with a situation may be gauged by the observed-to-expected ratio. The expected value is derived from the marginal totals of the table.

Whether or not there is any association of sound patterns with the circumstances of vocalizing is tested, without reference to the ordering of the axes, by measuring the overall heterogeneity of usage of sound types in a class. For the classes of sound with highly significant heterogeneity of usage among the measured utterances ($p < 0.001$), pairwise comparisons of pattern variants were performed. Only those patterns used with significant differences ($p < 0.05$) from others of the same class are described in Section III.

The distribution statistic employed for these tests is based on the logarithm of the likelihood ratio; it is the G statistic recommended by Sokal and Rohlf (1969) in preference to the similarly used Pearson chi-square. In addition to other advantageous properties, one major reason for selecting log-likelihood ratio statistics in behavioral research is that they may be used for large contingency matrices where there are cells with zero observations yet collapsing of categories is not warranted (Fienberg, 1970). The value of G is directly compared to the critical values of the chi-square distribution. Degrees of freedom are determined as for the Pearson chi-square tests.

Strength of association is measured by the contingency coefficient C, but using the value of G in the formula given by Siegel (1956) instead of the Pearson-series value for the chi-square variable. Predictability is measured by the index of predictive association of Goodman and Kruskal as given in Hays (1963).

A final analysis examines usage of the entire repertoire of ten classes ignoring fine distinctions within each class; each sound class is treated as a unit. At this stage of the investigation the complete set of notes and recordings was again reviewed, and the classes of sound were ranked by their extent of use in each of the circumstances in which vocal activity occurs. The circumstances are grouped into ten situations with common social elements: first by one hypothesis suggesting which elements of social behavior are important determinants of vocal behavior, and then by another.

The concordance of preferences for sounds used within these situations is measured using Kendall's (1970) coefficient of concordance, W; probabilities for gauging the statistical significance of each W are determined by methods given in Kendall and also using Rohlf and Sokal's (1969) tables. Two overall measures of concordance, labeled $\overline{W}$, are computed by averaging the ten W's pertaining to each of the hypotheses. Each W is weighted by the number of circumstances incorporated in deriving it, and the weighted mean is calculated. The ten independent probabilities associated with concordance measures of each hypothesis are combined by Fisher's (1970) method to give a probability estimate associated with the overall concordance.

A linear regression on ranked data is performed to examine the significance of a trend toward orderly covariation between sound pattern preferences and social situations. Even though the values of the coefficients are not meaningful (and hence are not reported), significance tests of the linear relationship are valid (Fretwell, 1969). A t-test for significance is used (Sokal and Rohlf, 1969).

III. THE REPERTOIRE

A. Discussion of the *Coo* Sounds

From the earliest observations in the field, it was clear that use of certain audibly recognized sounds is strongly tied to particular circumstances. It was possible to predict readily, for example, that growled sounds would accompany a threat whereas warbled sounds would not. These audibly distinctive utterances, which are closely tied to a narrow range of uses are among the patterns, acoustically defined as types, surveyed in Section III, C.

In the field, it was possible to predict only poorly, however, the characteristics of situations in which other patterns of vocalization occur. An examination of the phenomenon of sounds given in unpredictable or unrelated circumstances was conducted in two ways: by refining the concept of what is a distinctive sound and by reexamining the notions of what constitutes a similarity in behavioral circumstances.

Construction of a lexicon codifying these observations, by defining the acoustic patterns and scoring their behavioral distribution, generated an attempt to understand the rules by which particular vocal patterns are tied to situations. This was attempted at two levels. First, the *coo* pattern of sound, which was considered homogeneous in the field, and which is used in a large variety of situations, was subjected to a more refined analysis in the laboratory, part of which is reported in this section. Second, some of the insights garnered by this analysis of the relation of details of acoustic shape to function of the sounds were applied to the repertoire as a whole. The results of this examination of the organization of the repertoire and its uses are given in Sections IV, A and B.

Although the *coo* sounds are both audibly and spectrographically fairly homogeneous, they still show some variation. Some features of this variation might be of biological significance even though they do not yield distinctive patterns to the human ear or the sound spectrograph. The potential significance of such minor variations is examined in this section.

In addition to looking closely at signal variation, the notion of what constitutes a behavioral situation was reexamined. In judging which situations are the same or different, the subtler complexities of social structure and organization were examined more closely. The animals themselves reveal clues, from their point of view, about underlying similarities in superficially different circumstances.

It was difficult in the field to predict the occurrence of the simple tonal *coo* vocalizations. The various situations in which they were used seemed to have little in common. They were given by animals who were calm and in circumstances which had some element of contact or spacing-related function, but no overall pattern of use was clear. These sounds were also not recognized in the field as being composed of a variety of distinctive patterns, but rather were considered a single variable kind of pattern, each utterance a minor variation on the *coo* theme. An initial acoustic analysis seemed to confirm this impression of lack of distinctive heterogeneity in the sound patterns, each utterance being a simple clean tracing of a tonal band with some of its overtones.

A closer look at the *coo* sounds, designated Class II of the ten classes, is reported here. Some of the relations discovered about form and function of these tonal sounds are used in Section IV to examine possible rules that may characterize use of the repertoire.

1. Class II: Coos

In sounds of Class II all of the acoustic energy is in harmonically related bands (Fig. 1). The fundamental tones appear sonographically distinct and are usually the most intense bands. The harmonic structure is not rich, most of the energy being concentrated in only a few low overtones. Occasionally there is also a single intense higher overtone or region of higher overtones. The attack is never plosive. This feature, in addition to the lack of a rich overtone structure or noise overlay, separates these from some of the predominantly tonal types of Class IV or Class V sounds.

The tracings are continuous, with no abrupt shifts in pitch, represented as a displacement of the tracing along the vertical axis. There are no temporal discontinuities such as "clicks" or noise bursts. The number of inflection points, those portions of the fundamental tracing that can be observed to go through a local minimum or maximum, is at most 6. The continuity of the tracing and the limited number of inflections sepa-

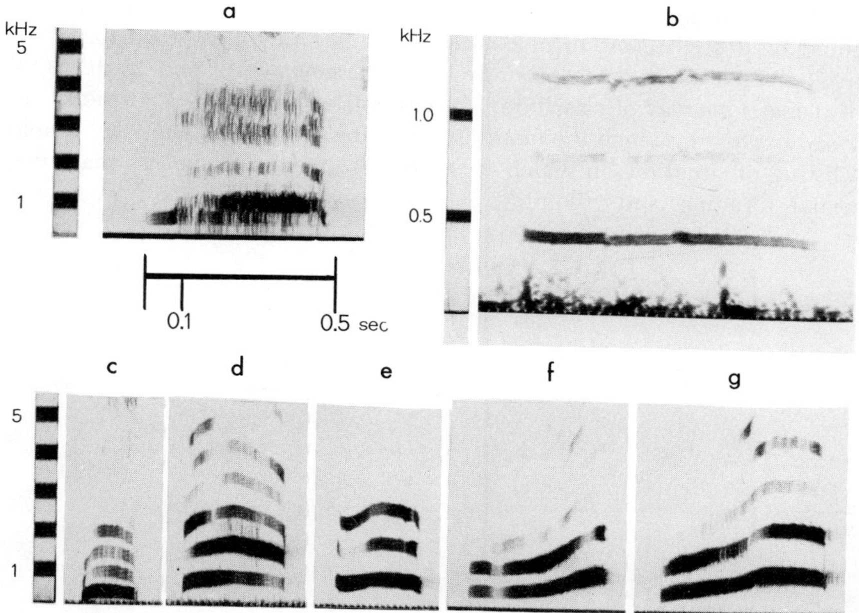

FIG. 1. Seven patterns of Class II *coos* differing in shape and used differently: (a) *double;* (b) *long low* (45-Hz analysis filter); (c) *short low;* (d) *smooth early high;* (e) *dip early high;* (f) *dip late high;* (g) *smooth late high.*

rate these sounds from those of Class III; the lack of articulatory modulations separates them from the *tonal girneys* of Class I.

The separation of this class into subdivisions takes into account the shape of the tracing of the fundamental tones. The tracings were measured for the attributes listed in Table II. A number of classifications of the *coo* sounds along single parameters such as duration, degree of slope, and maximum pitch were attempted. None of these yielded a typology that exhibits functional significance equal to that of a separation based on a combination of factors.

The typology used here incorporates the feature of a local minimum, called a *dip*. It also used the following criteria from among the measured features: the position of the highest point of the tracing relative to the total duration, the midpoint pitch, the total duration, and the occurrence of two simultaneous harmonic series.

The criteria for sorting these Class II sounds into types are shown in Fig. 2. The sounds are classified as to type by using this list as a key that is applied sequentially to the sound spectrograms. The key also defines the types.

All the utterances of this class of sounds occur under circumstances that could be described either as contact-yielding and solicitations for affinitive contact or as vocal coordination of movements and patterns of dispersal. One way of classifying the circumstances is to look at the broad social context in which the circumstances are enmeshed. This could yield a listing of situations in which these sounds are heard, such as maternal, sexual, foraging, and grooming.

COO TYPE	DISTINGUISHING CRITERIA*			
Name	Midpoint pitch $[P(\frac{1}{2})]$	Position of highest peak $[D(H)/D(E)]$	Duration $[D(E)]$	Other features
 Double	⩽ 510 Hz	N.A.	N.A.	Two overlapping harmonic series
 Long Low	⩽ 510 Hz	N.A.	⩾ 0.20 sec.	N.A.
Short Low	⩽ 590 Hz	≠ 1	⩽ 0.19 sec	N.A.
 Smooth Early High	⩾ 520 Hz	< 2/3	N.A.	No Dip
 Dip Early High	⩾ 520 Hz	< 2/3	N.A.	Dip
 Dip Late High	⩾ 520 Hz	⩾ 2/3	N.A.	Dip
 Smooth Late High	⩾ 520 Hz	⩾ 2/3	N.A.	No Dip

*N.A. = not applied for separation of types.

FIG. 2. Seven types of Class II *coo* vocalizations—distinguishing criteria.

The circumstances in which *coo* sounds are given may also be divided into situations taking account of many social and behavioral factors. Vocal events can then be classified and scored by a more refined concept of social behavior than the usual functional contexts.

Aside from the use of *coo* sounds by monkeys that are being provisioned, not included in this report, all the situations in which these Class II sounds are given may be separated into the ten groups shown in Table III. The concept of social behavior yielding this particular classification of situations for scoring occurrence of *coo* sounds will become clear as each of the sound types is discussed below.

TABLE III

SYNOPSIS OF CIRCUMSTANCES IN WHICH *coo* SOUNDS ARE UTTERED, ARRANGED INTO TEN CATEGORIES

None of the contact circumstances are agonistic; none of the directed utterances accompany gestures or expressions of threat. All animals appear relatively calm as compared with the agitated demeanor and arousal observed in situations characteristic of other sound classes.

a. *Separated male.* An adult male alone, calmly following the main troop concentration at a distance of at least 50 meters, directing vocalizations toward the troop or its straggling members.

b. *Female minus infant.* 1. An adult female that has not returned to normal intrafamilial clustering after the death of her infant; shows demeanor of lethargy and depression; vocalizations directed at the body, or as apparently searching. 2. A mature female, nulliparous or nonparous, showing similar demeanor, vocalizing while alone rather than at or within a family clustering typical of the birth season.

c. *Nonconsorting female.* In the copulatory season, a sexually active female that is neither consorting nor soliciting at the moment. She sits or lies calmly alone while vocalizing. She may have just completed a consort relationship, or abandoned pursuit of an unresponsive male, or was herself solicited but is unresponsive.

d. *Female at young.* 1. A mother vocalizing to her youngster (or adult daughter) while with or near it. 2. A mother vocalizing while her youngster is not at hand as she moves from place to place apparently looking for it, and her behavior changes on its appearance. 3. A mature female without an infant approaching to join a huddled grouping in the birth season and vocalizing at it; the grouping is of a different matriliny and contains at least one infant. 4. A mature or immature female alone near an infant-containing family grouping vocalizing as her visual attention is focused on the infant.

e. *Dominant at subordinate.* 1. A dominant in proximity to a subordinate, approaching it, or in affinitive contact with it. 2. A dominant initiating activity that leads to affinitive contact or behaving in the fashion usually leading to such "friendly" contact, e.g., grooming. 3. Within consortship, sitting calmly with female or grooming between mounts.

f. *Young alone.* A yearling or juvenile sitting very calmly, looking around; not near or with its mother, siblings, or playmates.

g. *Dispersal.* Scattered individuals or subgroupings out of visual contact with the main part of the troop, e.g., during troop progression or while in foraging parties.

(Continued)

TABLE III (Continued)

h. *Young to mother.* Calm youngster to its mother as near her, with her, or following her.
i. *Subordinate at dominant.* In calm approach or during affinitive contact or during behavior usually yielding such contact; not including youngster to mother or sexual solicitation.
j. *Estrus female.* 1. During earliest stages of solicitation, i.e., long-distance following of a dominant male. 2. During later stages of solicitation of a dominant male or female as closely following or as seated or lying nearby after a close approach. 3. During consortship as pursuing closely a male (copulatory) or female (pseudocopulatory homosexual) partner or between mounts as the partner leaves.

These groups of situations form one axis of a contingency table. The acoustic type of the sounds forms the other axis. Each vocal event of a Class II sound utterance is tallied in a cell of this contingency table shown in Fig. 3. Each event is scored independently by its acoustic pattern and by its social circumstance and is then tallied in the appropriate unique cell according to sound type and situation of use.

The types of sounds of this class were sampled equivalently in that all acoustical differentiation was performed after the field sampling, during laboratory measurement and classification. So, within each row representing groups of situations, the relative numbers of the different types of sounds uttered are proportional to their occurrences in these situations.

Examples and composite descriptions follow of some of the circumstances in which the seven types of *coo* sounds are uttered. Discussion of each sound type will concentrate on the situations in which it is the predominant type of sound and with which it shows the strongest associations as indexed by the observed-to-expected ratios in the contingency table (Fig. 4).

The discussion may serve to point out the unifying aspects underlying the circumstances in which the same type of sound is heard. This conception provides the basis on which the groups of situations were selected and is also reflected in the interpretation presented on the use of each type.

We begin with a sound with somewhat specialized usage, the understanding of which was something of a turning point in the analysis. It serves as an introduction to the general approach.

a. Double. Adult females give this type of *coo* (see Fig. 1a) in conjunction with the death of their infant. They may be either carrying the rotting or mummified corpse or may have recently abandoned it. The typical behavior of a female emitting this type of sound was not com-

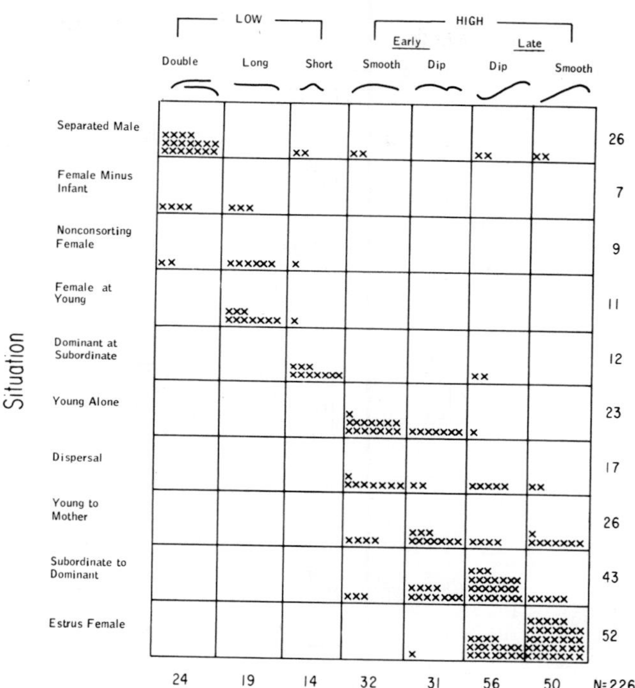

FIG. 3. Occurrence of Class II sounds in different situations. A sample of 226 utterances of *coo* sounds, not distinguished by acoustic morphology in the field, were separated into the listed types by spectrographic analysis. Each occurrence of a type is scored in a behavioral situation as indicated by the marks. For description of labeled situations, see Table III and text.

monly observed since the major circumstance in which it is given, the death of an infant, is rare.

After the stillbirth or death of an infant, the mother was observed to carry the body for a period of 2 to 16 days in this study. Early in this period the body is treated like a newborn infant: carried ventrally with one hand under it for support as the mother walks tripedally; held at the breast with one hand whenever the mother sits down. Sometime later this behavior changes and the mother begins to set the infant down upon the ground and then moves up to a meter or so away from it as she is foraging or is groomed. She continues at these times to treat it like an infant, both meticulously cleaning and grooming it on occasion and also

Type of COO vocalization

Situation	LOW			HIGH			
				Early		Late	
	Double	Long	Short	Smooth	Dip	Dip	Smooth
Separated Male	18 69.2% 2.8 *6.5*	 2.2	2 7.7% 1.6 *1.2*	2 7.7% 3.7 *0.5*	 3.6	2 7.7% 6.4 *0.3*	2 7.7% 5.8 *0.3*
Female Minus Infant	4 57.1% 0.7 *5.4*	3 42.9% 0.6 *5.1*	 0.4	 1.0	 1.0	 1.7	 1.5
Nonconsorting Female	2 22.2% 0.96 *2.1*	6 66.7% 0.8 *7.9*	1 11.1% 0.6 *1.8*	 1.3	 1.2	 2.2	 2.0
Female at Young	 1.2	10 90.9% 0.9 *10.8*	1 9.1% 0.7 *1.5*	 1.6	 1.5	 2.7	 2.4
Dominant at Subordinate	 1.3	 1.0	10 83.3% 0.7 *13.5*	 1.7	 1.6	2 16.7% 3.0 *0.7*	 2.7
Young Alone	 2.4	 1.9	 1.4	15 65.2% 3.3 *4.6*	7 30.4% 3.2 *2.2*	1 4.4% 5.7 *0.18*	 5.1
Dispersal	 1.8	 1.4	 1.1	8 47.1% 2.4 *3.3*	2 11.8% 2.3 *0.9*	5 29.4% 4.2 *1.2*	2 11.8% 3.8 *0.5*
Young to Mother	 2.8	 2.2	 1.6	4 15.4% 3.7 *1.1*	10 38.5% 3.6 *2.8*	4 15.4% 6.4 *0.6*	8 30.8% 5.8 *1.4*
Subordinate to Dominant	 4.6	 3.6	 2.7	3 7.0% 6.1 *0.5*	11 25.6% 5.9 *1.9*	24 55.8% 10.7 *2.3*	5 11.6% 9.5 *0.5*
Estrus Female	 5.5	 4.4	 3.2	 74	1 1.9% 7.1 *0.14*	18 34.6% 12.9 *1.4*	33 63.5% 11.5 *2.9*

IN EACH CELL:

Upper left number of utterances with attributes of acoustic morphology and situation indicated by row and column headings

Upper right percentage of total Class II sounds given in situation

Lower left number expected from marginal totals if sound and situation were independent

Lower right strength of association expressed as observed - to - expected ratio

BLANK ENTRIES REPRESENT ZEROES

FIG. 4. Association of Class II sounds with various situations.

protecting it, by retrieval, from close approach of human beings and from nearby fracases.

At these times the *double* type of vocalization is emitted either when she holds the infant or when she is close to it. Her activity level is abnormally low, her posture sags, and she appears depressed. The mother does not resume normal intrafamilial relations when she has a dead infant. Such relations during the birth season would include spending most of her time in association with the other adult females of her matriliny and with her own and their progeny. The vocalization is given when she is alone, then, except for the body. She is typically seated and looking slowly around in different directions as she vocalizes, but not apparently at any nearby animals.

Later, the timing being quite variable, the mother begins making longer duration and more distant excursions from the body, with these vocalizations now given when she is away from it as well as with it. At this time she may begin slowly walking toward females of other matrilinies either while carrying the body or after leaving it some distance away. She does not seem to orient her face toward these females or otherwise give evidence of directing the sound at them even though she may approach them. The impression is of an animal visually scanning about while moving lethargically toward others.

These females toward whom she is walking will, if the body is carried, stand up and slowly walk away from her line of motion. In any kind of formalistic scoring of the situation, this would be the lowest intensity of agonistic encounter, a supplantation. It would be a misleading representation of the social circumstances, however, much as it might be correct technically, since leaving the vicinity of a mother and dead infant occurs regardless of relative rank.

(Supplantations occur frequently when the supplanted animal is subordinant. Dominance status is ranked on a test independent of the event, such as priority of access to a peanut. Somewhat less frequently, supplantations occur in a direction that is the reversal of the usual hierarchical status of the animals, especially when the lower-ranking one has a temporarily conferred dominance by virtue of a higher-ranking ally being nearby.)

Human observers also move away from her to avoid the stench associated with the putrified body, and this may affect the monkeys' behavior as well.

During this later stage the mother is much less protective of the body than earlier, and continues to be still less so. Instead of carrying it infant-fashion, she holds it as monkeys are seen to do with novel inanimate objects such as a box of recording tape. When seated, she holds the body with only one hand and often lets it drag along behind her when she moves. While foraging she also sets it down in heavily trammeled areas rather than in the out-of-the-way locations selected for a live infant. When holding the body, she allows food particles to fall and remain on it, not at all like the scrupulous care taken to keep a live infant clean. She even allows the cloud of flies which has been hovering over it for several days to begin settling and crawling.

Some females who abandon the body after a few days do not seem to go through this stage of neglect. They are seen one day treating it similar to a live infant, then are not observed with the body from the next day onward. In other cases, gamekeepers remove the body at this stage by snatching it before the female can race over to retrieve it.

Some care is still exhibited at these late stages in that the body is retrieved and dragged off to a new location, then left there, whenever a human observer approaches it. Abandonment occurs at this stage by the female simply neglecting to go back and retrieve the body before beginning one of her movements to a new location. This appears to be an inevitable endpoint of the described progression through greater distances and longer times away from it. It is during these longer and longer absences from the body that the frequency of occurrence of these *double* vocalizations falls to zero after the body is permanently abandoned from its earlier peak when the female was engaged in behavior reminiscent of searching visually and calling vocally.

Only in the copulatory season, and only at Koshima, was the *double* type of vocalization recorded from males. It was heard there from 2 adult males, both of whom were solitary at that time. Failure to hear it from males of the other troops may be an anomaly since solitary males were culled at Miyajima and only rarely observed at the Arashiyama troops. *Doubles* were heard from adult females in each troop.

The typical behavior for either of the two solitaries emitting this vocalization resembles that of females. The male usually gives *doubles* while seated or walking slowly, intermittent with visually scanning. He is alone and apparently calm or even depressed. This isolation was more dramatic than the few meters of socially unoccupied space surrounding the females.

After the bulk of the troop at Koshima is observed during the day on Odomari Beach, it usually moves in the early evening back up the contiguous Odomari Valley. Occasionally the monkeys climb one of the slopes bordering the beach or continue around the shore on the rocky coastline adjacent to the sandy beach. In these cases, when the troop does not return up the valley, quite often the last animals to leave are the rogue males, the band of adolescents and subadults that form a distinct social unit of the troop. Not usually associated closely with the bulk of the troop, these monkeys may have been observed occasionally in the valley during the day and now descend to the shore and follow the troop as it vacates the beach. Then, behind these rogue males, at a distance of 50 or 100 meters, might follow one of the solitary males that has been shown to maintain a spatial association with the troop, especially during the copulatory season (Kawai, *et al.*, 1968).

As the last stragglers of the troop disappear behind the shoreline or up the valley walls, the solitary male begins vocalizing. Sometimes he follows them some of the way, vocalizes, and then returns to sit quietly on a vantage point such as an abandoned dinghy. At other times he walks toward the troop, then returns to the vantage point and vocalizes. Between vocalizations, he is inactive, appears depressed, and looks about quietly.

As dusk settles into nightfall, the male ceases vocalizing and returns up the valley away from the troop's line of progression. Responses from any of the troop are rare. Sometimes male or female stragglers return toward the beach and approach within 50 meters, but without any obvious interaction. Rarely an antiphonal vocalization is heard from one of the stragglers. On two such occasions when the straggler was an adult female, the vocalizing male seemed to be orienting toward her. He interspersed his vocalizations on these occasions with bouts of masturbation, finally eating the ejaculate, and ceasing to vocalize.

There is no striking superficial similarity in the situations in which males or females give *double* sounds. The uses described here, and other circumstances in which they are given, do have in common the lack of some sort of social contact or activity which might reasonably be expected to attend the emitting animal considering its age, sex, the season, and the typical social milieu of other monkeys at the time. This includes either responsiveness or presence of an infant in the birth season, the presence of the troop in its usual sleeping area in the evening, or a consort relationship in the copulatory season.

The situation might then be described in terms of this lack of contact plus the demeanor of the monkeys, inactive or lethargically searching. The social variables that affect the animal at the time and distinguish its behavior from others may also be considered. For example, the context in which a female with a dead infant utters this sound may include her feeding at the moment, but her behavior is of the pattern typical of females lacking live infants rather than one associated with females feeding. The lack of a live infant as the distinguishing social variable is then emphasized as part of the description of the circumstance in assigning the event to a social and functional situation.

b. Long Low. Only adult females were recorded giving *long low coos* (see Fig. 1b). One circumstance in which it is given appears to be a mother calling to her yearling or juvenile. In such a case, a female is seen to arrive at a location alone and look around. While seated she gives this sound once. After a pause of ½ minute or more, she may get up and walk slowly, emitting it again while moving or after seating herself at a new location. If her yearling or juvenile appears, the vocalizations cease. When a female is giving these sounds intermittent with this peripatetic motion, the young animal that eventually responds has not been seen nearby. On four occasions it arrived from a location definitely out of visual contact with the vocalizing mother. After the youngster appears, the mother walks toward it and then holds or grooms it. Only these *long low* sounds are given in a situation such as this.

There are other times, however, when a mother gives this sound when near and in sight of her youngster. On these occasions she does not move to a new location or repeat the vocalization. Response by approach in either of these circumstances is restricted to her progeny.

Adult females without a youngster, their progeny fully grown or no longer alive, also utter *long lows*. Such females give them toward a group of females with young animals among them; other types of vocalizations (Class I) are also given in this situation.

A quite different context in which these sounds are heard occurs only during the copulatory season. They are given by a female seated quietly alone and who is in estrus but not oriented toward or soliciting a male. Most often, she is in between copulations, resting after culminating one mating and not yet soliciting further sexual activity.

Long lows are also given in the copulatory season by a female that is oriented toward or near a male but that is not engaged in the typical solicitous behaviors which usually occur in female–male behavior in this season. After giving a *long low* tonal sound in the vicinity of a male, the vocalizing female does not attempt to maintain a spatial relationship with him by slowly following him or engaging in the active pursuit which characterizes sexual solicitation.

Although a mother calling her youngster has little in common on first glance with an estrus female sitting quietly and resting between matings, these situations have an underlying similarity. All the circumstances in which *long lows* are given by females are in the appropriate seasonal context for the kind of affinitive social contact which is prevalent and might be expected of them. They are given in circumstances with respect to males in the copulatory season, with respect to their own young, or, for females without young, with respect to other families with young in the birth season when clustering of females and youngsters is the dominant mode of social aggregation.

The sounds seem to be employed as a low-intensity indicator of desirability of contact. Even if they are very unspecific indications of lack of affinitive contact and receptivity toward it, the context in which they are emitted may supply enough additional information that the sound and circumstances together may be an event quite specifically interpreted by respondents.

The response by a young monkey to its mother's vocalization might then be an assessment of the utterance of an "alone, contact desired" indicator plus the contextual knowledge on the part of the youngster that these are exclusively adult female sounds. If, furthermore, its own mother is not in sight, it might then walk to a place within view of the vocalizing female, and if it is indeed the mother, she could then approach the youngster as has been observed. This sort of interpretation eliminates the necessity for suggesting individual vocal recognition or a narrow and specific meaning to a call. Yet it still allows for a specificity in function as fine as "mother calling her infant."

c. *Short Low.* Although there was no single situation for vocalizations of this type (Fig. 1c), all occurrences did involve some common social elements which will be described following specific examples. No composite of a typical circumstance may be constructed, however.

Example 1. W, the alpha male of the Arashiyama-A troop, is in consortship with female N'62. They are sitting quietly close together between mounts. W scratches himself underarm and vocalizes. There is no change in the behavior of N'62.

Example 2. Akakin, the senile, but beta male of Koshima, lies down in front of Kaki, a juvenile female, in a grooming presentation. She begins to groom him and vocalizes (a *smooth late high coo*). Akakin then gives a *short low* vocalization in antiphony. There is no change in Kaki's behavior after Akakin's utterance; she continues to groom him.

These two examples involve a close spatial association of two individuals. The contexts differ: consortship versus a temporary grooming relationship. The behaviors of the vocalizing animals differ: sitting and scratching versus lying down. In both cases the vocalizing animal is a high-ranking male and is dominant to his associate. Spatial conjunctions of high-ranking males with juvenile or adult females are unusual, unless the male's behavior encourages proximity, and he actively participates in maintaining the association.

Example 3. A play group is near 2 old males, Akakin and Yon. Play groups are frequently seen to stay close to old males. As Yon, the third-ranking male of Koshima, participates in a play-fight with them, Akakin vocalizes at them. Play continues around the 2 males.

Example 4. In the birth season at Miyajima, female No. 126, which is holding her infant, is together with female No. 125 and her infant. Female No. 126 is the third-ranking female, and No. 125 is low-ranking; they are also from different matrilinies. Other animals are in the area. Females Nos. 126 and 125 are rifling the investigator's pockets for peanuts. Without any directional component to her behavior, No. 126 utters a *short low*. There is no change from any of the animals in her immediate vicinity, but adult female No. 18, low-ranking, which is walking by at a distance of 10 meters stops and looks back at her.

In these two examples the utterances of a *short low coo* are given in different associative situations, but again by the most dominant animal. Tolerance of old males to play groups near them is sporadic; in this case proximity is not discouraged by Akakin. Tolerance of a subordinate animal at close quarters next to a food source is rare, but No. 126 shows no sign of threatening the subordinate female No. 125.

In these four examples, then, the sounds are given when a dominant individual is close to others as part of a behavioral complex allowing or encouraging this proximity.

Example 5. Ego, the fourth-ranking adult female on Koshima, vocalizes at her niece, Ine, the seventh-ranking one, who is a few meters away. This *short low* initiates a vocal interchange, with Ine girneying in return. The vocal interchange continues during a mutual approach which culminates in grooming.

Example 6. Enoki, a high-ranking adult female, has been following an unrelated subordinate subadult female, Shiba, across the beach on Koshima. She follows slowly, then sits 1 meter away and vocalizes while looking at Shiba. Enoki is ignored in this case, and there is no record of her subsequent behavior. On other occasions similar behavior results in her approaching to contact and groom the animal she so follows.

The females giving *short low* sounds in these two examples are initiating a chain of events that generally culminates in affinitive contact. In both cases they are high-ranking and also dominant to the monkey to which their behavior is directed.

In all six of the examples, the utterance of a *short low* sound occurs in a circumstance of spatial conjunction which is initiated, encouraged, or maintained by the appropriate behavior of the vocalizing monkey who may be promoting affinitive social contact. The kind of contact encouraged or solicited here includes play, mating consortship, and grooming. The common feature of these examples lies in the relative dominance of the vocalizing monkeys and in the promulgation of an association, rather than in the broad functional contexts of the kind of association.

To examine whether this concept of the situations in which *short low* sounds are given may be extended, two counterexamples are discussed in which the circumstances might ordinarily be judged agonistic. The vocalizing monkey in these could be considered aggressive rather than friendly if only the withdrawal or defensive reaction of the respondent is considered.

Example 7. Pe'64, a subadult female of the Arashiyama-A troop, is accompanied by and associating with a subadult male, N'64. She approaches in affinitive fashion Mo'62, another subadult, the lowest-ranking male. Even though she exhibits no threatening gestures, he trots away. She vocalizes, using a *short low coo,* and he continues running away.

Example 8. Thirty seconds after her mother, Sasa, has bitten her, Sasage, a yearling female, returns the 30 cm she had backed away while screaming. Now next to Sasa, she utters a *short low* while facing and prancing toward another yearling. This other yearling grimaces in response. Two adults 1 meter away look up at Sasage as the vocalization is given.

In example 7, male Mo'62 continues to run away after being supplanted by the vocalizing female, Pe'64. Male Mo'62 is dominant to her but subordinant to the male who accompanies her, N'64. The female may then have a temporary status of conferred dominance over Mo'62 because of her association with N'64. If so, the observed supplantation is then in the same direction as usually occurs in agonism: the subordinate animal withdraws. This supplantation is consistent with scoring the event as a straightforward low-intensity agonistic encounter.

It must also be noted, however, that Pe'64 does not show any of the gestures, postures, agitation, facial expressions, or any of the other aspects of demeanor which in themselves are threats or are otherwise associated with offensive behavior. Quite the opposite, she is walking toward Mo'62 in a fashion typical of contact-yielding approaches.

In the absence of evidence of offensive behavior by the vocalizing animal, Pe'64, and since her demeanor is suggestive of nonaggressive approaches, consider the possibility that her approach to Mo'62 is an attempt to encourage an affinitive association with him. Even though the response of yielding ground by Mo'62 is one usually included within a context of low-level agonism, according to an alternative interpretation of the circumstances in which this *short low* is uttered, the manner of Pe'64's approach is considered. In this particular instance Mo'62 continues to move away from her and from N'64, a male dominant to him. Female Pe'64, however, may be uttering the *short low* as a component of an approach which indicates a willingness to associate, rather than its opposite, aggressivity. The circumstances may be formulated as a vocalization given by a dominant monkey, with conferred status in this case, as approaching a subordinate one in a fashion usually yielding affinitive contact. Such a description takes account of past associations with the demeanor of the vocalizing animal rather than considering only the response evoked in the present instance. It is also consistent with the description of common features of the circumstances of the earlier examples of *short low* utterances.

Similarly, a close look at the entire situation, rather than only at the response, affects considerations of the eighth example also. Sasage is returning to a calm state after a short withdrawal while screaming, following the administration of a grab and bite by her mother, Sasa. In such situations, a rebuffed youngster usually resumes normal social contact very quickly afterward. She approaches in this case an unidentified yearling to which she is dominant because of their proximity to Sasa—again a case of conferred dominance.

The vocalization is given during an approach without any threat component and in a fashion which is usually followed by the young animals engaging in play. The other yearling responds to Sasage's approach with a grimace. The grimace is normally a component of withdrawal in the aftermath of a threat or other offensive behavior; it is so closely associated with being threatened that it is sometimes labeled a "fear grimace" (cf. Kaufman and Rosenblum, 1966).

This grimacing response is rare from one young animal when approached by another in the circumstances in which play normally ensues. Although formally this instance could be considered a low-intensity agonistic encounter if the grimace is given weight, it is also possible to interpret the situation as an unsuccessful attempt at initiating play. In examining other features of the situation, including Sasage's demeanor, and noting the usual outcome of similar circumstances, the situation can also be phrased as an animal in a dominant status promulgating contact with another, although unsuccessfully.

Unrelated adults generally ignore the activities of nearby yearlings. Since the response of looking up at Sasage given by 2 nearby adults is unusual, it is likely that utterance of a *short low* is an unusual occurrence among yearlings. This possibility is consistent with interpreting such an utterance as associated with monkeys of dominant status.

All eight of these examples may then be interpreted as use of a *short low* sound in a situation in which a dominant monkey is promoting affinitive contact with a subordinate.

d. Smooth Early High. Utterance of a sound of this type (see Fig. 1d) by one monkey is frequently associated in time with occurrence of other *coo* sounds by other individuals. The momentary activity of animals giving these sounds may be included in a number of different contexts including feeding, foraging, huddling in a matrilineal group, allogrooming, and autogrooming. Regardless of their immediate activity, animals that are alone or in small subgroupings away from the troop or are part of a spaced progression during troop movement utter sounds of this type when looking at and calling to one another. Animals giving these *smooth early highs* may also do so without any apparent orientation to another monkey. Yet other monkeys, including those out of sight, frequently respond, therefore apparently only to the vocal aspect of the behavior of the emitting animal. These antiphonal vocal responses, or vocal coordination (Itani, 1963), may at times be so extensive throughout the troop or a subgroup that they form a predominent undercurrent of noise. Not only are the sounds more likely to occur when the troop is spread out, but also they are more prevalent among the farthest dispersed animals and in parties moving somewhat independently of the main group.

Example 1. Kumoi, an adult female, is eating. While picking at scattered grain she utters a *smooth early high.* Other animals of Arashiyama-B troop are dispersed nearby since they are also eating the same provisioned food. Some respond in vocal antiphony.

Example 2. During a very stretched-out troop progression on Miyajima, an unidentified pregnant female stops and sits alone momentarily. She utters a *smooth early high*; a vocalization follows a few seconds later from a group of 3 animals that have been moving in the same progression 10 meters from her and are the only other monkeys nearby.

Example 3. A subgroup of seven animals is foraging more than 100 meters from the rest of the Koshima troop. Satsuki, the alpha female, utters a *smooth early high* in antiphonal response to a sound from an unrelated subadult female, Shiba, which is part of this same temporary foraging party. She vocalizes while continuing to forage and without looking up.

Other utterances of *smooth early high* sounds are given in more complex circumstances. To gauge the important social and functional aspects of these events, the momentary behavior of the monkeys in these situations must be noted in addition to the other aspects of the context.

Example 4. Adult female No. 8 and her adult daughter No. 308 have lingered behind during a troop movement toward the feeding arena on Miyajima. Both are late in pregnancy and not prone to associate with other monkeys at this time. The bulk of the troop is out of sight beyond the crest of a hill 40 meters away. A vocalization is heard from that direction, after which No. 8 pauses in grooming her daughter, then cranes her neck and peers up the slope while uttering a *smooth early high.*

Using the cue provided by No. 8's behavior, the social and functional situation which is considered predominant can be scored as that of an individual of a dyad within a progression vocalizing toward the main group. Other aspects of the context do exist, of course, but are not pointed out by the vocalizing animal's demeanor as being significant to the circumstances. The context in Example 4 is also a grooming relationship, a dominant monkey with a subordinate, a maternal association, and a prepartum social separation.

Monkeys giving *smooth early highs* in all circumstances are inactive and do not appear aroused. The occasions when the sounds are a dominant undercurrent of noise are those times of low overall activity by members of the troop; there is a greater likelihood of hearing these in the early morning and late evening or at night than at other times. They are also more frequent on stormy or foggy days when the monkeys are dispersed in the various nooks and crannies where the terrain offers shelter from the cold wind or rain.

The occurrence of these sounds is labile in the sense that any shift to greater activity among any of the members of a grouping in which these vocalizations are being exchanged leads to their abrupt cessation by all members. Such shifts occur at the appearance of a human being in the area, at the outbreak of a fight within sight or hearing, at the sudden motion of any monkey in the group, or at any other event that intrudes on their serenity.

Yearlings or juveniles who are alone within the heart of the troop, but not near relatives or playmates where they are usually found, also utter *smooth early high coos* while sitting calmly and looking around intermittently. Often they are autogrooming and may even be dozing lightly just prior to or after these sounds are given. Although these young monkeys are often gregarious toward human beings, it is especially difficult to record them in this situation since the approach of an observer is enough to alter their vocal activity even though they remain outwardly unchanged in demeanor. These unwary youngsters seem sensitive to disturbances by humans when uttering this type of sound. It is, therefore, possible that low activity is not the only characteristic of monkeys uttering *smooth early highs*. Some extremely calm internal state, perhaps readily disturbed, may be required as well.

Utterance of *smooth early high coo* sounds may be interpreted as a position marker. Weather, terrain, or dispersal patterns can yield conditions in which visual contact is difficult. During periods of low activity, few foliage movements or vocalizations serve to pinpoint monkeys' locations. The *smooth early high coo* is the principal sound used at these times of low activity or poor visibility. The exact function is not clear, but these sounds are probably involved in maintaining group cohesion.

e. Dip Early High. The *dip early high coos* (see Fig. 1e) are given in circumstances of low activity with the vocalizing monkey apparently unaroused. Many utterances are given in the circumstances of dispersal and position marking as are the *smooth early highs*. The monkeys also utter them, however, against a background of higher activity than with the preceding type, and not during the extreme calm characterizing use of *smooth early highs*. Generally, *dip early highs* are given in situations of greater social involvement, either as part of some ongoing relationship between two individuals or directed to a specific individual. The directional component of the vocalizing monkey's behavior is usually a head turned or thrust toward the other monkey. All the directed utterances of this type were given by a vocalizing subordinate toward a dominant.

The situations in which this sound type finds its most extensive use lead to the interpretation that it is both a component of subordinate status-marking and also part of indicating the desirability of affinitive contact. Younger monkeys are more disposed to use this type than older ones, especially toward their mothers. Adult males were not recorded uttering it at all.

Example 1. The yearling female No. 729 still maintains a close relationship with her mother, No. 29, during the birth season on Miyajima, probably because her mother is nonparous this year and No. 729 has not been weaned. After walking up to No. 29, the youngster sits in front of her mother and vocalizes, uttering a *dip early high.* She then moves to No. 29's breast and takes her nipple. The mother holds and suckles her.

Vocalization from a youngster directed to its mother, or as it is near or with her, is a special case of the general situation in which *dip early high coos* are given. In general, the situation comprises a dyad in which the subordinate monkey vocalizes toward the dominant one at a transient point in their social interaction. Characteristically, the alteration in the relationship is evidenced by a change in distance at the time the *dip early high* is uttered.

Example 2. Ikaru, a juvenile male, is with his cousin Goma, an adolescent female dominant to him, eating near the shoreline on Koshima and away from the concentration of the troop. She stands and starts to walk away; he looks up at her uttering a *dip early high.* She continues to leave.

Example 3. In the birth season, subadult female No. 311 is grooming Miyajima's alpha male, No. 36. This pair maintains a consortlike spatial association sporadically throughout the birth season and are frequent mating partners in the copulatory season. As she grooms him, he stands and moves about 1 meter looking toward some disturbance in the troop. She stands and utters a *dip early high* as she moves toward him. He lies down and is rejoined by her.

In many ways the use of this type of sound is the converse of that of the *short low coos.* Although these sounds are uttered within different functional contexts (e.g., grooming, feeding, mother–young relations), many circumstances have in common the subordinate status of the vocalizing monkey and the initiation or maintenance of affinitive contact.

f. Dip Late High. These sounds (Fig. 1f) are used by subordinate animals toward dominants. One interpretation of the circumstances in which they are used allows one to consider them as part of active contact solicitations. A readily identified situation included in such a general social format is the sexual solicitation of dominant males by subordinate females.

Dip late highs accompany the very earliest stages of female-initiated consortship solicitations and are rare at the later stages. The next type, the *smooth late high coos,* has the converse usage. They are more frequent accompaniments of the somewhat advanced stages of solicitation.

These early stages are often difficult to detect. The first indication of solicitous activity may be nothing more than a female sitting near or within a concentration of animals, perhaps alone among them perched on a limb, and vocalizing occasionally. Infrequently, during and just prior to the vocalization she may be looking at an adult male. Irregularly, some small motion by the monkey on whom she appears to focus seems to trigger her vocalizations. This early pattern of attention was observed only between vocalizing estrus females and mature dominant males.

If a male leaves the area, the female may follow, but at a considerable distance behind (ca. 100 meters). She gradually reduces the distance to as little as 5–10 meters after he stops. Vocalization, usually with *dip late high* sounds, continues at irregular intervals during this period when she once again approaches his vicinity. This maintenance of a fixed spatial relationship between the two by active pursuit on her part is difficult to detect, unlike later stages of solicitation.

It is also difficult to confirm that this activity is directed toward a particular male because she may switch orientation many times in one day, returning to solicit the same males again amid bouts of following others.

Sometimes the solicitation is successful at this early stage.

Example 1. In the copulatory season, Fut'64, a subadult female of the Arashiyama-B troop, has been following subadult male Mo'64 to which she is subordinate. She sits 5 meters from him and utters a *dip late high coo.* He stands and walks away, and she follows in immediate pursuit. One minute later she calmly approaches him again by a circuitous path, finally moving up to his side. He holds still, as she completes an approach to him and then begins grooming. After 3 minutes of grooming, mounting begins.

Early indications of success may be more subtle, however.

Example 2. Odamaki is 8 meters up in a tree uttering *dip late high* sounds at Arashiyama-B troop's alpha male, Zola, in the copulatory season. He is on the ground near the base of the tree. Each time he moves, she vocalizes. Zola shows no obvious notice of her, yet he remains close to the tree for over 1 hour except for occasional forays to break up fights. He then climbs into the tree after she jumps into a neighboring one. The following day a more obvious, but occasional, consort relation is apparent, and 2 days later they are seen copulating for the first time.

Continued solicitation without success in forming a consortship or in initiating copulations is also observed. As unsuccessful solicitations continue, the relative frequency of occurrence of different types of vocalizations changes. *Dip early highs* gradually give way to the next *coo* sound, *smooth late highs.* Much later these are, in turn, replaced by sounds of other classes.

Dip late highs are the predominant type of *coo* only in the earliest stages of sexual solicitation. The other circumstances in which this type of sound predominates are so various that characterization of a typical sequence is not possible.

They are used in situations where the behavior leads directly to establishment of an affinitive social contact.

Example 3. Between the birth and copulatory seasons, small clusters of females are the typical social grouping of the Arashiyama troops. Ume'67, a yearling female, approaches a cluster of 4 older females including her juvenile sister, aunt, and cousin. She is subordinate to the aunt, cousin, and an unrelated subadult. As she walks toward them she utters a *dip late high* and continues moving during the vocalization until she joins them. There is no detectable change in the behavior of the 4 monkeys joined during this approach and utterance.

Example 4. Tsuga, one of the lowest-ranking females of Koshima, runs away from Kaminari and Yon, the alpha male and the third-ranking one, respectively, which have approached her in the copulatory season; possibly these approaches are part of mating solicitations. She continues moving away from them and walks 15 meters to a dominant female, Ine. At the last portion of this approach to Ine, Tsuga utters a *dip late high.* Ine remains in place and is then groomed by Tsuga.

The *dip late high* sounds are also used by monkeys not necessarily at a transient point of spatial change in a relationship. In addition to their use in consortship solicitations, they are also given by those monkeys who are enmeshed in a variety of other sequences generally leading to affinitive contacts.

Example 5. In the birth season at Miyajima, the typical social grouping is a huddled cluster of members of one family. Juvenile female No. 612, an orphan, is 3 meters from an unrelated family of mother, juvenile daughter, and infant. The juvenile daughter, No. 604, had left her mother, No. 4, and been with No. 612 but has now returned and huddles with No. 4 which is holding infant No. 804. Between hesitant approaches to within 3 meters and while looking at the family, No. 612 occasionally utters vocalizations including this *dip late high coo.* Seven minutes later she joins the huddle.

These other examples include approaches and contacts within the different contexts of grooming, matrilineal clustering, and other kinds of affinitive contact. Yet in these uses, as well as in sexual solicitation, there is the common element of the vocalizing monkey's subordinate status with respect to those approached or contacted. An animal vocalizing in these situations is likely to be more active than in situations in which the previous types are heard. This greater activity may occur near the time of vocalizing or may be seen as locomotion during the utterance.

Dip late highs appear to be used as part of a complex of behavior patterns by which one monkey makes affinitive contact with another or tenuously attempts such contact. Noting that use is exclusive to subordinates, perhaps it is not unreasonable to assign a common substratum of apprehension to some of these contact circumstances. This could also be inferred from the lack of straightforward approaches and the consequent long time courses until contact is achieved.

g. Smooth Late High. Like the preceding type, these *coos* (see Fig. 1g) are used chiefly in contact circumstances by subordinate monkeys toward dominants. Youngsters use them extensively toward their mothers much as was described for *dip early high* sounds.

Example 1. At Koshima, Kiji follows about 1 meter behind his mother, Enoki, in the first copulatory season subsequent to his birth. He is not weaned, but is just beginning to spend extensive amounts of time away from her. He utters a *smooth late high coo.* Enoki immediately stops and then grooms him as he reaches her. He remains held by her and is in oral contact with her nipple several minutes later.

These *smooth late highs* are the principal type used in sexual solicitation by estrus females when consortship has begun, i.e., proximity is mutually maintained, but copulation has not ensued.

Example 2. Fut'64, a subadult female of Arashiyama-B, is seated near Pe'60, an adult male of the same troop who is also seated. She vocalizes with a *smooth late high.* Male Pe'60 immediately moves closer, to about 30 cm from her, then ½ minute later puts his hand on her back. A sporadic copulatory consortship ensues over the next few days with each of them also mating with other monkeys.

The same sounds are also used by estrus females within a consortship during some lapse in copulatory activity, generally as the male has left the female's vicinity for a moment.

Example 3. Adult female No. 35 is in mating consortship with the alpha male of Miyajima, No. 36. There has been a series of mountings, but this mount is not the ejaculatory one of the copulatory sequence. Just after he dismounts her, he walks away from her and goes toward the sounds of a fight. She vocalizes with a *smooth late high* just after he leaves and he continues moving away. He returns and mounting begins again a minute later.

Smooth late high sounds come to predominate over *dip late highs* as the earliest stages of consortship formation with their long distances and slow, gradual pursuits are replaced by later stages in which the female decreases distance and follows closely, actively maintaining proximity.

Similar behavior patterns of more active solicitation are given by females in preludes to homosexual consortships and pseudocopulatory behavior. At this time the *smooth late high coos* are directed by a subordinate female to a more dominant one. The short distance between the pair is typical of the closer spacing in the later stages of female-to-male use. The early stages of female homosexual solicitation, analogous to the early stages of ordinary sexual solicitation in which *dip late high* sounds are used, were not observed.

It is possible that the earlier stage of gentle and long-distance pursuit does not exist between females or that it is much more brief than the typical development of female-initiated solicitations of males.

Example 4. A subadult female of Arashiyama-B, Co'65, has been closely following alpha female, Mino, all day and has appeared increasingly agitated. Just as Mino stands and begins to walk away, Co'65 utters a *smooth late high* which is the beginning of a series of intergraded sounds. Mino continues walking away, and Co'65 pursues her. Female Co'65 is beginning to employ the approach and immediate withdrawal locomotory pattern of sexual solicitation. This pursuit develops into a homosexual consort relationship over the succeeding 4 days. Mounting is observed on the fifth day hence.

In all of these uses, subordinate monkeys direct the vocalizations to a dominant. The circumstance in each case is part of a situation of affinitive contact, or prelude to affinitive contact, with a monkey of the age and sex class appropriate to the typical seasonal and social milieu of the one vocalizing. This includes youngsters toward their mothers and estrus females toward prospective or actual male consorts. In the Arashiyama troops where female homosexual behavior is prevalent, the sounds are used in this kind of solicitation.

In many cases in which *smooth late highs* are heard, the vocalizing monkey is more aroused or agitated than when uttering the other types of *coos*. This is particularly evident in many of the sexual solicitations where the vocalizing female pursuing her prospective partner is very sensitive to its movements, especially any locomotion away from her.

Smooth late highs are also used in a number of other dyadic circumstances, including juvenile males vocalizing at rogue males with whom they frequently associate on the periphery of the troop, and juvenile females at one of the older adult males around which they are seen to cluster and play. In such circumstances, the situation is that of a vocalizing animal engaged in affinitive contact or its initiation. If there is a relatively greater degree of tension on the part of the vocalizing animal when the social circumstances include proximity to a very dominant individual then, in these other cases, agitation or arousal may not be observed, but perhaps may be inferred.

2. *Summary of Coo Sounds*

In summary, the seven types of *coo* sounds, forming a single class of the vocal repertoire, are given in a wide variety of circumstances. The social elements of their use include lack of affinitive contact, but desirability of achieving it, vocal contact, attempts to make contact, maintenance of affinitive proximity, and solicitations for specific kinds of contact, especially maternal or sexual.

One formulation of behavioral categories in which these *coos* are used could impress on all these circumstances a common designation of "affinitive contact" situations. Another way of looking at the circumstances would be based on the overall contexts in which the vocalization is enmeshed. This could include the social contexts of maternal, sexual, agonistic, and grooming behaviors as well as troop movements and foraging.

A different conception of social behavior is suggested here. It takes account of both the overall functional context and also the immediate social environment as well as the demeanor of the vocalizing animal. The behavior accompanying an utterance indicates which aspects of the circumstances may be paramount to the vocalizing animal. Since, as in the fourth example of *smooth early highs*, a vocalizing animal can be involved in a troop movement, in a maternal interaction, and in a grooming relationship all at the moment of an utterance, its total behavior at that moment is used as a clue to classify the circumstances into a single but compound situational category.

Clues include any orientation component to the behavior, such as looking in the direction of others in the troop progressions, and timing, such as a pause in grooming during the vocalization. In other cases, account is taken of behavior patterns that are stereotypically assignable to certain kinds of interactions. Monkeys engaged in the motor patterns that are generally preludes to copulation are said to be engaged in courtship or solicitations for sexual activity even though the entire sequence through copulation may not be observed or may not be culminated.

Responses by other monkeys are often valuable clues to the social and functional aspects of a situation which includes a vocalization. Although cognizance of the responses is desirable, the nature of the response is not relied on as the exclusive determinant of the categorization of the circumstances. Withdrawal from a dominant animal or submissive behavior toward it is not considered an agonistic encounter, and the dominant one is not considered as engaged in aggressive or threatening behavior unless its demeanor is also of the nature usually provoking withdrawal, i.e., it is employing a threat gesture, facial expression, body posture, or locomotory pattern.

B. Association of *Coo* Sounds with Social Situations

1. Analysis

When Class II utterances are scored by acoustic pattern and also by the social and functional circumstances conceived as we have defined them, then a relationship between the types of *coo* and the situation in which they are used becomes discernible.

Grouping the circumstances into ten groups of situations by their functional, social, and behavioral attributes, it may be seen that in each group the types of *coo* sounds are given to different extents. These preferences for the sound types differ according to different situations (see Fig. 4). The significance of these differences will now be examined.

Because the sampling in the field was opportunistic by situation, the different behaviors may have been sampled for vocal activity to different degrees. Therefore, although it is clear that there is a relationship between the type of sound and the situation in which it is given, the strength of this relationship must be examined using statistics based on marginal totals.

The first statistical test on the contingency table examines whether the observed heterogeneity of use of the types of sounds is likely to have arisen by chance from a homogeneous distribution of usage. The computed G (Table IV) indicates that this array of tallies is highly unlikely to arise by chance if there is no association between the acoustic patterns of Class II *coo* sounds represented by the rows and the indicated categories of situations in which *coo* sounds are used ($p < 0.001$).

This demonstrates that it is possible to derive acoustic characters from a sample of very similar sounds which can then be used to sort that sample of sounds in a way that relates them in a nonrandom fashion to their use in various social and functional situations. It is especially interesting that the shape criteria for the acoustic patterns used in this differentiation are not very pronounced features of the sounds and were not distinguishable as natural units in the field or even after an initial appraisal of the spectrograms.

The acoustic criteria were derived from examination of an initial sample of 180 Class II utterances. It is conceivable that the sort of differentiation achieved here, if both the derivation of criteria and their application are related to the same sample, may not exhibit any general validity. To examine that possibility, this initial sampling was supplemented by a subsequent sampling of 46 more Class II utterances. The selection of these supplemental *coo* utterances was not influenced in either form or function by the results on the first sample; they comprise the 46 *coo* patterns that had already appeared accidentally as background noise on spectrograms which had been produced for analyzing utterances of other sound classes.

The result of a test on the G-statistic comparing the observed distributions in separate contingency tables for the two samplings indicates that they are not significantly different ($p \approx 0.3$). A correlation coefficient may be calculated in which the number of tallies in a given cell for the initial sampling is paired with the same score for the second sampling. This coefficient, $r = 0.62$, is significantly greater than a value expected on the basis of chance correlation ($p < 0.01$).

These tests confirm that, after classifying the *coo* utterances using criteria derived from the first sample, the second sampling does not yield significantly different results from the first sampling; the data are strongly positively correlated. These two samplings are, therefore, combined to yield the contingency table based on all 226 *coo* utterances of the sample.

Since overall heterogeneity is established by the G-test, pairwise comparisons are permitted amongst all seven of the sound types to test for differences between their usage. The G statistics for these comparisons are given in Table IV. These statistics are tested for significance by a conservative formulation which considers each test to have 9 degrees of freedom from its two rows and ten columns even though some of the columns for certain pairwise comparisons may show zero observations. This procedure is conservative with respect to yielding spurious significance, because a higher value of G is required for the same level of significance with more degrees of freedom.

Looking at all the pairwise comparisons in Table IV, it may be seen that each sound type is used significantly differently from all of the other types of Class II sounds.

Knowing that a distribution is significantly different from that expected if there was no association between the attributes heading the columns (sound type) and the rows (categories of situations), does not reveal the strength of such an association. The strength of overall association may be measured by the contingency coefficient which we compute from these data as $C = 0.81$. This value is ca. 86% of its theoretical maximum for a 7×10 array.

We may also look at the contingency coefficients calculated for each pairwise comparison based on separate 2×10 arrays. These results are displayed in Table V where, to ease comparison, the numbers have been normalized with respect to the maximum observed contingency.

Notice that for those pairs with the greatest degree of separation by the acoustic attributes into different social and functional usages, the numbers are highest. Where the patterns of social usage are most similar between two forms, the numbers are lowest. In general, these values serve as a numerical index for the degree of difference in distribution of occurrence contingent on the acoustic form of the vocalizations (or vice versa).

A contingency coefficient equal to its theoretical maximum implies complete contingency, i.e., if one of the attributes of an event is known then the status of the other attribute can be best predicted. With complete contingency, once there is knowledge of either type of vocalization or situation for an utterance, then it is possible to make a least-error guess as to the converse attribute of the utterance, either the situation in which it is given or its acoustic pattern.

The degree to which the error may be reduced with different levels of contingency is measured by an index of predictive association. When there is complete association between attributes in such a way that knowledge of one yields zero uncertainty in predicting the other, then this index is 1.00, i.e., perfect prediction is possible.

This measure is not equivalent to either the G-statistic or the contingency coefficient. To paraphrase Hays (1963): it is quite possible for some statistical association to exist even though the value of the index of predictive association is zero. In this situation, the attributes are not independent, but the relationship is not such that giving the status of either causes one to change one's bet about the status of the other. The index is other than zero only when *different* categories of one attribute would be predicted from different information on the status of the other attribute.

The two asymmetric and the symmetric indices of predictive association for the seven Class II *coos* and the ten social and functional categories in which they are uttered is given in Table IV. Given information on either sound type or category of behavior, we may predict the other aspect of the utterance with about 42% fewer errors, on the average, than otherwise.

This reduction in uncertainty achieved by being able to specify the

TABLE IV

STATISTICAL RESULTS ON USE OF SEVEN TYPES OF CLASS II *coo* VOCALIZATIONS [a]

1. Overall heterogeneity of usage of seven types in ten situations: $G = 421.7$; $df = 54$; $p<0.001$
2. Differences in distribution of use are measured for the 21 pairwise comparisons and yield the tabled values of G. These, and the p associated with each (from $df = 9$: two types $\times$ ten situations), are in the cell determined by the intersection of the types being compared.

	long (LOW)	*short* (LOW)	*smooth* (HIGH, EARLY)	*dip* (HIGH, EARLY)	*dip* (HIGH, LATE)	*smooth* (HIGH, LATE)
double (LOW)	40.5 / 0.001	33.2 / 0.001	63.5 / 0.001	75.4 / 0.001	84.7 / 0.001	80.2 / 0.001
long (LOW)		32.5 / 0.001	67.4 / 0.001	66.4 / 0.001	84.9 / 0.001	81.2 / 0.001
short (LOW)			51.0 / 0.001	55.8 / 0.001	53.7 / 0.001	61.7 / 0.001
smooth (HIGH, EARLY)				18.5 / 0.05	55.1 / 0.001	68.3 / 0.001
dip (HIGH, EARLY)					30.8 / 0.001	48.6 / 0.001
dip (HIGH, LATE)						24.5 / 0.005

3. Overall contingency coefficient $C = 0.81$ (86% of theoretical maximum)
4. Indices of predictive association

   Given type of vocalization: $\lambda_A = 0.420$

   Given nature of situation: $\lambda_B = 0.415$

   Given either (unspecified): $\lambda_{AB} = 0.417$

[a] Data are the tallies in Fig. 3, a contingency table.

situation in which a vocalization is used from knowledge of its type is not just a measure of the information content of sounds but it is also one pertinent to the nature of social intercourse. Optimizing the reduction in uncertainty is the very heart of the communication process in social animals. If there is available additional knowledge of the situation, age and sex of the emitter, for example, then a communicant can eliminate more of the potential situations and further reduce the range of possibilities. In this way the uncertainty as to what is the precise social and functional circumstance in which a sound is emitted may be minimal for an animal that can both hear and see the emitter of an utterance without necessarily seeing all of the events surrounding the vocalization.

2. *Inferences*

For the animals to utilize these associations of sound type with situation effectively, they must have some kind of built-in set of rules for encoding and decoding. The results presented, all based on order-independent statistics, could apply to any kind of rules such as a simple lexicon. They do not in themselves require or suggest any set of generalized rules by which the syllabary of sounds is related to social usage.

Certain aspects of the order-independent results are strikingly similar, however, to order-based conclusions indicating existence of a single rule rather than only an arbitrary lexicon. These conclusions derive from a consideration of the position of cells showing strong associations with respect to the row and column orderings. The strong associations, as indexed by the high observed-to-expected ratios, fall roughly on a diagonal within the contingency table (see Fig. 4). This suggests a rule by which acoustic pattern of *coo* is related to its social usage, the rule reflecting the ordering of the headings on the table's axes.

The order in which the row headings naming the categories of situations appear down the side of the table is not arbitrary. In the first three categories—separated male, female minus infant, and nonconsorting female—the animals are quiescent, move slowly, and appear depressed. If they are seeking contact, they are doing so with a minimum of activity. If apprehension and some degree of tension or arousal in internal state accompanies many kinds of intimate social interaction, this seems lacking in these animals vocalizing in situations characterized by lack of dyadic sociality.

The next two categories—female at young and dominant at subordinate—also include animals that are relatively inactive. They generally appear calm at the time of vocalizing; their dominance relationship in these situations may account for their being unaroused. Confidence, rather than apprehension, could be imputed to the internal state of animals that are in a calm social relationship or seeking affinitive contact with subordinates, including their own progeny, rather than with dominants.

In the two following groups of situations—dispersal and young alone—the monkeys are maximally calm in appearance. The two groups of situations are similar in that the monkeys employ utterances apparently as position markers. Also, even though they appear least aroused in these situations, they are more readily disrupted in their vocal behavior than in others in which Class II *coos* are given. This lability in the behavior of monkeys vocalizing in these circumstances suggests a sensitivity that is belied by other aspects of their demeanor.

The remaining three categories of situations are of subordinate animals vocalizing at dominant ones including young to mother, estrus female soliciting sexual activity from dominant animals, and other instances of vocalizations directed from subordinate to dominant. In these, the vocalizing monkeys are active and appear sensitive. The sensitivity in the case of estrus females is evidenced as a triggering of vocal activity by movement on the part of the solicited animal. The monkeys also appear to be more aroused in these circumstances than in the previous ones and may become even more aroused with the passage of time if the maternal or estrus contact solicitations do not develop into affinitive interactions. Because these all are circumstances of vocalization by subordinates within dyadic encounters, there may be some degree of apprehension or at least lack of confidence that distinguishes the internal state of these monkeys from those that are dominant in a dyad or that are not socially interactive as they vocalize.

There is then a gradation in demeanors, which can be observed, and in internal state, which can be inferred, as the list of situations is traversed. The listing progresses from a group of situations in which the monkeys are inactive, depressed in appearance, and lacking contact to those in which they appear agitated and aroused, are sensitive, and are actively soliciting contact from dominant monkeys. Intermediate states are of calm animals, dominant and seeking or maintaining dyadic interaction with a subordinate, and very calm ones that are sensitive but whose social interactions are principally vocal and not directed to specific individuals.

This ordering of one axis of the contingency table has two principal features. The headings closest together are of situations in which the demeanor of the vocalizing monkeys and certain social facets of the circumstances surrounding the utterance are most similar. The groups of situations are listed in an order that grades to the greatest observed activity and arousal and to the social circumstances in which one might expect the monkeys to be most apprehensive and tense.

The ordering of the columns, the types of *coo*, results from placing the types with acoustic patterns that are similar nearest to each other and those with least alike patterns the furthest apart. Although this ordering is readily accomplished by visual comparison of the simple tonal patterns (once they are delineated from the homogeneous mass of *coos*), it can also be confirmed with a multivariate analysis performed on those few acoustic parameters that vary within this class (Green, 1972). Dealing with only minor variations on a single structural pattern, this ordering is accomplished without difficulty. Sounds of high pitch are at one end of the table and those with low pitch at the other. *Doubles* are similar in shape to *long lows* except for the presence of the twin series of resonances. *Short low* sounds have low pitch in common with *long lows* but are similar in shape to *smooth early high coos*, so are placed between them. *Smooth early high* and *dip early high* are quite similar, being separated only by the "dip" feature, and the same is true for the two kinds of *late high* sounds. These pairs are neighbors—the two types with a "dip" differ only in the position of their peak, and these are placed as neighbors.

The meaning of the diagonal pattern of the strong associations in the contingency table is now clear. Sounds that are most alike are used in situations in which the demeanor of the monkeys and the social and functional aspects of the circumstances are most similar. Sounds that are least alike are used in situations which are least similar by this conception.

Examining once again the numerical indices derived from contingency coefficients measuring differences in distribution of use of all pairs of Class II sound types (Table V), it should be remembered that they are independent of the ordering of the rows and columns of the contingency table. Yet note that they fit closely this same notion as to how differently the several kinds of sounds are used, that is, different in the sense of the degree of dissimilarity or behavioral separation in the nature of the situations in which they are employed, not the statistical differences in the distribution of occurrences from which the indices are derived.

The *smooth late high* and *dip late high* sounds, for example, show

TABLE V

RELATIVE DIFFERENCES IN DISTRIBUTION OF USE OF SEVEN TYPES OF CLASS II *coo* VOCALIZATIONS

	LOW	LOW	HIGH: EARLY	HIGH: EARLY	HIGH: LATE	HIGH: LATE
	long	*short*	*smooth*	*dip*	*dip*	*smooth*
LOW *double*	92	90	96	100	94	95
LOW *long*		93	99	99	96	97
LOW *short*			95	98	87	92
EARLY HIGH *smooth*				63	82	89
EARLY HIGH *dip*					67	81
LATE HIGH *dip*						57

Use of each sound is compared to all others and contingency coefficients calculated for each pairwise comparison. The 21 results have been normalized with respect to the maximum calculated $C = 0.76$ and appear in cells determined by the intersection of the named types.

similar distributions of occurrence and both are used most extensively in the categories at the end of the table (by animals most aroused and appearing least confident among the situations in which *coo* sounds are used). The index of 57 is the lowest in Table V, reflecting the similar distributions of occurrence of these two types. It is also true that these two types closely resemble each other in acoustic pattern, differing only in the presence or absence of a local minimum of pitch.

Looking at the other entries in the same set of results in Table V, this phenomenon is repeated. The low figures tend to occur between a pair where both acoustic pattern and situations of use have been judged to be similar. The high numbers tend to occur between pairs where there is both great acoustic differentiation and also where the situations in which each finds employment are towards the opposite extremes of the array of categories given as row headings in the contingency table (see Fig. 4).

This is the same phenomenon exhibited in the contingency table as the trend line of observations from corner to corner. It is a manifestation of a tendency for acoustic similarity in sound type to covary with similarity of demeanors and social and functional facets of the situations. The same degree of contingency or association of sound type with situation as in these results could be achieved without covariation. The data in that case could be the same as reported here, but could not be arranged by independently ordering the axes into a table exhibiting a regular trend. This hint of covariation indicates, then, that there may be one overall rule that relates sound form to function, a general link between patterns of sound and the vocalizing animal's behavior and demeanor, rather than a set of encoding and decoding rules separate and specific for each vocalization.

Not only does this covariation suggest the existence of such a rule, but also it suggests by the direction of the trend what the rule is. The chief acoustic character of these Class II *coos* is the tonal band that represents the fundamental frequency of phonation. This frequency rises as the ligamental tension in the region of the larynx is increased and with increase of transglottal pressure (Lieberman, 1967). These changes in tension and in the subglottal breath force are accompaniments of arousal such as can be observed in the crying of newborn human infants (Lieberman *et al.*, 1971).

The three types of low-pitched *coo* sounds are used in situations that have in common a low level of apprehension or agitation of the vocalizing monkeys. In contrast, the four high-pitched ones are used much more in situations where the animal is sensitive, agitated, and probably apprehensive. Those observed to reach the highest pitch are the *smooth late highs*—these are used by the most agitated animals.

The relation between the pitch characteristic of these *coo* sounds and their uses may then be a direct reflection of some aspect of the internal states of the monkeys. When the monkeys are in those situations in which they appear more aroused, there is postulated some corollary phenomenon in their internal states such that they produce sounds of higher pitch according to the dictates of phonatory physiology.

Given only the pitch of a *coo* sound, then, this rule allows one to predict at least the arousal tendency of the internal states of the vocalizing animal. Knowing this arousal level alone will permit a more accurate prediction of the one of the many "contact" circumstances in which the *coo* is uttered. This increase in predictive ability furthermore requires no specific knowledge of the demonstrated associations of the detailed shapes of *coo* types with their situations of use.

Vocal signals indicate only whatever aspects of internal state are reflected in sound morphology. A communication system without rules other than those inherent in the physiology of sound production and of arousal can still be quite complex and yield very specific information.

A limited range of behaviors and demeanors indicates each kind of internal state. Monkeys foraging do not show the extreme arousal of those threatening most intensely; those involved in attack are not calm. Since generalized internal states are related to demeanors and also govern acoustic morphology, then knowledge of a sound pattern implies by itself improved predictability of demeanor.

Furthermore, since demeanors segregate among the various social situations in a predictable way, then knowledge of a sound pattern also implies by itself certain limitations of the social circumstances in which it is given. A large repertoire of vocal signals can then be socially exploited, each pattern indicating its own nuance of internal state, without a complex lexicon. Classes of sound may indicate internal states associated with a large number of situations much as the *coo* sounds may indicate internal states associated with affinitive contact. Minor variations are free to reflect even more subtle differences as do the seven types of *coo*. Knowing only age, sex, and perhaps reproductive state of the vocalizing animal limits uncertainty still more. Added to this may be all the increased specificity afforded by the various pantomimed gestures, facial expressions, individual identities, and so on. Together the vocal patterns, visual information, and social heritage may combine into a communication system of great complexity.

C. Review of the Repertoire

Basic patterns other than simple tonals are recognizable in the repertoire. Ten groupings of such patterns are employed to organize the typological classification of the repertoire. Each grouping or class of sounds contains sound types that have patterns showing acoustic features in common or which are modifications of some shared features.

Even though the ten classes are derived strictly by examining spectrograms for common features, each includes those patterns that are judged by ear to be similar. The patterns within a class are relatively homogeneous to the ear when compared to patterns of other classes; this is not surprising in light of the design characteristics of the Kay sound spectrograph (Joos, 1948). For this reason it is possible to assign each

TABLE VI

SOUND CLASSES IN THE VOCAL REPERTOIRE OF JAPANESE MONKEYS

Class	Name	Some distinguishing features [a]
I	Girneys	Articulations alone or superimposed on voiced sounds
II	Coos	Simple, nonplosive, uniformly tonal
III	Whistles and warbles	Nonplosive, modified tonal and predominantly tonal; duration $>$ 0.18 sec
IV	Squawks and squeaks	Duration $\leqslant$ 0.18 sec, plosive tonal; rich harmonic structure; complex, predominantly tonal
V	Chirps and barks	Duration $\leqslant$ 0.18 sec or multiunit; plosive or noisy, predominantly tonal compound, some with clicks or dropouts
VI	Squeals and screeches	Duration $>$ 0.10 sec, may be multiunit; noisy, predominantly tonal compound or complex, predominantly tonal; dominant frequency above 2400 Hz
VII	Shrieks and screams	Duration $>$ 0.10 sec; simple or modified atonal, or predominantly atonal; dominant energy above base line (ca. 80 Hz) and below 2400 Hz
VIII	Whines	Complex, predominantly tonal; duration $>$ 0.18 sec; overtones more intense than fundamental; major energy $\leqslant$ 2400 Hz
IX	Geckers	Simple atonal, or complex, predominantly atonal, or plosive modified atonal; duration of units $\leqslant$ 0.10 sec
X	Growled sounds and roars	Pulsed and/or uniformly atonal, or complex, predominantly atonal with dominant energy at base line

[a] This list does not serve to classify sounds unambiguously; it indicates characters shared by most sounds within each class. Descriptive terminology from Table I.

class a mnemonic from the common names given to animal sounds, including where possible names used by other workers in describing spectrographically similar monkey vocalizations.

These names are assigned to the classes so as to be close to the audible impression usually associated with them by English-speaking people; they are not, of course, an acoustic definition or description of the patterns in the classes, but are less unwieldy for quick reference than a physically descriptive term. For easy reference names are also given to the types of sounds into which classes are divided.

The names of the sound classes and a brief phrase describing their salient acoustic features are listed in Table VI. A few names refer to only some of the patterns within classes which span a broad range to the ear. The "girney" mnemonic for Class I sounds refers, for example, to only two of the three major types of sounds found in that class; it does not describe the unvoiced articulations such as teeth-chattering or lip-smacking included in the same class.

The ordering of these classes from 1 to 10 is not arbitrary. It is the result of placing classes near each other if they have been observed to be uttered in unbroken transition, as will be discussed in Section IV, B.

Each type of sound described here, although acoustically defined, is not necessarily a minimum acoustic unit of a single continuous element of sound. Some utterances, such as of the "trill," a type of Class V sound, are composed of separate repeated elements.

The actual classification of the individual spectrograms is made on the basis of measured values of a number of variables reflecting spectral distribution of energy, the temporal properties of the sounds, both relative and absolute, and the presence or absence of various features affecting the morphology of the spectrographic pattern such as inflections and clicks. A taxonomic key was developed for classifying the spectrograms of *Macaca fuscata* vocalizations based on the acoustic features previously indicated in Tables I and II. This key also served to define the patterns of sounds (Green, 1972).

A small portion of the key used to sort the *coo* sounds of Class II was given in Fig. 2. The full range of uses of these *coos* has also been given, with analysis revealing that each *coo* pattern is used differently from the others. A similar analysis has also been performed on the types of sounds in the other classes. The declarations of predominant use made in the following section are based only on results statistically validated (Green, 1972). In addition to the typical uses indicated here, a summary table showing the range of uses of sounds of these classes accompanies the concluding results (Section IV).

A presentation of the types of sounds comprising nine of the ten classes of vocalizations is given in this section. For each type only the strongest associations with social usage are mentioned.

1. *Class I: Girneys*

The sounds of Class I are produced by articulation or by articulatory modulations of vocalizations. Three types of sounds comprise this class; examples of each are shown in Fig. 5.

The first type, *articulations* (Fig. 5a), includes lip-smacking, teeth-chattering, and other sounds produced when mouth or tongue are manipulated without any concomitant vocal process. The principal use is by males in the copulatory season toward subordinate females whom they are courting. They generally give *articulations* at the end of a stylized trot toward the female which is one of the preludes to consortship formation.

The *atonal girneys* (Fig. 5b) are sounds with predominantly atonal structures that are modified by a process such as lip movements or teeth-scrapings during the vocalization. The characteristic use of this type is by an adult female late in the birth season which does not have an infant and which is approaching a cluster of females in which infants are present.

Nulliparous females, those nonparous for the season, and those whose infants of the year were either stillborn or died, all gave *atonal girneys* in such situations. On occasion they uttered these sounds while orienting specifically toward an infant as it moved within or near a group. The vocalizing female often postures submissively, and she may eventually sidle in backward to join the grouping.

Tonal girneys (Fig. 5c) are the extremely variable, predominantly tonal sounds in which the vocalization is superimposed by an articulation such as tongue or lip movements. These are the characteristic vocalizations of an adult female approaching an unrelated dominant female that is not in a huddled cluster in the birth season. These approaches, also performed hesitantly and with submissive postures, usually result in a grooming relationship.

Before contact is established, *tonal* and *atonal girney* vocalizations may be interchanged extensively. Females in huddled face-to-face contact may engage in mutual rocking back and forth while holding each other and reciprocally girneying; girneys are the only types of sound that accompany this activity. The *tonal girneys* are the only sounds given by males that are mounting one another in situations described by Itani (1963) as a greeting.

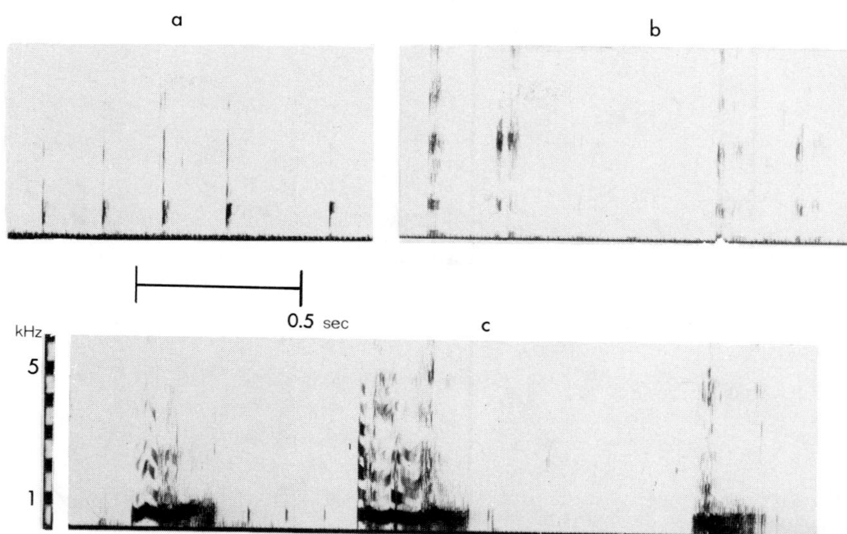

Fig. 5. Class I sounds produced by articulatory tongue and lip movements alone or modifying vocalizations. (a) *Articulation*—lip-smacks. The formant frequency of the mouth at ca. 1 kHz rises in each successive smack as the mouth configuration changes. (b) *Atonal girneys*—the vocalizations appear to be doublets where they are interrupted by the tongue flicking in and out of the mouth. (c) *Tonal girneys*—the sounds are modified by rapid tongue flickings and lip-smacks (seen separately between the first two vocalizations). The sounds are of labile morphology because a slightly new vocal tract configuration may be assumed after each articulation.

2. *Class II: Coos*

Class II is composed of the simple, uniform, and continuous tonal sounds whose typology and use was presented earlier.

3. *Class III: Whistles and Warbles*

These are the lengthy, predominantly tonal utterances that are not uniform and continuous. Examples of sounds of this class are given in Fig. 6. Included are sounds in which an initial simple tonal segment is compounded with a briefer atonal one or with some other kind of structure in which the tonal band becomes modulated or discontinuous. Also included are patterns of closely spaced units each of which might be a simple tonal sound. All the simple variations on the basic tonal theme of this class have been subdivided into seven structural types.

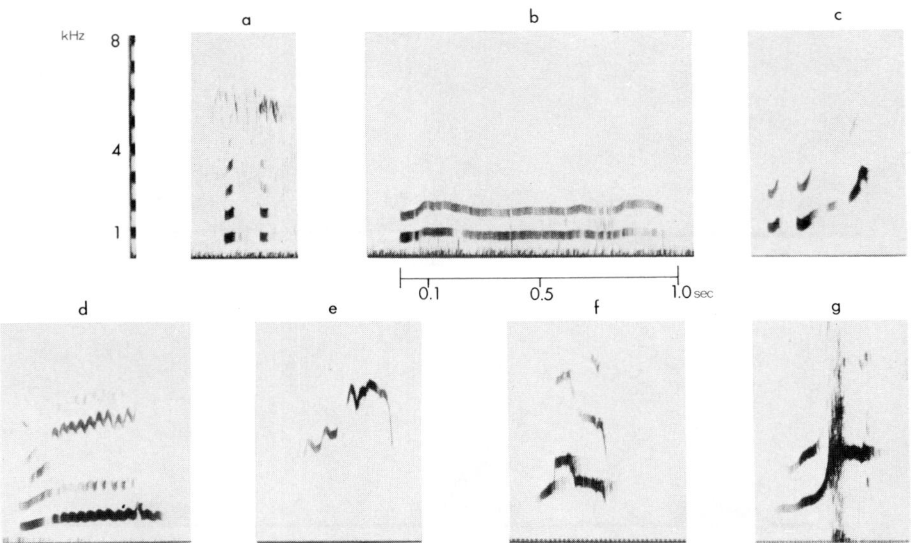

FIG. 6. Modified tonal sounds of Class III. (a) *Toots*—this example is a pair of tonal units of complementary shape. (b) *Stops*—long tonal sound with dropouts. (c) *Musical*—tonal units in sequence. (d) *Warbly*—frequency-modulated tonal sound. (e) *Whistly*—multiply inflected and ranging more than one octave. (f) *Oui*—sharp rise in pitch evidenced as a vertical discontinuity. (g) *Rise*—a burst of greater intensity accompanying a rise in pitch.

Toots (Fig. 6a) are two-unit sounds in which the units are either congruent in shape or time-reversed mirror images. Generally they are given by young subordinate individuals while approaching other youngsters or adults or while in affinitive social contact with them. Their principal use is highly specific to young animals in situations followed by play or grooming.

Stops (Fig. 6b) are used similarly by young animals at the last stages of approach to groom a dominant adult. They are also given frequently by young animals or others very subordinate to an approaching animal in situations that generally yield affinitive contact. These are long tonal sounds with a very brief cessation in the otherwise continuous tracing.

Multipart sounds with each discontinuous unit a simple or modified tonal band are grouped together as the *musical* type (Fig. 6c). These are overlapping in usage with the *stops*. They have additional prevalent usage by young animals toward their mothers. The usual circumstance is part of a prelude to a youngster being retrieved by its mother after it has been calling at her and as she approaches it.

Warbly sounds (Fig. 6d) show a temporally regular and small magnitude frequency modulation of the basic tonal structure. They are given by an aroused yearling or juvenile that has been pushed or shoved by its mother from a huddled group of females and young. These sounds are given both as the youngster is hurriedly backing off and then as it once again reapproaches the cluster.

Tonal bands showing irregularly timed, large pitch excursions, often more than an octave, are called *whistly* (Fig. 6e) sounds. They are given by young animals that have appeared aroused in a situation revolving around maternal contact. A youngster that has been lost, temporarily abandoned by a mother that has gone elsewhere, perhaps to forage or to mate, or that has been pushed away or otherwise rebuffed by its mother gives these as it is calming once again after regaining maternal contact. Both as the mother retrieves it and for a few moments afterward while it is in contact with her, usually held snuggled at her breast, it may emit these *whistly* sounds. The sounds sometimes continue after the youngster appears completely calm and is following along near her after such an event.

The *oui* type (Fig. 6f) show such an abrupt upward shift of pitch that the tracing is displaced vertically with an apparent discontinuity. A lost or abandoned yearling, after climbing to the end of a twig of shrubbery or to the edge of a rock precipice or to another locale with an unobstructed view, gives these while scanning visually. If its mother is in sight, it orients toward her while calling. Sometimes an older sibling, especially a juvenile sister, may join it, after which these calls may cease. If the mother does not respond, the vocalizations may become screeches; if she approaches the yearling, its vocalizations change to the *musical* type.

These *oui* sounds are also used within sexual solicitations by subordinate females toward and while pursuing potential copulatory partners and also toward other females with which nonreproductive, pseudocopulatory homosexual activity occurs.

The last type of Class III sound, *rise* (Fig. 6g), exhibits a dramatic increase in intensity accompanying a rise in pitch. This additional energy is evidenced either as a burst of noise crossing the tonal band or as a sudden enrichment of the overtone structure. Its use overlaps completely with the last type although its distribution of occurrences is statistically different.

4. Class IV: Squawks and Squeaks

Class IV sounds are short and contain a predominance of harmonically related energy in a rich series of overtone bands (Fig. 7). Many have plosive attacks, some have an overlay of noise, and in all of them the

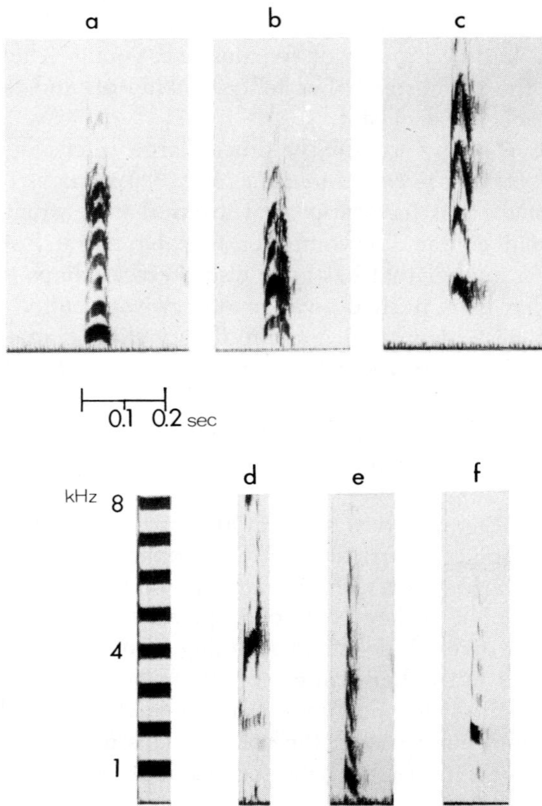

FIG. 7. Short predominantly tonal sounds of Class IV. (a) *Low squawk*—maximum frequency of fundamental band less than 750 Hz. (b) *Medium squawk*—maximum frequency of fundamental band between 750 Hz and 1250 Hz. (c) *High squawk*—maximum frequency of fundamental band above 1250 Hz. (d) *Squeak*—note high pitch, click at onset, and sharp slope. (e) *Ech*—noisy with monotonically negative slope; click at onset. (f) *Uh*—similar to *ech* without heavy noise overlay.

tonality is continuous from the attack to the end of the units. Six types within this class have been identified. The three *squawks* (Fig. 7a,b,c) are similar in appearance but differ in the pitch of the fundamental tone.

Low squawks (Fig. 7a) are mildly plosive, short sounds in which the low-pitched fundamental is intense. These are used mostly by adult females in sexual solicitations, both hetero- and homosexual. They are directed toward a dominant individual while following closely on its heels or as darting toward it then backing away. Often the utterances seem triggered by a sudden motion or other unpredicted activity of the solicited animal.

Medium squawks (Fig. 7b) are used by females in the same way as *low squawks*. In addition they are used by animals engaged in active wrestling and chasing play and by males that are soliciting females by grabbing them as part of what Tokuda (1961–1962) has termed a "check-attack."

High squawks (Fig. 7c) are given by both males and females with copulatory activity. Their use in this situation does not overlap with the uses of the other types of *squawks*. The major uses of *high squawks* are in nonreproductive situations termed defensive; in these an animal is supplanted or withdraws from the approach of a dominant or cringes at its approach.

Squeaks (Fig. 7d) include the tonal sounds of high pitch that are initiated plosively and have a sharp upward slope. The predominant use is by subordinate females closely following other females in the copulatory season and during agitated approach to and withdrawal from them after lengthy and so far unsuccessful homosexual solicitation. They are also used with this same approach and withdrawal behavior pattern in the analogous heterosexual solicitations, but are a more characteristic accompaniment of the female homosexual variety.

Ech sounds (Fig. 7e) are short, plosive, and noisy with a negative slope to the low point at their end. Of similar shape are the noise-free *uh* sounds (Fig. 7f). Both occur in such a wide variety of social circumstances that generalizations of typical usage are impossible. They are both characterized, however, by their employment at the point in time when a very transitory arousal of the vocalizing monkey decays into a more calm state. Included, for example, would be the moment when a struggling infant manages to grasp its mother's nipple or when an adult male culminates copulation. He ceases thrusting, ejaculates and, just prior to dismounting, *uhs*!

5. *Class V: Chirps and Barks*

This region of the repertoire includes those short sounds in which tonality is still the basic theme but atonal noise may be prominent. Figure 8 shows examples of the six types into which this class of irregular nonuniform sounds is divided. Most of the utterances of this class have a barklike quality, especially the very plosive or multiunit ones.

A *trill* (Fig. 8a) is composed of multiple, closely spaced units, each of which is a rich series of overtones to a computed fundamental which is

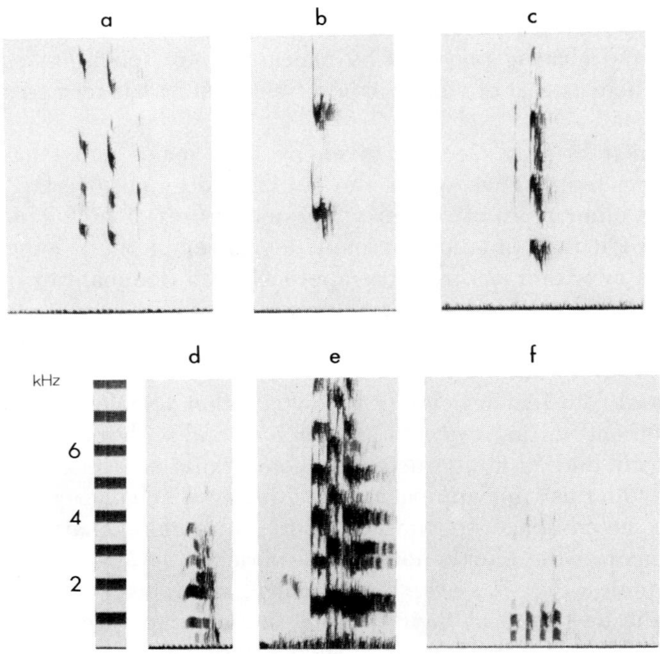

FIG. 8. Short non-uniform sounds of Class V. (a) *Trill*—multiunit bark; fundamental band at 1.2 kHz is not present. (b) *Chirp*—initial tonal structure is followed by atonal sound at the same frequency. (c) *Chirrup*—two-part bark of dissimilar units. (d) *Nweh*—harmonic structure, emphasizing even overtones, is less intense at beginning. (e) *Yap*—vertical lines separate tonal segments of slightly different frequencies. (f) *Chutter*—lapses in acoustic output separate tonal segments of slightly different frequencies.

absent or very weak. Use of these sounds is concentrated in weaning situations. Yearlings or juveniles give them seated a meter or so from their mother after having been rebuffed from attempted breast contact. Most of these sounds occur within bouts of utterances during a long sequence of attempted contacts and subsequent rejections.

The *chirp* sounds (Fig. 8b) are short high units with initial tonal banding followed by a noise burst or other marked irregularity in tonality. These sounds are uttered in situations of defensive withdrawal. As a subordinate animal is supplanted by the approach of a dominant, or after some movement by a dominant animal that precipitates the exit of a subordinate from a huddled grouping, the individual withdrawing gives a sound of this type and may continue it for a short while.

The *chirrup* (Fig. 8c) is a multiunit sound in which the successive units do not exhibit the same morphology, as do those of the *trill*. Each unit shows the kind of modifications of tonal structure that characterizes this region of the repertoire—a plosive attack preceding a rich harmonic structure overlaid by noise. Most of these sounds are two-part "barks" such as in Fig. 8c. They are used by an animal that is wincing and pulling its upper body away from a nearby dominant monkey who may have threatened it. They are also used by an animal that is cringing away from, and grimacing toward, an approaching dominant whose pathway brings it close.

Nweh (Fig. 8d) is a single-unit type of sound used under similar circumstances as the above *chirrup*. The *nweh* shows its harmonic enrichment after an unemphatic tonal initiation; the even overtones are enriched more than the odd ones. They are particularly common in one kind of situation in which cringing behavior is seen, that in which an unreceptive female remains seated and pulls her upper torso away from a male who is approaching or grabbing her in the manner of sexual solicitation. In unhabituated troops of Japanese monkeys, this cringing behavior is exhibited to an approaching human being. It may also be seen directed toward human beings in the habituated troops if the animals are approached in a portion of their range where humans seldom appear. *Nweh* sounds may accompany both of these situations.

Yaps (Fig. 8e) are distinguished by the occurrence of multiple clicks or vertical lines separating segments of tonality within a continuous tracing. This type is used as a component of approaching and threatening, then withdrawing, by subordinate females toward dominant monkeys or human beings. These sounds are given intermittently during both the threat and withdrawal phases of this defensive threat pattern of behavior.

Defensive threat is seen in a variety of circumstances, but only in two kinds of situations is it accompanied by *yaps*. Both involve a female evidently protecting a young monkey, not necessarily her progeny, from a potentially endangering event. The first situation is that of a mother that yaps and defensively threatens as she dashes to retrieve her infant or yearling that is in the path of an approaching adult, particularly an adult male, to which she is subordinate. This occurs when the young animal is near a commodity, such as food, toward which the dominant is approaching; it also happens if the youngster is anywhere on the route to a commodity or an event toward which the dominant's locomotion is oriented.

The second kind of situation occurs when a human being comes very close to a youngster away from its mother in a provisioned and habituated troop, or, in an unhabituated troop, comes anywhere near youngsters not being carried. The human being is greeted by an advancing then retreating adult female that *yaps* and defensively threatens; this behavior may become contagious and result in a mobbed attack on the intruder.

Chutter sounds (Fig. 8f) are similar to the preceding *yaps* but with momentary lapses in acoustic output, rather than the *yap's* clicks, as a disruptive acoustic feature. The *chutters* are given predominantly in the copulatory season by adult males and females in the vicinity of a male not usually associated with the troop. Such a male may be an adolescent or subadult from a rogue band, a solitary, or from a different troop. The females who give these may be in estrus, and it is possible they are soliciting the males. Males may *chutter* either on sight of a strange male or after a female's *chutter*; in either case a mobbed attack by males on the intruding animal may follow. In other seasons, *chutters* were used by animals orienting toward feral dogs; such dogs are monkey predators. Most instances of this sound were uttered in unknown circumstances, but they, and the ones appraisive of dogs, resulted in many nearby animals scattering into the nearest shrubbery or quickly ascending trees and looking at the vocalizing animal.

Other unrecorded "bark" sounds judged by ear to belong to this class were used in alarm responses to human beings, dogs, raptors, a snake, and nontroop monkeys.

6. *Class VI: Squeals and Screeches*

Squeals and screeches include a large variety of acoustic shapes (Fig. 9). All of them show some evidence of tonal banding, major energy of high pitch, and extensive noise and disruption of the tonality. Four kinds of sounds are common and distinctive in morphology, but intermediates are so frequent and intergradation so complete that attempts failed at typological separation into patterns with functional differences.

The first kind of pattern, a squawky screech (Fig. 9a), resembles a group of *squawk*like elements strung together into one long unit. The second kind, a squeal (Fig. 9b), looks very much like a Class III *whistly* sound with irregular modulations, but also includes internal clicks and sometimes heavy noise overlays. The third, a geckered screech (Fig. 9c), appears similar to a string of tightly spaced or continuous *echs, chirps,* or *trills* with very heavy noise overlays. The fourth, a bridged squeal (Fig. 9d), is a high pitched, noisy sound bridged by a narrow band. It is not clear from the spectrograms if this band is sharply formant-compressed or tonal energy.

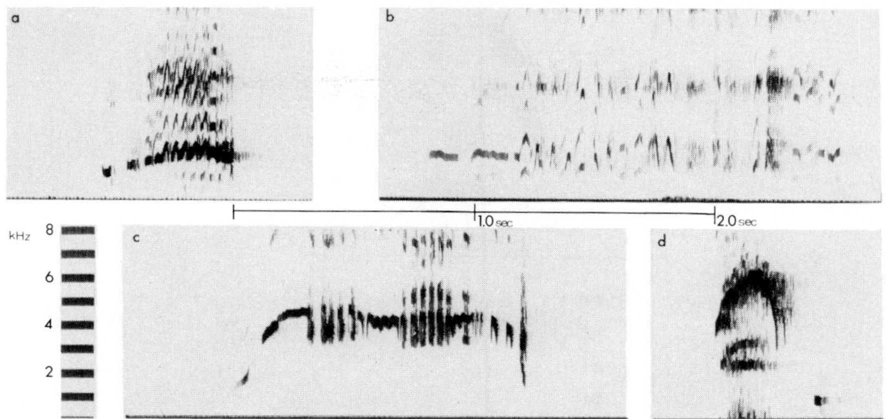

FIG. 9. Varied structure of Class VI squeals and screeches: (a) squawky screech; (b) squeal; (c) geckered screech; (d) bridged squeal.

These patterns are used in the behavioral extremes under circumstances in which the acoustically related types of other classes are heard. Their occurrence is strongly associated with the most advanced stages of unrequited estrus solicitation, with a continuing series of rebuffs from the breast by a youngster being weaned (weaning tantrum), and with infants or yearlings who have been calling without success as they are lost, abandoned, or merely away from their mother and suddenly frightened. Squeals and screeches are also commonly used by monkeys in a "stand-off," a kind of agonistic encounter in which animals face each other, spar, and periodically look back over their shoulder, a gesture which often recruits aid.

7. *Class VII: Shrieks and Screams*

Shrieks and screams are long and mostly atonal. They are composed of a wide band of noise with poorly demarcated regions of emphasis (Fig. 10). The shorter sounds within this class also tend to be lower pitched; they are audibly distinct from the longer harsher sounding and higher pitched ones although a complete continuum exists. The two ends of the continuum are not defined as separate types. The shorter lower ones are called screams (Fig. 10a) and the others shrieks (Fig. 10b). Both kinds are often used in a single bout of vocalizing in one situation with the lower ones beginning and the higher ones ending the bout.

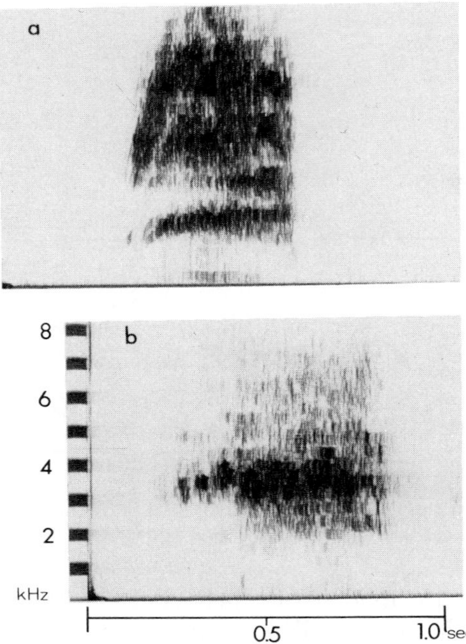

FIG. 10. Long atonal sounds of Class VII: (a) scream; (b) shriek. Lowest emphasized region is at higher frequency than in screams.

These sounds are given most often by a monkey that has lost an agonistic encounter and flees or is chased some distance from the site of the original threat or squabble. As it turns to face its antagonist, the sounds are directed toward this opponent by the loser that may stand bipedally to look over the heads of others located between them. Screams and shrieks are also given during flight, especially while a monkey is being chased by many others recruited into a widespread fracas.

8. *Class VIII: Whines*

The typical form of whining sounds is a structure of narrowly spaced harmonic resonances of a missing, low, warbled fundamental. This rich series is emphasized by a formant process into two or three broad regions of greater intensity within the wide bandwidth (Fig. 11).

Whines are used by monkeys that have been very aroused and agitated, particularly after a weaning tantrum, after being lost or abandoned, and after losing a fight. They are given as the animal once again comes into affinitive contact. This may be at its mother's breast, as it joins a con-

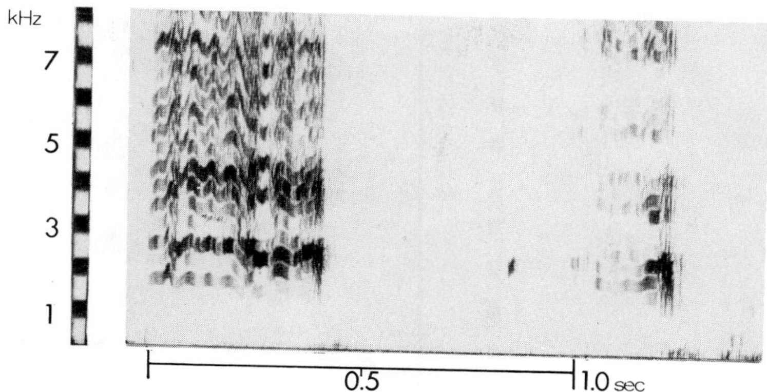

Fig. 11. Class VIII *whine.* The closely spaced series of resonances may indicate a nasalized vocalization.

gregated grooming party, or in whatever activity is appropriate to its typical social milieu.

9. *Class IX: Geckers*

Geckers are composed of units which are short, predominantly or exclusively atonal in structure, often pulsed, and usually plosive. The two types are differentiated by their occurrence in a series either with or without markedly accentuated inspirations between the exhalational units (Fig. 12).

Hacks (Fig. 12a) are those units given without marked sounds of inspiration. An infant gives them, usually singly, when suddenly manipulated by its mother that may be shifting it while grooming or adjusting its position at the nipple. An infant or occasionally a yearling may also *hack* when approaching but not yet in contact with its mother. In these cases a convulsive racking of the body may be seen in synchrony with these vocalizations. *Hacks* are also used substantially in consortships by a female just as she is pushed or positioned into standing for a mount or as she is mounted.

Cackles (Fig. 12b), utterances with intervening loud inspirations between the units, are typically given by females immediately following the ejaculatory mount of a copulatory series of mounts. They are also given by both males and females during the penultimate or antepenultimate mountings of a series, especially as the female is reaching backward and grasping the male's leg, thigh, or scrotum, or as she reaches upwards and tugs at his beard.

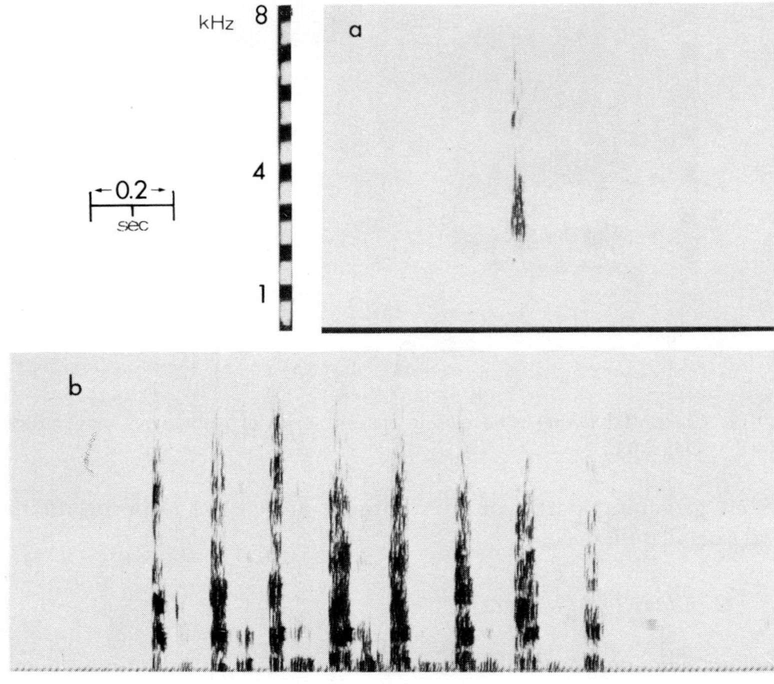

FIG. 12. Short atonal geckers of Class IX: (a) a single *hack;* (b) *cackles.* Lower-intensity sounds of inspiration are seen between the exhalational units.

10. Class X: Growled Sounds and Roars

The last class of sounds contains those units that are predominantly or completely atonal, are pulsed, and/or have their major energy at the pitch base line. As in the previous class, these units may be given in a bout structure with intervening inspiration sounds (Fig. 13).

The *gruff* type (Fig 13a) of Class X sounds show little base line energy; fine vertical striations or pulsing occur as well as a longer-period amplitude modulation which appears as bursts or pulses. To the ear *gruffs* resemble a pant which is trilled and voiced simultaneously. They are given with a great range of intensities that correspond unfailingly to the apparent vehemence of the accompanying facial expressions of threat. At the least intense extreme, these are barely audible pufflike sounds or very airy pants given by an animal with a mildly puckered mouth. With increasing intensity they become more gruntlike. The spectrographic

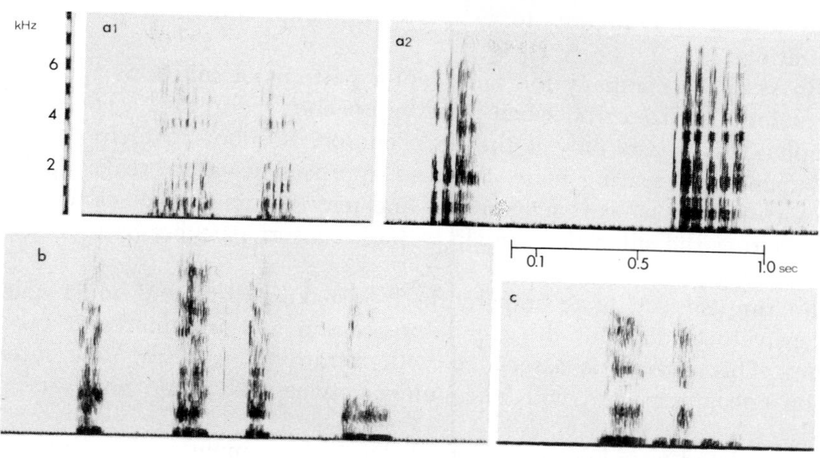

FIG. 13. Growled sounds and roars of Class X: (a1) *gruff* that sounds like an airy pant; (a2) *gruff* with a more grunt-like quality; (b) *growls* showing a continuation of the trend to more energy at the base line and less pronounced pulsations; (c) *roar*—three growl-like units of decreasing intensity and two intervening loud inspirations.

change includes loss of the microscopic pulsations and increasing emphasis of the base-line energy. The casual direct stare of the less intense versions is replaced by a direct beetle-browed stare. The mouth is opened more widely than the preceding slight pucker and the threatening animal may lunge rather than merely stare.

With still further increases in intensity and likelihood of attack, the *gruff's* fine pulsations diminish and the proportion of energy at the base-line is greater. At the point where the base-line energy is predominant, we have drawn an arbitrary line dividing them from the next type, the *growls* (Fig. 13b). Whereas *gruffs* are given during mild or moderate threats and within stand-offs, *growls* accompany the most intense threats. Dominant animals usually *growl* while they lunge and attack. Their opponents in these more serious encounters may be chased at full run and mauled if caught.

Units similar to the *growl*, but given in a series with juxtaposed and intervening sounds of inspiration, are called *roars* (Fig. 13c). They are occasionally used within agonistic threats by adult animals of either sex. Other uses of *roars* are exclusive to adult males. *Roars* are given most frequently in the "check-attack" sexual solicitations by a dominant male as he approaches, grabs, leaps over, shakes, or otherwise manhandles a female. They are also given, in addition to these preconsort solicitations, after a pair has already formed a consortship and as a prelude to mounting.

Roars also accompany the same motor patterns of solicitous approach by males, a stylized trot, when these approaches are toward a consorting couple. This occurs only if the male consort is subordinate to the approaching and roaring male. Such an approach usually breaks up the consort activity at least temporarily and may terminate the relationship since either the subordinate male or both consort partners precipitously withdraw.

Roaring is heard to accompany the tree-shaking display of adult males extensively throughout the copulatory season and irregularly at other times. This activity is associated with intratroop agonism, with intertroop encounters, and with encounters between the troop and solitary males.

It is related to unknown aspects of dominance among males. Only the higher-ranking adult males *roar* with the tree-shake, but tree-shakes without *roars* may be given by immature males, low-ranking males, or females, as well as by young animals in a play version. Occasionally a number of adult males may sequentially shake the same tree in an order corresponding to the dominance hierarchy, with a high-ranking male last—sometimes after running hundreds of meters to reach that particular tree even though others suitable and frequently used by him may be nearer at hand.

IV. RESULTS AND HYPOTHESES

A. Sound Preferences and Situations: Is There Concordance?

The strongest associations of sound patterns with usage were reported in the previous section. Table VII is a synopsis of ninety social circumstances in which utterances occurred indicating the classes of sound recorded on such occasions. Excluded are circumstances associated with provisioning and those in which sound production is concomitant to some asocial physiological process, e.g., coughing. All other vocal events of this study may be assigned, when the observation of the event is adequate, to one or more of the circumstances in the table.

A vocal event is assigned to a single circumstance by taking account of age and sex of interactants, kinship, dominance, their demeanors, gestures and facial expression, the history and context, and social spacing. Clues offered by the vocalizing animal may be used to determine the classification of ambiguous and multipotent events as was discussed in

Section III, A. Assignment is made to the circumstance that most specifically describes the event; for example, a young animal vocalizing to its mother is the appropriate circumstance of such an event rather than the more general, but also correct, circumstance of subordinate vocalizing at dominant.

If only a single sound class occurs within one circumstance, e.g., Class X sounds in tree-shaking displays, then a "+" is marked next to that circumstance in the proper column indicating the one sound class. For those circumstances in which sounds of two classes occur, the predominant one is scored with a double mark, "++," and the other with a single "+" mark. By similar extension, if more than two classes of sound are used in one circumstance, the class that identifies the preponderance of utterances is given the most marks, the one with those next most used is given one less, the third two less, etc., until that class that contains the least-used sounds is given a single "+."

It is clear from Table VII that in no circumstances is the full range of sound classes employed. It is also apparent that the preferences for sound patterns exhibited by the monkeys and noted by the marks differ depending on the circumstances.

In assessing any general agreement or concordance of preferences for sounds of different classes among monkeys in related circumstances, it must be realized that the degree of any such agreement depends on the particular conception of "related." Two ways of grouping circumstances for examining concordance are to be compared. Each way groups all ninety circumstances into ten broader situations, but they do so by different hypotheses of relatedness.

The first hypothesis states that monkeys vocalizing in functionally related contexts use similar sounds. This is the hypothesis that implicitly underlies many efforts at tabulating kinds of vocalization by usage. When only one kind of sound is heard to serve a particular function, the hypothesis is readily confirmed and the listing of a single sound for a given function is accurate. Since this kind of vocal stereotypy does not characterize use of the Japanese macaques' repertoire, the hypothesis will be tested in a more general form by taking account of the observed variability. The more general statement predicts that monkeys in circumstances related by functional context will show the same preferences for uttering the various sound classes.

To test the agreement of the results with this hypothesis, the circumstances in Table VII are divided into ten groups. Within each group the vocalizations are given in contexts related by social function. The assignment of each circumstance to one of these ten functionally defined situations is noted in the table by a letter.

TABLE VII

RELATIVE EXTENT OF USE OF SOUND CLASSES IN DIFFERENT SOCIAL CIRCUMSTANCES

Circumstances of vocalization and salient features distinguishing them	Sound class [a]										Assignment to two situations [b]	
	I	II	III	IV	V	VI	VII	VIII	IX	X		
Alone with respect to the dominant seasonal modes of interaction; not directing vocalizations to any nearby animals; not aroused												
1. Perinatal female, prepartum, or just postpartum and holding infant, not associating with kin	+										A	2
2. Solitary male, far from concentration of troop		+									A	2
3. Mature female without new infant (or whose infant has died) in the birth season, not gravid and not with family		+									A	2
4. Nonconsorting mature female, momentarily without a consort relationship in the copulatory season and neither actively soliciting nor being solicited; recently or subsequently active		+									A	2
5. Youngster sitting very calmly, looking around; not with mother, siblings, or playmates											A	2
Dispersed individuals, or subgroupings out of visual contact with main part of troop												
6. Feeding dispersal: individuals spread out as feeding or foraging, at least 2 meters between them		++	+								B	2
7. Troop movement: from an animal at one place along a progression to one at another, not part of the same band moving together		+									B	2

8. Separated group: from members of a foraging group to the main party or to another such group which is out of sight		+++	++		+	B	2
9. Joining or leaving separated foraging party or band in troop progression		++	+			B	2
Calm animals approaching or in contact with others, or acting in a fashion usually yielding calm contact; neither maternal, nor sexual nor agonistic contact							
10. Dominant approaching or with a subordinate		+				C	2
11. Female or young by itself, oriented toward an infant in a huddled family grouping with its mother or with its mother only	+++	++	+			C	1
12. Mature female to unrelated dominant female not in a group, in the birth season, directed from a distance with or prior to hesitant sidling approach	+++	++	+			C	1
13. Subordinate female joining family huddle of an unrelated dominant female; hesitantly approaching in short spurts, often backing in		+				C	1
14. Mature female without infant hesitantly joining group of unrelated females with infant in birth season	+++	+		++		C	1
15. Reciprocal vocal interchange arising from circumstance 12, 13, or 14, and initiated by a							

(Continued)

TABLE VII (Continued)

Circumstances of vocalization and salient features distinguishing them	Sound class [a]										Assignment to two situations [b]	
	I	II	III	IV	V	VI	VII	VIII	IX	X		
subordinate, or between females grooming or huddled	++	+									C	1
16. Other cases of subordinate approaching dominant, e.g., as prelude to play or grooming interactions			+								C	1
17. Subordinate approaching very dominant (i.e., great disparity in rank) as prelude to grooming, etc.			++					+			C	1
18. Subordinate as approached by very dominant; not showing defensive behavior	++++	+++	+++++	+				++			C	1
19. Male–male mounting not arising from agonistic encounter; mounted animal then mounts	+										C	1
Mother with respect to progeny												
20. With or near her young		+									D	2
21. Young is not visible; mother apparently looking for it		+									D	2
22. At dying infant or as carrying one which recently died; aroused and looking at it or around						+					D	5
23. After abandoning dead infant's body; aroused and scanning						+					D	5
Youngster with respect to mother												
24. To, with, near or following mother; calm		++	+								D	2

25. Infant or yearling as mother retrieves or approaches it after it was aroused and vocalizing at her	+							D	3
26. Infant as manipulated by mother, e.g., as she cleans it		++	+				+++	D	3
27. Infant or yearling alone, but mother in sight; aroused	++++		+	+++	++			D	3
28. Infant or yearling alone, scanning, but mother not in sight	+++			++	+			D	3
29. Calm infant or yearling suddenly aroused or frightened such as by close approach of human being; mother comes	+++			++	+			D	3
30. Weaning: juvenile has been repeatedly pushed away by mother and now reapproaches as it crouches and follows her; aroused			++	+++	+			D	5
31. Rebuffed: pushed or attacked–threatened by mother, backs off few centimeters; may begin reapproach	+	++	++++++	+++++	++++		+++	D	5
32. Agitatedly returning to mother after a fight with another monkey or in reaction to nearby squabble	++++++	++	+++++	++++	+++		+	D	5
33. Returning to or retrieved by mother after being lost, abandoned, or rebuffed	++++	+++	++			+++++	+	D	7

Agitated or aroused animals in affinitive contact;

(Continued)

TABLE VII (Continued)

Circumstances of vocalization and salient features distinguishing them	Sound class [a]										Assignment to two situations [b]	
	I	II	III	IV	V	VI	VII	VIII	IX	X		
neither sexual, nor agonistic, nor young to mother												
34. Active play, chasing, wrestling				+							C	5
35. Older female behaving as youngster in 30, 31, and 32 toward older sister or mother						+++	+		++		C	5
36. Youngster behaving as in 30, 31, and 32 toward an adult male						++	+	+++			C	7
37. Female, aroused after fight, joins unrelated one; demeanor as in 32						+		++			C	7
Animals terminating one kind of behavior and beginning another; dramatic change in demeanor is evident												
38. Locomoting from one situation to another, e.g., walking away from a terminated relationship				+							A	7
39. Relaxes tense, grimaced face at end of aroused vocal outburst and restores calm demeanor after fight, rebuff, unsuccessful sexual solicitation, etc.				+							A	7
Sexual behavior: copulations, the events preceding them, analogous behavior between females, and activities during mating consortships												

Males with respect to dominant females								
40. Subadult male mounting dominant female	+						F	2
Males with respect to subordinate females								
41. Trailing her from a distance, ca. 50 meters; very calm		+					F	2
42. Within consortship, sitting calmly with her or grooming between mounts		+					F	2
43. Soliciting by rapid stylized trot directly toward her	+++			+		++	F	9
44. Soliciting by grabbing or otherwise roughly handling her	+++		++	+		++++	F	9
45. Mount solicitation within copulatory sequence by pushing seated female forward from waist					+		F	8
46. Mounting and thrusting	++				+		F	8
47. Mounted by female			++		+		F	8
48. Ejaculating and dismounting or just afterward			+		+		F	8
Females with respect to solitary, subadult or subordinate males								
49. Following him from at least 10 meters				+			E	3
50. Calmly approaching then leaving vicinity				+			E	3
51. Engaged in sexual activity with another animal near such a male, whose approach or activity seems to trigger vocalization				+			E	3
52. Solicitation by close approach and pursuit				+			E	3

(Continued)

TABLE VII (Continued)

Circumstances of vocalization and salient features distinguishing them	Sound class [a]										Assignment to two situations [b]	
	I	II	III	IV	V	VI	VII	VIII	IX	X		
Female-initiated activity with respect to dominant partners, male or female												
53. Earliest stages of solicitation—long-distance following		++	+								E	3
54. During consortship, as closely pursues partner, or as male walks away between mounts		++	+						+++		E	3
55. Mildly agitated close following, or seated or lying nearby and triggered by activity of followed animal; still a solicitation		++++	++++++	+++++++	+++++	+++	++		+		E	5
56. Repeated agitated close approach and withdrawal; late in unsuccessful solicitation			+++++	++++++	+++	++++	++		+		E	5
57. Extremely agitated close approach to contact and then immediate fleeing and/or presentation posture; late in unsuccessful solicitation; extremely aroused			++	++++		+++	+++++		+		E	6
Female as being solicited or in sexual contact												
58. Retreats from male's stylized trot toward her					+						E	4
59. Dips shoulder, grimaces, and looks back over it as soliciting male approaches or grabs her; same manner as defensive (see 77)				+	++						E	4
60. Grabbed or pushed in solicitation, demeanor other than as 59				+							E	5

61. Activity prior to, and which is first sign of, the impending mount, e.g., as she gives a hand-slap gesture which often seems to signal copulatory readiness	+				E	5
62. Being mounted or pushed into copulatory posture, or as mounting		+++	++	+	E	5
63. Change from intermount lassitude to sudden alertness, apparently triggered by nearby activity		++	+		E	8
64. Reaching back over shoulder and grabbing beard of mounting male		++	+		E	8
65. Immediately after being dismounted by male		+	++		E	8
66. At the end of ejaculatory mount or immediately after it as animals part, or analogous end of pseudocopulatory sequence with another female		+	++		E	8
Female toward a consorting heterosexual couple						
67. Harassing: approaching and withdrawing similar to solicitation			+		E	8
Couple toward harassing female						
68. Activity triggered by harassing female: either partner behaves in threatlike fashion, but attack not carried through and often changes abruptly to copulatory activity with partner				+	E	9

(Continued)

TABLE VII (Continued)

Circumstances of vocalization and salient features distinguishing them	Sound class [a]										Assignment to two situations [b]	
	I	II	III	IV	V	VI	VII	VIII	IX	X		
Male toward a consorting heterosexual couple												
69. Runs or trots directly at them in fashion akin to solicitation of female; usually results in breakup of consort if interloper is dominant										+	F	9
Alarm: other animals scattering or mobbing, while vocalizing animal looks from a conspicuous vantage point at stimulus												
70. Man					+						G	4
71. Solitary male interloper					+						G	4
72. Monkey from another troop					++					+	G	4
73. Raptor					+					+	G	4
74. Snake					+						G	4
75. Dog					+						G	4
Agonism (other than maternal rebuffs, intertroop encounters, and consort breakup)												
76. Defensive withdrawal: nonagitated retreat from a threat, push, or approach of a dominant, or just afterward as normal activity resumes				+++++	++++++	+++	++		++++	+	H	4

77. Defensive subordination: wincing or cringing from, and grimacing toward, approaching or threatening dominant	++++	+++++	+++	+		++		H	4
78. Defensive threat: momentarily holding ground, stare-threatening and eye-flashing over shoulder in recruitment gesture before turning and fleeing; may be repeated in approach–withdrawal series		+++++	+++	++		+	++++	I	5
79. Two individuals facing each other, stare-threatening, employing flashback gestures which usually recruit aid, and swiping and grabbing at each other in a standoff			++	+		+++	++++	I	5
80. Fleeing, or being bitten, grabbed or mauled after being caught in a chase			++	+++		+		H	6
81. Fled: having run from encounter, now facing fight site or antagonist, still highly aroused			++	+++	+			I	6
82. Mild stare-threat: direct stare with head slightly thrust forward and jaw dropped							+	J	9
83. Stare with lunge: as in 82, with motion of upper body toward directee (moderate threat)							+	J	9
84. Lunge or dash toward standoff, chase, or full fight, which may be distant; demeanor as in full threat, also including bobbing or bounding on ground, but not directed toward nearby individual							+	G	9

(Continued)

TABLE VII (Continued)

Circumstances of vocalization and salient features distinguishing them	Sound class [a]										Assignment to two situations [b]	
	I	II	III	IV	V	VI	VII	VIII	IX	X		
85. Full threat: stare with jaw dropped, full piloerection, tail erectile, and dash toward other monkey with head lowered and thrust forward						+	++			+++	J	9
86. Chase: full run; pinning and biting monkey if catches it										+	J	9
87. Full fight: mauling, biting, ripping										+	J	9
Bounding on a branch or shaking a tree												
88. By chasing animal, during a chase										+	G	10
89. Triggered by a fight or a prior tree-shaking										+	G	10
90. Not triggered, or prelude not seen										+	G	10

[a] + indicates extent of use by monkeys in each circumstance relative to other sound classes; more marks indicate greater use; no marks indicate no recorded use.

[b] Letters indicate assignment to one of ten groupings of circumstances by similarity in functional attributes of context (see Table VIII); numbers indicate assignment to one of ten groupings of circumstances by arousal and demeanor similarities in vocalizing animals (see Table IX).

Table VIII lists these ten situations and notes how the circumstances comprising each are related by functional context. The concordance among the sound class preferences in each situation is given by Kendall's coefficient of concordance, *W*. Its probability of being at least this large by chance, i.e., *W*'s significance, and the consensus first preference are also indicated for each situation.

For this way of grouping circumstances, there are only two categories that show statistically significant concordance ($p < 0.05$) among the sound class preferences. In the first, dispersed animals call among each other with a pronounced preference for Class II vocalizations. In the second category, animals that are threatening show a distinct concordance in preferring to utter Class X sounds within all five of the agonistic circumstances; they use Class X sounds exclusively in four of them.

It is thus concluded that the first hypothesis successfully predicts significant concordance between sound preference and situation for only two situations defined by functional context, namely, agonistic threats and dispersed calling.

TABLE VIII

CONCORDANCE BETWEEN SITUATIONS AND SOUND CLASS PREFERENCES: SOCIAL FUNCTION-RELATED CONTEXT

Situation[a]	Circumstances grouped by functional context	Consensus[b]	W[c]	$p\approx$[d]
A	Asocial: monkeys lacking contact	II	0.111	0.55
B	Contact calling: feeding and troop movement	II	0.905	0.01
C	Affinitive contact (not mother–young or sexual)	I	0.095	0.25
D	Mother–young contact	III	0.110	0.20
E	Solicitation and copulation: females	IV	0.092	0.10
F	Courtship and copulation: males	I	0.042	0.90
G	Action provoked at a distance: alarm, male dominance display, etc.	V	0.010	0.75
H	Agonistic: defensive or submissive	V	0.383	0.25
I	Agonistic: ambivalent or recruiting aid	VI and X	0.219	0.70
J	Agonistic: threatening or aggressive	X	0.792	0.025

Overall concordance for circumstances grouped by functional context: $\overline{W} = 0.171$, $p < 0.05$

[a] Circumstances included are indicated by these letters in Table VII.

[b] Sound class with least sum of rank-ordered preferences.

[c] *W* is Kendall's coefficient of concordance.

[d] Approximate level of probability of obtaining by chance a *W* of at least the observed value; actual *p* is bounded by number in table and the next lowest from a series: 0.95, 0.90, 0.85, . . ., 0.05, 0.025, 0.01, 0.001, less.

The figure indicating an average concordance ($\overline{W} = 0.17$) is a significant measure of the agreement of the results to the hypothesis ($p < 0.05$). This number indicates that disregarding any lack of agreement on sound preferences among the ten situations, the overall concordance of sound preferences with social circumstances grouped by functional context is low.

Table VII also indicates into which group of ten the circumstances are assigned by the second hypothesis. This hypothesis predicts concordance between sound pattern preferences and those circumstances related by the conception of situations introduced in Sections III, A and B. This concept places into the same group all circumstances in which the demeanor, arousal, social spacing, and orientation of the vocalizing animals are similar. These ten polythetically defined situations are listed in Table IX.

The concordance of sound pattern preferences among the various circumstances is statistically significant ($p < 0.05$) for eight of these ten situations. It is thus concluded that this second hypothesis predicting concordance between sound class preferences and situation is verified in more situations assembled by this hypothesis than by the first. The overall concordance ($\overline{W} = 0.34$) is twice that of the previous hypothesis and is highly significant ($p < 0.001$). [There is no test for directly assessing the significance of a difference between two such concordance measures (Kozelka and Roberts, 1971). The maximum $\overline{W}$ achieved by a systematic partitioning of the 90 circumstances into 10 arbitrary groups was 0.50; the grouping with maximum $\overline{W}$ did not yield any sociobiological insights.]

Consideration of the overall concordance of preference rankings within these situations and of the indication that the strongest preferences vary depending on the nature of the situation, leads naturally to an investigation of the manner in which this variable preference for sound classes is associated with situations.

B. Sound Preferences and Situations: Association through Internal States

1. *Arraying the Repertoire in a Natural Order*

The vocalizations of Japanese monkeys include intermediate gradations among all the basic sound patterns in the repertoire. The sound patterns differ by the presence or absence of acoustic features, by qualitative

TABLE IX

CONCORDANCE BETWEEN SITUATIONS AND SOUND CLASS PREFERENCES: INTERNAL STATE-RELATED CONTEXT

Situation[a]	Circumstances grouped by demeanor and arousal	Consensus[b]	W[c]	$p \approx$[d]
1	Subordination: submissive gestures and hesitant approach in nonagonistic, nonsexual contact	I	0.285	0.05
2	Calm, confident: alone and not engaged in active interaction except for dispersed vocal exchange or close contact with subordinate or with mother	II	0.434	0.001
3	Contact uncertainty: calm or mildly aroused subordinate seeking contact initiation or restoration as part of maternal or sexual relationship	III	0.086	0.50
4	Alert avoidance: supplanted or wincing/cringing from approach of dominant or as appraisive of potential danger, e.g., looking at a dog from a vantage point	V	0.450	0.001
5	Aroused imploration or contact: labile behavior or ambivalence in demeanor, usually toward dominant, and often expressed as alternating approaches and withdrawals, e.g., weaning and late unrequited sexual solicitation	VI	0.144	0.05
6	Extreme imploration and avoidance: fleeing or having just fled from an attack or fight, or estrus female approaches solicited male and touches him, then flees; teeth-bared grimace is most intense, mouth widest open	VII	0.661	0.05
7	Sustained arousal: at end of lengthy aroused avoidance or lack of contact following weaning, separation from mother, fleeing from fight; as normal sociality resumes	VIII	0.203	0.45
8	Sensitive and attack-prone: between mounts or as copulating, animals are very agitated and sensitivity to disturbance is greatest; disturbance greeted by threat	IX	0.365	0.05
9	Confident agitation: threat or rapid direct frontal approach by dominant	X	0.424	0.001
10	Maximum agitation: as vigorously shaking tree or branch	X	1.000	0.01

Overall concordance for circumstances grouped by demeanor and arousal: $\overline{W} = 0.339$, $p < 0.001$

[a] Circumstances included are indicated by these numbers in Table VII.

[b] Sound class with least sum of rank-ordered preferences.

[c] W is Kendall's coefficient of concordance.

[d] Approximate level of probability of obtaining by chance a W of at least the observed value; actual p is bounded by number in table and the next lowest from a series: 0.95, 0.90, 0.85, . . ., 0.05, 0.025, 0.01, 0.001, less.

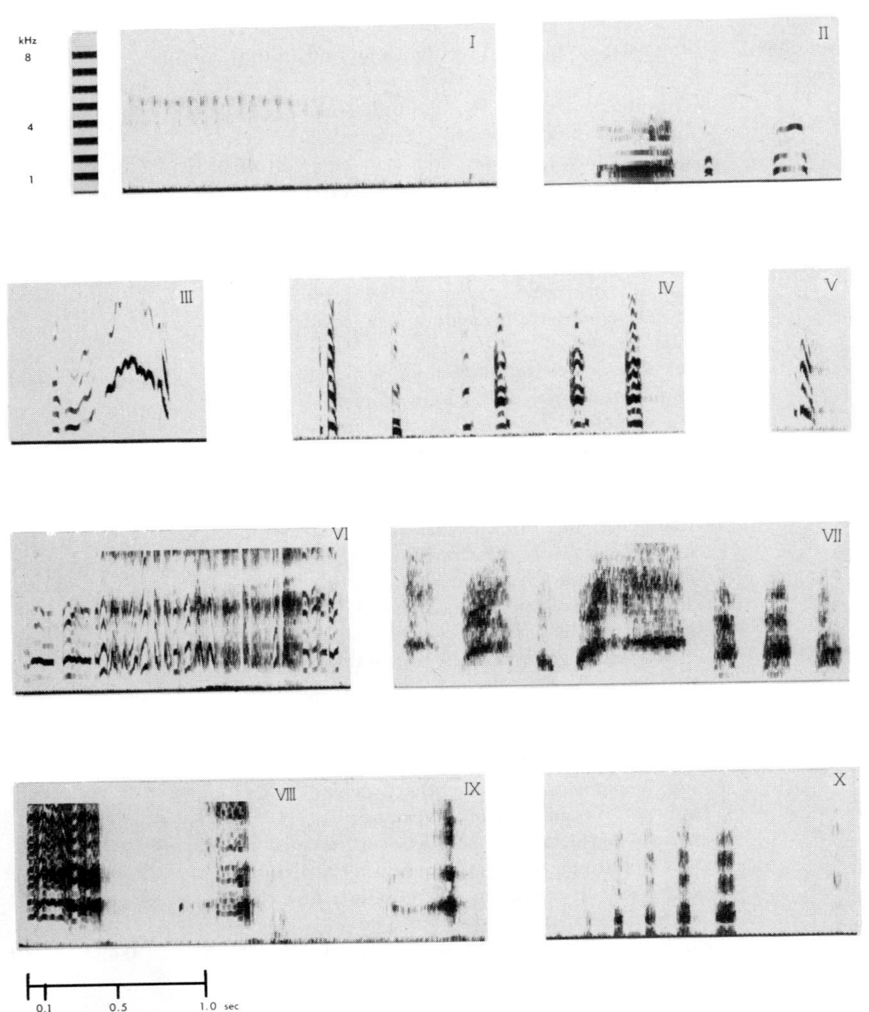

FIG. 14. Examples of each class of the repertoire. The occurrence of variation along many dimensions independent of others can be seen; namely, in frequency, duration, degree of tonality, and modulations.

factors, and along quantitatively measurable parameters. When the entire repertoire is considered, sounds vary in many dimensions (Fig. 14).

Arranging and rearranging sound spectrograms without reference to their order of utterance, like a taxonomist classifies specimens, does not reveal discrete divisions of the repertoire. Judgment of pattern similarities cannot be made reliably by eye. Such judgments are possible only for a limited portion of the repertoire representing minor variations on a stable basic pattern, e.g., the seven types of Class II *coo*. The repertoire as a whole does not divide naturally into a syllabary of dissimilar patterns on morphological bases. But considering instead the dynamics of vocalization does aid in identifying and ordering related classes of sounds.

a. Division into Classes. In Fig. 15A five sequences show transitions between juxtaposed vocal patterns of different morphologies. Each of these is given by a subordinate estrus female unsuccessfully soliciting sexual activity. The earlier solicitations have gone unheeded. Those which are solicitations of a male began tens or hundreds of meters from him, while looking out from a branch and softly *cooing*. Preceding these recorded sequences, the early, long-distance pursuit activities oriented toward the male, or the closer following of a homosexually solicited female, may have been going on for hours or days beforehand.

Then the soliciting females followed more closely and in their active pursuit no longer appeared unaroused as in the earlier states of calm and slow following. At the times these sequences were recorded, the females approach within a few meters and periodically dart toward and agitatedly scuttle away from the solicited monkey. Vocal bouts may be triggered by slight body movements of the solicited animal. The soliciting estrus female is sensitive not only to these motions, but also to nearby activity of other monkeys or of people. She may look about frequently, casting her head from side to side. Facial expressions with grimacing components are seen during the approach–withdrawal locomotion and are also triggered by the solicited animal's movements, just as the vocalizations may be.

Early in such sequences of solicitation behavior, estrus females' sounds are primarily of the simple tonal types, such as initiates the utterances in Fig. 15A (b, c, and d). Then, during more active pursuit, the sounds are principally modified tonals. Modifications include breaks in the continuity of the tracing as can be seen in Fig. 15A (a and c). Later the sounds are squeaks and squawks, the high, harmonically rich, noisy, and plosive sounds; these are given especially during approach and withdrawal [Fig. 15A (a and e)].

After the point in time when these sounds were recorded, if the females still remain unrequited, their behavior will progress to frequent and extremely agitated direct, close approaches accompanied by screech and scream vocalizations. Vocal outbursts are even more readily triggered in these visibly very aroused and sensitive females.

Points of transition between sound patterns are seen in these and similar vocal sequences, and in other circumstances of long duration, e.g., in a tantrum by an abandoned or weaned youngster. While the functional significance of these rare utterances showing juxtaposed, different pat-

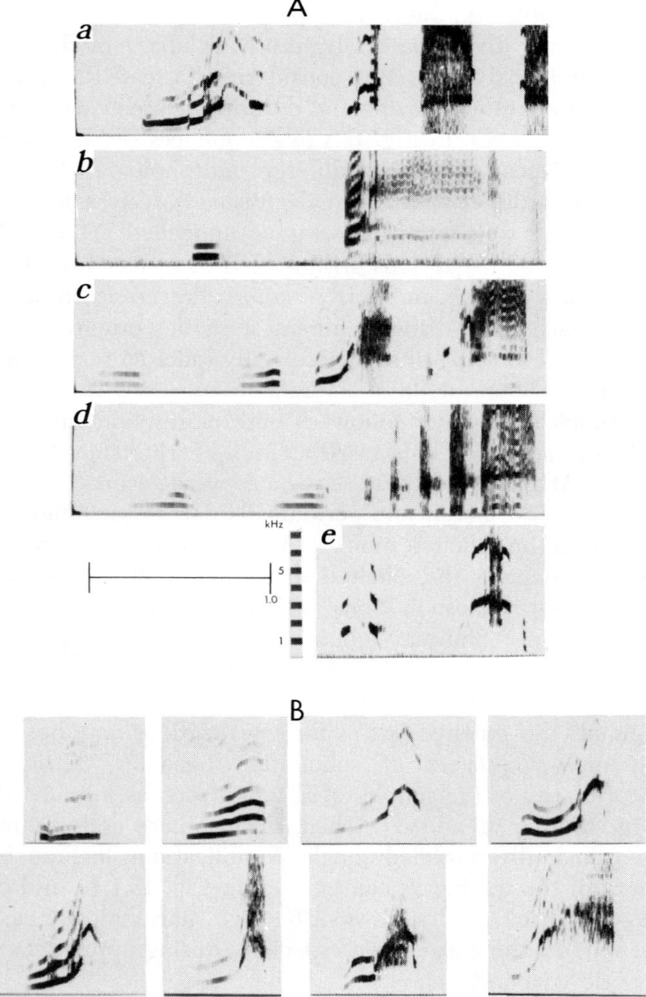

Fig. 15. (A) Five sequences showing dynamic transitions in vocal patterns as given by subordinate estrus females becoming more aroused during lengthy, unrequited sexual solicitation. (B) Sequence of vocalizations from different animals arranged to show a gradation in patterns for *dip late high* and *smooth late high,* through modifications of tonality classified as Class III *oui* and *rise,* and finally predominance of the atonal segment in a related Class VI compound sound. Reading left to right from the top, the first, second, fourth, fifth, and seventh utterances are by different adult females that engaged in progressively more agitated versions of sexual solicitation behavior. The third and sixth are by a yearling that is being weaned. The seventh is by a very aroused infant that has been ignored by its nearby mother.

terns is unknown, these points of transition from one acoustic shape to another mark a natural dividing line. It is here proposed that the dynamic changes in pattern can be used to divide a repertoire which shows morphological intergradation into "natural" units, i.e., those units demonstrable by the animals' usage (cf. Altmann, 1962, 1965). These divisions are natural not only in the sense that they demarcate different kinds of sound morphologies but also in that they illuminate the monkeys' phonatory division points separating production of different patterns. They can thus be employed in achieving a natural classification from an intergraded group of patterns without requiring *a priori* knowledge of uses.

Since the monkeys themselves, by their structures of vocalizations, indicate which features are distinctive in separating one morph from the next, these same features revealed in transitional utterances are taken to serve as boundaries between classes of sound. When the repertoire was divided into the classes noted in Section III, C, this division was accomplished by assembling into the same class those patterns sharing in presence and magnitude the distinctive features revealed by the monkeys in these vocal transitions.

b. Linking Patterns Together. The same dynamic considerations that helped demarcate natural divisions in the repertoire are also used to deduce their natural arrangement. If the same distinctive patterns employed to categorize the repertoire into classes are placed into an order, then this order can be applied to the entire repertoire. Such an order places as neighbors the classes related by occurrence of dynamic transitions between patterns comprising them.

After examining contiguous sound patterns included in transitional sequences of vocalization, the distinctive acoustic shapes were arrayed into a linear order. The patterns observed to neighbor in transitions are placed close together in this order. Those never observed juxtaposed are placed furthest apart. The exact order is determined by the same procedure by which a geneticist arrays hereditary units onto a map on the basis of occurrence of phenotypic linkages or a biochemist places amino acid units of an unknown polypeptide into an order on the basis of fragments showing different peptide linkages. In the case of these sound patterns, just as in some of the more complex cases encountered in genetics and biochemistry, a strictly linear array did not give the best arrangement, but is preserved here for the sake of lucidity. It is this array that serves to indicate the order into which the classes as a whole are placed.

Closely linked classes are defined as those containing patterns with distinctive features which are produced in temporal juxtaposition. Those classes containing sound patterns that are never produced in continuity are the least closely related. The order arrived at is now represented by the sound class numbers I through X.

The order which has been imposed on these sound classes has, in summary, been derived independently of the social or functional attributes of vocal events. It is based solely on linkages between neighboring sound patterns observed when an animal alters its vocal output from sounds of one type to those of another.

c. Inferences from Dynamic Linkage. The order of sequential patterns that occur in transitions hints strongly of a relation between variation of vocal pattern and alterations in internal state of the monkeys. Important to note is that the temporal patterning of vocal behavior on the gross scale, as sketched for estrus females over a period of hours or more, is reflected in Fig. 15A in a fraction of a second or a few seconds of vocalizations. The females vocalize progressively with patterns of sounds, first given alone and then together with those that will follow. At this in-between stage, transitions are seen between patterns showing the same morphology as sounds also given separately both earlier and later.

The female alters the acoustic pattern at a point in time when the intensity of her solicitous behavior is also changing. From calm and slow following she has progressed through increasingly agitated behaviors directed and oriented toward the solicited animal; she has also exhibited increasing sensitivity to its behavior. Later she will be extremely sensitive, appear very aroused, and engage in frequent and agitated locomotory movements toward and away from it. Females exhibit increasing arousal during this progression of unrequited solicitous behaviors. Those vocal patterns given earlier by less aroused females precede in the brief transitional sequences those given later by more aroused ones.

Those sounds produced by the females when they are more aroused can be accounted for by a more intensely excited vocal mechanism. Looking again at the transitional series of utterances in Fig. 15A, the pattern morphology changes first from simple tonals to sounds that show abrupt rises in pitch or vertical discontinuities indicating very rapid pitch shifts, and/or bursts of atonal noise. This change occurring in transition, and also representative of the longer-term changes, shows that the patterns from more aroused animals are those variations on the earlier pattern that may be expected from a more forcefully excited phonatory apparatus (as reviewed in Lieberman, 1967; see also Lieberman *et al.*, 1971). Other changes in sound morphology also occurring in these sequences follow this same pattern: the temporal order of changes reflects a sequence consonant with greater vocal excitation in successive patterns. There is a progressively increased harmonic enrichment, loss of pitch stability, and then aperiodic excitation of the fundamental phonation until the final relaxation at the end of an utterance.

This phenomenon of covariation of acoustic pattern with the internal state of the vocalizing animal was encountered earlier in examining the association of Class II *coo* sounds with situations. A static analysis of the variation of certain features of the sounds with small shifts in circumstances led to the same conclusion that the physiological states underlying social behavior are also directly reflected in sound morphology. The *smooth late high* sounds, which achieve a high pitch, are given later and at a more excited stage of solicitation than are the *dip late high* sounds. In fact, a continuous gradation of sound patterns can be arranged from *dip late high* to *smooth late high* to Class III *oui* and *rise* sounds (Fig. 15B). Such gradations occur actively in a dynamic transition as the individual becomes more aroused. In addition, utterances of these patterns from different individuals can be similarly explained: the *dip late high* sounds are given by animals engaged in the earliest and calmest following, the *smooth late high* sounds by those at later stages, and the Class III types in solicitations by fully aroused animals.

It is hypothesized that all sound patterns, not just those within utterances given in transition, are linked to some aspect of the internal state of the vocalizing animal and vary with changes in behavior. Vocalizations with acoustic features in common are predicted to occur from animals with similar internal states. Certain patterns are thereby expected to occur more frequently from aroused animals, that is, are "preferred" by them, and others by those which are calm. In general, we predict that sounds of those classes that are most strongly linked will be used in situations of similar internal state. This prediction of covariation of vocal pattern preference with internal state may be tested if first the situations are arranged into an order corresponding to some gradation of internal states.

2. *Ordering the Situations*

In Section IV, A the circumstances of vocalization were grouped into ten situations placing into the same group those circumstances in which the monkeys look as though they are behaving similarly and in which the immediate social environments are similar. Disregarded or unemphasized for this way of grouping the circumstances is the broad functional context which may be served by the observed behavior.

Since there are common features of observable behavior within the circumstances grouped into any one situation, one may postulate shared aspects of the internal states of monkeys in these circumstances. For each of the social situations delineated from this study in Table IX, it is likely the vocalizing animals have in common a similar underlying physiological substratum, resulting in similar demeanors.

Even without specifying the exact nature of this internal state, it is, nevertheless, possible to array the situations into an order based on gradations of its arousal component. Situations in which the monkeys are agitated (judged by locomotory activity, intensity and frequency of motor patterns, etc.) or highly aroused (judged by piloerection, defecation, and other autonomic indicators as well as sensitivity to disturbance) are conceived as different from those in which this is not the case. One may, in general, infer in which situations the monkeys have similar underlying internal states and place them closer together in a classification than situations contrasting in the salient features. Those situations indicating a calm state are placed far from the situations indicative of excitation of the vocalizing animals. Inferences of arousal based on relative dominance and context also influence this order. A monkey suddenly approached by a very dominant one is postulated to be in a dissimilar state from one that has been in an ongoing association with a dominant.

The order into which these situations are placed by such considerations is the order in which they are presented in Table IX. This order locates at the extremes those situations in which the behavior observed and the internal state which is inferred differ most from the "neutral" state of an animal quiet and alone. The first extreme deviates from neutral in a direction of animals appearing least aggressive and most hesitant in their approaches to others. It is in these circumstances that supplicating gestures such as sidled approach and lack of eye contact are employed, long time courses of approach are observed, the initiating animals are subordinate, and the context and responses indicate that the behavior generally yields or maintains nonagonistic contact.

At the opposite extreme are placed those circumstances in which the monkeys appear most aggressive and aroused. Their behaviors include rapid frontal approaches to others, stares and other facial expressions of threat, and/or adopting postures that maximize their visibility by piloerection, exaggerated gait, and bounding on limb or ground. At this same extreme are those circumstances in which the observed behavior is similar except for the lack of expressions of threat and the even more intense postural and gestural components that maximize visibility. At this extreme the monkey is least "self-effacing," announcing by vocal and visual display not only its presence but also its dominant status.

3. Association of Sound Class with Situation

It is now possible to test the hypothesis associating in a graded manner vocal patterns with social circumstances. This association was suggested by analysis of the Class II *coo* sounds, and then again by considering the dynamics of the rare transitional utterances and their covariation of morphology with arousal.

A contingency matrix with tallies reflecting preferences of the monkeys for the ten classes of sound patterns is shown in Table X. The sound classes are arrayed along one axis in the order derived by observing dynamic linkage between the patterns. The situations along the other axis are arrayed in the order already indicated which judges gradations of internal state of the vocalizing animals independent of the pattern of their utterances. The tallies are the combined "+" marks which were introduced in Table VII to indicate relative extent of use.

Many of the possible associations of sound class with situation do not occur at all and the cells are blank. In cells with tallies there is also an index of deviation from chance association. The index is computed as a ratio of observed preferences to those expected from marginal totals. The greater is this number above 1 for any cell, the stronger is the association between situation and sound class labeling its column and row.

One can discern a regular pattern of covariation although there is a great deal of variability as to which sound is preferred under what circumstance. The strongest associations are not scattered throughout the table in an irregular manner but rather fall roughly on a diagonal of the matrix. This diagonal trend reflects the predicted covariation of sound morphology (indicated by class) with internal state (indicated by situation) when each is ordered.

The trend toward orderly covariation hints that the most preferred classes of vocal patterns may be a function of the arousal level of the vocalizing animals. This may be tested by assigning two independently determined indices to each circumstance—one is the number representing the order of the situation in which it is included and the other is the sound class most extensively used. A regression analysis indicates the sound classes preferred in these circumstances can be accounted for by a monotonic dependence on arousal level to a degree highly unlikely to have arisen by chance ($t = 10.7$; $df = 88$; $p < 0.001$). The results, therefore, indicate the likelihood of a process suggested by the hypothesis: there exists a functional relation linking vocal morphology to arousal. They do not confirm arousal to be the principal determinant since it is only one aspect of internal state that can change in a regular fashion as the list of situations is traversed. It is concluded that one of the major determinants of vocal pattern in any given circumstance is that aspect of the internal state of the vocalizing animal reflected in its nonvocal behavior.

TABLE X: Association of Sound Class Preferences with Situations[a]

Situation of vocalization (increasing arousal[c] ↓)	Sound class[b] I	II	III	IV	V	VI	VII	VIII	IX	X (→ Later)
1	5.7	2.6	1.9	0.2				2.1	0.5	
2	1.1	7.7	1.5							0.5
3		1.0	2.4	0.4	1.0	1.3	1.0		1.4	
4				1.5	3.1	0.9	0.7		1.2	0.8
5	0.1	0.3	1.0	1.2	1.3	1.6	1.3		1.0	0.8
6			0.6	1.1		2.0	4.5	1.1	0.7	
7			1.2	1.4	0.6	0.9	0.4	11.2	0.4	
8	1.4			3.0					4.8	
9	3.1			0.5	0.5	0.3	0.8			7.0
10										12.7

[a] Tallies are scores indicating extent of use of each sound class. They are sums of tallies indicating relative extent of use of sounds in the circumstances constituting each situation (Table VII). Assignment of circumstances to each numbered situation is noted in Table VII. Numerical cell entries are the nonzero observed-to-expected ratios indexing the extent of deviation from chance association (=1.0) of preferences.

[b] The ten acoustically differentiated classes of the repertoire are arranged by the temporal order in which distinctive patterns of each class occur in transitions (see text).

[c] The situations are derived by arranging all ninety circumstances of vocalization into ten groups, each with similarities in typical demeanor and arousal of the vocalizing animals.

V. DISCUSSION

A. Pattern Variation

Rowell and Hinde (1962; Rowell, 1962) revealed that intergradation of sound patterns characterizes the vocal repertoire of rhesus monkeys. They arbitrarily divided the repertoire into acoustically different varieties and reported that each has its own characteristic uses. Variation in acoustic parameters was thus demonstrated to be communicatively significant for a repertoire composed of graded vocal signals. Left unresolved was the degree to which fine details of structural variation among sounds of one kind might be utilized socially.

The variability of one kind of intergraded vocal signal of the Japanese monkey, the *coo* sounds, was examined in Sections III, A and B. It was shown that a few details of structure suffice to divide this one variety of vocalization into a number of separate but continuously related types. The monkeys use each type of *coo* differently from the others, thereby revealing significance of pattern variation in one portion of the repertoire.

The repertoires of squirrel monkeys (Winter *et al.*, 1966) and vervet monkeys (Struhsaker, 1967) are characterized as discrete, that is they are composed of acoustically nonoverlapping classes of sounds. The social use of a repertoire of this kind is believed to rely on the acoustic and, hence, perceptual discontinuities between kinds of sounds rather than on variations of the sound patterns of any one kind (Altmann, 1967b; Marler, 1961, 1965; Moynihan, 1964). Winter's (1969) description of the variability within each of two readily recognized classes of squirrel monkey vocalizations does not ascribe any communicatory significance to the variations. Struhsaker's (1967) statistical analysis of some vervet calls also reveals variation of basic patterns but again without any suggested significance for social communication. Winter (1969) and Altmann (1967b) suggest that vocal signal variation within a discrete category provides the prerequisite raw material for evolutionary development. Marler (1974) suggests the degree of stereotypy observed within the discrete vocal categories of sympatric, African forest *Cercopithecus* monkeys may be a function of specific distinctiveness—a type of behavioral character displacement which he earlier (Marler, 1957) demonstrated in birds.

An additional possibility is that pattern variation within a discrete vocal category may be of communicatory significance. A discrete repertoire does not preclude acoustical variation within individual categories from systematically reflecting social distinctions. The *coo* sounds described earlier are used in circumstances which could be labeled "affinitive contact." It is only because the details of social behavior are so readily available for the Japanese monkey that it was possible to discover the nuances separating use of the *coo* types. Variations within a discrete category of a discontinuously organized repertoire could reflect subtle nuances of use, just as the variations in an intergraded repertoire demonstrated here, but this possibility has yet to be examined.

B. Repertoire Size

Some of the logical and procedural difficulties in establishing repertoire size have been reviewed by Altmann (1967b) who concludes that the appropriate behavioral elements are the natural rather than physically arbitrary ones. Natural units of classification are revealed by usage. Marler (1965) suggests that, to avoid circular reasoning, consideration of function be postponed until a typological or physical classification of signals is completed. These approaches coincide for a discretely organized repertoire in which discontinuities between patterns can be easily discerned.

Because the number of signal elements in a graded repertoire composed of a continuum of patterns is potentially without limit, one must inevitably turn to functional considerations when discussing how to divide the continuum and enumerate the categories. Struhsaker (1967) suggests employing for both discrete and graded repertoires a synthetic attack combining physical description of a signal with the circumstances in which it is given and any response it evokes. The difficulty still obtains, however, as to when to split and when to lump, such decisions becoming a function of the emphasis put on similarities in signal form, situation, or response. For repertoires in which natural demarcations and discontinuities in signal form are not evident, the approach of most investigators has been to treat a signal continuum as if it were discrete. By selecting in any arbitrary fashion examples of vocal signals which are far removed from each other in physical form, it has proved possible in many species with graded repertoires to construct a listing of easily characterizable sounds. The uses of each of these sounds abstracted from a continuum have been reported to differ in baboons (Bolwig, 1959; Hall and DeVore, 1965), macaques (Itani, 1963; Rowell, 1962; Rowell and Hinde, 1962), gorillas (Schaller, 1963; Fossey, 1972), chimpanzees (Marler, 1969a; Reynolds and Reynolds, 1965; Goodall, 1965), titi monkeys (Moynihan, 1966), and red colobus (Hill and Booth, 1957; Marler, 1970), for example. A lower limit is thereby set to the number of meaningful elements in a graded repertoire. It is this number that is presented, along with the number of discrete components of repertoires lacking intergradation, in summary reviews of species' repertoire size (e.g., Struhsaker, 1967).

The continuum of vocal patterns of the Japanese monkey was categorized in Section III, C into ten classes which are an exclusive and exhaustive division of the repertoire. Although the number of divisions is arbitrary, the demarcation points reflect phonatory distinctions that the monkeys themselves make, as described in Section IV, B. This categorization is thus similar to the usual analytic treatment of a graded repertoire by a discrete approximation but differs in that the entire continuum, not just nodes along it, is unambiguously classified with each pattern assigned to a class by explicit criteria. As suggested by Marler (1965), classification was accomplished without reference to functional significance, yet it is both natural and rigorous as Altmann (1968) urges.

Further subdivision of each class, also acoustically exclusive and exhaustive, was performed in the fashion exemplified in the discussion of one class, the *coo* sounds. The acoustic differentiation within each class is among sounds that are quite similar. It is, therefore, not surprising that the uses of the patterns defined by this process overlap. When examples of vocal patterns are cited from a graded repertoire, they are usually acoustically distant and their characteristic use is consistently distinct from other examples. Because all uses, not just the characteristic ones, were scored for this study and because close relatives from the continuum of sounds were demarcated for examination, statistical heterogeneity is necessarily employed here to determine differences in use. Only those patterns so far shown to be employed by Japanese monkeys in a fashion significantly different from all acoustic relatives are enumerated in this report. The thirty-seven types of sound described in Section III represent a lower limit of repertoire elements, each a portion of the vocal continuum which is both acoustically defined and differentially employed. There is no indication of the factors determining an upper limit. The number of types can be expected to increase substantially when the repertoire is analyzed in greater detail by performing finer acoustical divisions and relating these to more refined conceptions of behavioral circumstance.

C. The Meaning of Vocal Signals

1. *Experimental Study*

The most satisfactory determinations of information transfer rely on experimentally controlled conditions. Miller (1971) reviews some techniques explored in studying nonvocal communication in captive rhesus monkeys. No studies have been reported that experimentally examine primate vocal communication in the laboratory by measuring the responses of conspecifics to different signals.

In pilot experiments in Uganda, playback of intertroop roars triggered like calls from the leader adult male in a troop of *Colobus guereza* (unpublished data of the author with N. Acheson, A. Leskes, and K. Minkowski; intertroop roars of this species are described by Marler, 1969b, 1972). In attempting the same techniques with Japanese monkeys during this study, it was discovered that only *coo* sounds evoked a response other than approach to and interest in the playback apparatus. The monkeys normally use *coos* in marking their positions, and they antiphonally exchange them; similar calls were uttered in reply to the experimental *coos* as well. Human imitation of *coos* could also elicit a *coo* in response. Human renditions of a call pattern not reported in this paper, one exchanged between distant individuals in provisioning circumstances, also elicit calls in kind. Two other calls often responded to at a distance by animals out of sight, barks used in alarm and roars used in tree-shaking displays, were not experimentally investigated. Playback of other kinds of vocalizations, those used at close range, resulted only in nearby monkeys congregating around and investigating the playback equipment.

Both the appropriate vocal and also the unusual investigative responses are predicted by Marler's (1965, 1967) hypothesis differentiating the function of sounds usually received with as opposed to without corollary visual signals. For those vocal signals that function principally at long ranges and in intertroop use, when visual information is lacking, experimental playback can be expected to elicit an appropriate response. Both the black and white colobus and Japanese monkeys responded in their usual fashion to the experimentally produced audible signals which are usually generated out of their sight. For sounds used at close range, however, once the Japanese monkeys came within sight of the broadcast apparatus, their attention turned to scanning visually and otherwise searching for the "hidden monkey," including lifting and looking under the speaker apparatus. After the monkeys auditorially localized and approached the source of playback vocalizations, an unusual response, searching, was then elicited from them. This response appears to be governed more by the unusual context of vocalization occurring without a monkey visible than by the nature of the sound. Field techniques more sophisticated than simply a playback of recordings will be required to elucidate the function of the audible portion of composite signals for those parts of vocal repertoires employed principally in close-range intratroop communication.

2. *Inferences from Observational Study*

In both experimental and observational studies, the information utilized by the recipient of a signal can be gauged only by assessing response, i.e., the shift in behavioral probabilities which can be ascribed to exposure to the signal. Altmann (1965) constructed for the rhesus monkeys of Cayo

Santiago the first- through fourth-order transition probabilities between elements of the behavioral repertoire which he (Altmann, 1962) compiled. Because audible signals are but one component of the composite signal patterns that regulate social behavior at close range in many primates (Marler, 1965), Altmann's achievement does not allow ascribing any informative function to the vocalizations per se. In Japanese monkeys, composite signals are fully available to most communicants because their close-knit troops, living in relatively open habitats, have few signal barriers between members. Their vocal behavior is inextricably tied to simultaneous olfactory, tactile, and visual signals, hence consideration solely of evoked responses cannot disentangle the roles of the concurrent signals available by different sensory modalities.

Although there are difficulties in interpreting situations and responses in terms of any separate role of the audible portion of composite signals, the results of this study, nevertheless, allow examining vocal patterns for probable social significance. If all the environmental, physiological, and social factors surrounding each vocal event were listed, a baffling array of data would result. The inherent variability in biological processes could yield a limitless list with no discernible structure. What an observer does, of course, is to interpret data even as it is gathered by abstracting and grouping those variable factors that experience with the animals indicates are possibly relevant to the issue (cf. Schneirla, 1950). The splitting and lumping compromises, which must inevitably be faced, are discussed by Altmann (1967b) who also notices that independent observers of the same species often make the same choices. The search for contextual factors in common among different signal events is treated by Smith (1965, 1969) who emphasizes the utility of examining the referents of the signal rather than the responses to it. Those referents that are relatively consistent for a signal pattern he calls the "message" of the signal. The message is thereby an indicator of at least the identity, circumstances, and activity, ongoing or likely, of the signaller.

Although Smith's approach has been adopted in this report, many of the consistent referents of vocal events have not been made explicit. Some are implicit to all events, e.g. the signaller is a *Macaca fuscata*. Other referents, while consistently present for certain signals, do not serve to distinguish messages of similarly used signals from each other. The sounds serving in courtship and copulation, for example, all have as consistent referents, and therefore as part of their message, the season. Only those parts of the message that are necessary to distinguish use of one sound pattern from that of another similarly used one have been indicated in the results.

In many primate species, if the repertoire is discrete, or if the limits defining one pattern are a broad portion of a graded repertoire, then, as Carpenter (1942) notes, the signals "characteristically occur in definable situations and . . . stimulate a definable range of responses in associates." Listing these characteristic stimulus situations and functions alongside vocal signals has been accomplished in many field and laboratory primate studies. Each list abstracts invariant factors surrounding the typical or composite description of a vocal event. Conceptions of which factors are relevant differ, however, among investigators (see also Smith, 1968). As Andrew (1962) notes, "Systematic studies of the vocalizations of Primates (and indeed of mammals and birds in general) always have classified the different calls into such functional categories as 'warning call' or 'threat call.'" In many recent studies, however, the social parameters of the stimulus situations have been emphasized as invariant principles around which the repertoire is organized and presented (Bertrand, 1969; Carpenter, 1934; Chalmers, 1968; Furuya, 1961–1962; Hall and DeVore, 1965; Jolly, 1966; Moynihan, 1964, 1966; Poirier, 1970; Reynolds and Reynolds, 1965; Schaller, 1963; Struhsaker, 1967; Van Lawick-Goodall, 1968; Winter *et al.*, 1966). Most of these studies also note the social function believed served by the characteristic responses. Other studies emphasize the emotion or mood presumably communicated (Bolwig, 1959; Itani, 1963; Jay, 1965; Rowell, 1962; Rowell and Hinde, 1962).

Such lists are necessary prerequisites to understanding the significance of vocal signals, but, as Sebeok (1967) notes, for comparative work, "as in linguistics, not inventory but system is the base of typology, and to comprehend a system a mere listing of its components is insufficient." He urges adoption among primate workers of some consistent framework.

Andrew (1962) attempted to find such a framework for primate vocal signals and concluded that the form of vocalizations indicates the "contrast" in the stimulus situations. His search for a systematic relation was an attempt to account for the fact that, in his words, "In all species, the same call may occur in widely different situations within the same group or adjacent groups of situations."

By examining the subsidiary uses of vocalizations in addition to the principal or characteristic uses, Andrew was able to generate a hypothesis relating the common factors or messages to the form of vocalization. The conceptual importance of taking account of variability was also a precedent in Smith's work leading to his message hypothesis. The eastern kingbird gives a Kitter vocalization, he observed (Smith, 1963), "with a tendency to approach, irrespective of whether the approach is toward a perch, potential perch, mate, or occasionally, some other bird."

To discover a framework by which signals and their use are systematically related, as distinct from merely being part of the same list, is to uncover their linguistic *meaning* in Quine's terminology as employed by Bastian (1965). The optimal indication of meaning is to be found in that relationship that best accounts for variability in uses of signals as well as characteristic usage. A seminal attempt to examine variability in use of vocal signals of one species was reported by Winter *et al.* (1966) They estimated the degree of subsidiary usage of the elements in the squirrel monkey's repertoire but did not discern any systematic organization to the variable association of sounds and situations. The problem is long-standing in animal communication; Darwin (1872) also remarked that "The cause of widely different sounds being uttered under different emotions and sensations is a very obscure subject."

After listing the characteristic uses of vocal signals by Japanese monkeys in Section III, all of the messages and the variability in classes of vocalization associated with each were introduced in Table VII. The existence of systematic differences as to which sound occurs in what situation was taken as presumptive evidence of communicatory significance. Discovering meaning relies on identifying the organization underlying the observed variability. The concordance analysis undertaken in Section IV, A does not detect any unique meaning of vocal signals. It does indicate which of two hypotheses of meaning fit the data better.

Organization of circumstances by attributes revealing their sociofunctional message, as is traditional, was shown to characterize less adequately the meaning of classes of vocal signals than organizations by other attributes indicating demeanor and internal state messages (Tables VIII and IX). The relationship between sound class and this conception of situation better accounts for the variations, as well as the principal associations, of sound with circumstance. It thereby yields a set of social meanings distinct from the traditional ones. Uniform tonality, the distinctive feature of Class II, means, for example, a calm state and the desirability of affinitive contact whether maternal, sexual, or otherwise. Pulsed atonality, a distinctive feature of Class X sounds means, in this sense, dominance, excitation, and the likelihood of bounding or rapid frontal approach, whether in attack, tree-shake display, or courtship. The semantic use of such meanings, to yield new, yet predictable (decodable) meanings by blending or sequencing acoustic features, will be described in a future report.

D. Determinants of Signal Form

Itani (1963) noticed in the Japanese monkey that "sometimes a certain sound . . . gradually slides to another sound . . . This may be regarded as a phenomenon corresponding to a change in the psychological condition of the utterer." These transitions are not predicated on changes in the social circumstances eliciting them but rather on changes occurring within the vocalizing animal. The examples of transitional utterances discussed in Section IV, B were given as arousal was increasing; at other times transitions occur during decreasing arousal. As affinitive contact is restored following an excited parting, weaning-rejection or fight, for example, screams (Class VII) are dynamically linked with whines (Class VIII) just when the vocalizing animal is visibly calming. In general, transitions are heard when there is a change in the utterer's degree of arousal; this observation has been taken to point up the likelihood that arousal is one of the major components of the internal state determining signal form. Two independent lines of evidence, deriving from vocal physiology and from neuropsychology, are suggestive in this regard.

Most studies of the physiology of the vocal process in primates emanate from an interest in human speech sounds. The underlying principles which have been illuminated should apply equally well to all primates of related anatomy and physiology. The precise form of sounds produced by homologous processes will, of course, differ depending on the acoustic constraints related to each species' anatomy (Lieberman *et al.*, 1972). Most studies on human speech production are not easily related to monkey vocalizations, however, since human speech is concentrated in only one part of the total vocal repertoire of which human beings are capable. Some nonspeech vocalizations of man are similar to some monkey vocalizations and hint of the relation of acoustic form to arousal. When a human infant is aroused and cries, its subglottal air pressure exceeds a critical limit beyond which the medial compression of the vocal folds is not maintained and aperiodic noise results (Lieberman *et al.*, 1971). Noisiness thereby indicates higher arousal than does pure tonality of structure.

There is no theory yet to account for what acoustic features pertain to which aspect of internal states. None of the pertinent studies cited in Lieberman's (1967) review restrict themselves to examining a single aspect of emotion, such as arousal. Hence, Lieberman can only conclude, "There probably are some general aspects of intonation and stress that relate to particular emotional states." In the same volume, he details how perturbations of air pressure and of tension regulation in the laryngeal musculature are determinants of vocal patterns. Some findings have been mentioned above, e.g., a rise in tonal frequency is associated with increasing tension. It is yet known only in a very general way how breathing patterns and muscular tension are correlated with arousal. Hence, the rules relating sound pattern to arousal are not yet clear, but, so far, degree of atonality and frequency of tonal bands are among the features clearly associated with this aspect of internal state. These relations help explain the observations that only calm monkeys give low-pitched tonal sounds and only excited ones utter screeches, screams, or roars.

Arousal is but one of the components of "emotional" behavior that can be elicited by experimental neurological interference. A dominant role for the limbic system (Papez, 1937) has been implicated in most studies, particularly since Klüver and Bucy's (1937) classic experiments involving surgical lesions on monkeys. Surgical intervention has also demonstrated that the limbic system is directly involved in the behaviors of social interaction (see Pribram, 1962). Direct electrical stimulation of this same system and the hypothalamus can yield integrated emotional behavior patterns including demeanors and arousal characteristic of naturally occurring states (see synopsis in Thompson, 1967). Recently, it has been shown that electrical stimulation within these areas can also elicit vocalizations that are part of the normal repertoire in squirrel monkeys (Jürgens and Ploog, 1970; Jürgens *et al.*, 1967) and gibbons (Apfelbach, 1972). Since each different arousal level and demeanor represents a different state of neural activation, then any simultaneous vocal behavior is also expected to reflect the same state. Thus, the observed changes in demeanor and arousal in transitional sequences are accompanied by changes in the pattern of vocalizations. The same neuroanatomical structures functioning in emotional behavior are suggested to yield in general covarying changes in vocal activity depending on arousal.

E. Communication

1. Responses

The idea that animal vocalizations are linked to and determined by the emotional substratum of the emitter may be traced back to antiquity (see Plato's *Kratylos*). In modern times, Darwin (1872) emphasized that vocal utterances express emotions and, furthermore, suggested the manner in which vocal form may be tied to internal states, e.g., "Rage leads to the violent exertion of all the muscles, including those of the voice" (Darwin, 1872, Chapter IV). Some relationship of this kind has been assumed by many linguists attempting to trace the origins of human speech. de Laguna (1927), for example, states that animal cries are "an expressive movement made under the influence of the emotional state." The correlation between vocal sounds and internal states has been remarked upon by many investigators of primates. Rowell and Hinde (1962) introduce their study of rhesus monkey vocalizations by noting that "The noises made by monkeys express their mood, and are effective in communicating it to others." Itani (1963) states that for his characterization of Japanese monkey vocalizations, "The classification of . . . the vocal sounds may probably not [be] so far from a phonetic classification" although it is actually a "fundamental classification by the emotional overtones of the sounds."

If vocalizations encode internal states and thereby signal affect, they must be decoded by a listener in order to serve a social function. Decoding could occur via a learned lexicon. A monkey who fails to withdraw from a dominant's *growl*, a vocal threat, will be attacked and is more likely next time to decode a *growl* as indicating aggressive mood. The extent to which responses to vocal signals, i.e., a decoding lexicon, may be learned awaits experimental investigation. Learning need not be postulated, however, to account for consistent decoding. Since the monkeys possess an internalized set of physiological rules for encoding emotional states into sounds, these same rules of sound production can suffice for decoding sounds back to the states they represent. The listener may perceptually recreate the motivational state of the emitter by an unconscious reconstruction of the state of the phonatory apparatus which produced the vocalization. A similar view, that knowledge of an encoding process helps perceptual decoding, is embodied in a recent psycholinguistic hypothesis called the "motor theory of speech perception" (Liberman *et al.*, 1967). Darwin presaged this notion of using our physiological lexical rules to decode affective content of sounds. He (Darwin, 1872) quotes a Mr. Litchfield on the expressive content of vocal utterances, e.g., "Indeed it is obvious that whenever we feel the 'expression' of a song to be due to its quickness or slowness of movement —to smoothness of flow, loudness of utterance, and so on—we are, in fact, interpreting the muscular actions which produce sound, in the same way in which we interpret muscular action generally."

Although an analysis based on situations surrounding the emitter was selected for this study, responses also covary in parallel with the class of vocalization. Within the paradigm of aggregation and dispersal (Marler, 1968), typical responses suggest that the monkeys decode the vocalizations to yield a meaning similar to the ones reflected in our analysis. Sounds at the predominantly tonal end of the repertoire generally indicate desirability of affinitive contact; these are responded to by aggregation. Those at the atonal and noisy extreme indicate an excited and aggressive state; they are followed by precipitous withdrawal, i.e., dispersal. Intermediate sound classes have a more variable relationship with responses, the vocalizations perhaps reflecting a greater lability or uncertainty in internal states. These sounds are used in situations in which the animal's demeanors are most variable, and the widest range of sounds are employed at these times. Tendencies both to approach and withdraw, or to attack and flee, for example, may be manifest within a few moments of each other. There is a corresponding uncertainty in the responses elicited, such as fleeing or counterattacking, mobbing or scattering.

Although communication of affect may predispose the respondent to a certain range of responses, especially as regards aggregation and dispersal tendencies, the exact form is variable. The particulars of the context in which vocalizations are uttered affect the fine structure of response. A mother responds to a *coo* not by consorting with her youngster but by holding it; when a female uses that same *coo* in soliciting a male, he does not attempt to suckle her but rather consorts with her. Both responses are affinitive as is appropriate to the sound class and the internal state thereby denoted. They differ in detail as a function of context. In considering a total communication system, Smith (1968) uses *meaning* to characterize the consistently evoked response to a composite signal in light of its context. In this sense, there is an extremely precise meaning ascribable to the utterance of a *coo* sound by a youngster to its mother. One component of meaning, the animal's affinitive state, is contributed by the vocal pattern while the remainder is contributed by other information available to the mother, e.g., identity of the vocalizer as her progeny and its orientation toward her.

2. *Grading*

Vocal repertoires composed of intergraded signals can be expected among those primates with ready visual access to each other (Marler, 1965). Intergradation is more frequent in those species least dependent on audible signals for their social regulation (Moynihan, 1964, 1966). Conversely, when limitations are imposed on the visual mode, or sounds alone serve to mediate long-range interactions, discontinuous organization of the vocal repertoire occurs, e.g., in forest-living *Cercopithecus* monkeys (Marler, 1974). Discrete signals are believed to indicate an evolutionary premium on lack of ambiguity (Altmann, 1967b).

Similar principles can be seen operating in human language. Man sometimes exploits a rich multimodal system allowing nuances but at other times detail is sacrificed for intelligibility. The drum languages used by some West African societies, for example, employ relatively discrete signals. The information they convey is impoverished compared with spoken language, but not because they use a different code. The drummings are iconic representations of speech, reflecting many of its temporal and tonal properties (Stern, 1957). The signals are intentionally selected for their lack of ambiguity to the receiver. Richness of information and the number and kind of messages are necessarily sacrificed. It is readily acknowledged within these societies that when detailed information need be conveyed, a man is sent rather than a message drummed. The same kind of tradeoff is seen in the radio argot used by pilots or astronauts in those communications where precision is valued more than subtlety.

We have noted that exploitation of signal gradedness was predicted to characterize vocal communication by primates whose social organization and habitat allow them redundant information, particularly by seeing each other. Many studies reporting vocalizations have focused on the semiterrestrial monkeys which live in large and socially complex troops. Among the macaques, for example, the Japanese monkey (Itani, 1963), rhesus (Rowell and Hinde, 1962), pigtail (Grimm, 1967), and stumptail (Bertrand, 1969) are reported to have vocal repertoires in which the signals intergrade. The results and discussion in this report suggest some ways in which variability of vocal signals is utilized by the Japanese monkey as part of a complex communication system.

The subsidiary as well as the principal sounds uttered in each circumstance were analyzed for any concordance in the monkeys' preferences. A greater concordance was revealed by taking account of the vocalizing monkeys' arousal tendencies rather than by purely functional considerations. Broad classes of sounds within the intergraded repertoire reflect and may be determined by the same aspects of internal state which are manifest as arousal or emotional behavior. All vocal sounds are necessarily iconic representations of the state of the mechanism producing them. If the configuration and dynamics of the mechanism have strict neuropsychological determinants, then vocalizations are direct representations of the motivational status of the vocalizing animal. Decoding can make use of the same physiological relationships that govern sound production; no complex code nor any elaborate learning therefore need be postulated for the monkeys to decipher appropriately from a limitless variety of signals the neuropsychological state of the emitters. Vocalizations that make members of a group aware of the emotional status of others must help regulate social interactions. Relatively unspecific affect may be usefully reflected in vocal signals, but there is also the potential for variants to reflect detailed nuances. In the Japanese monkey, subtle aspects of motivational state may be communicated by small differences in sounds. The minor acoustical pattern variations within the broad category of tonal sounds were shown to covary with small differences in situations; responses to these sounds are generally similar in that they are affinitive. Respondents may modify their behavior according to both the details of vocal pattern and also the rich contextual information available to them. Signal gradedness may in general be exploited in this fashion within a composite communication system: the vocalization contains by its gross distinctive features an indication of unspecific affect, the fine structure reveals more detail of state, and context allows appropriate interpretation and fine-tuning the response.

VI. SPECULATIONS: ARTICULATED SOUNDS AND THE ANTECEDENTS OF SPEECH

Characteristic of human speech is the use of a variety of sounds in a labile sequence which may evoke extremely variable responses. In addition to articulated and voiced elements, however, humans also employ roars, cries, shrieks, screams, screeches, and a variety of other sounds. These sounds are not only acoustically homologous with those described here for the Japanese monkey, but they are also used in analogous situations by primates with similar inferred internal states. Roars are used by enraged people, cries by babies abandoned or otherwise distressed, screeches in tantrums of youngsters, and whines as they reach the comfort of a mother's embrace. Tonal sounds, including yodels and whistles, are used in hailing to make audible contact and to command attention to a location, especially as a prelude to initiating locomotion. Shrieks and screeches are employed in stressful situations, and screams are used by the victims of aggression. Characteristic responses to these sounds are far less variable than those to speech sounds: whistles evoke counter-whistles, cries of a baby evoke the approach of the mother, screams may yield both aggregation and avoidance, and the screeches uttered in a tantrum by a youngster are often greeted by a sharp rebuff from the mother.

In addition to the phylogentic homology, there are also ontogenetic parallels between the full repertoire of Japanese monkeys and development of human vocal repertoires. The sounds of human infants resemble sounds of nonhuman primates more than they do the speech sounds of adult human beings (Lieberman *et al.*, 1972). With maturation and perhaps the development of refinements in internal state, the infant repertoire is at first augmented by the babbling of articulated sounds and then by the speech sounds themselves. Babblings are thought to be early manifestations of speech sounds occupying the calm or playful young human who, nevertheless, relies on inarticulate sounds for communication. Later these articulate sounds are employed in speech and come to predominate in the vocal output. Speech is then an elaboration of only one portion of the vocal repertoire of human beings; it assumes its paramount communicatory role as control of its fine structure becomes ontogenetically refined and as the states represented by the inarticulate remainder of the repertoire become socially less significant with maturation, e.g., diaper distress.

It is intriguing to note that the communicatory significance of the full range of human sounds does not necessarily diminish with the emergence of articulatory speech. In addition to the occasional employment by adults of inarticulate noises such as screams, the acoustic features that define these sounds persist into adulthood as the paralinguistic phenomena known as prosodic aspects of speech. These may effectively communicate "mood" in conjunction with and sometimes independent of the linguistic content of speech. Speech may be roared in anger, screamed in anguish, or screeched in emotional extremity. These prosodic aspects of speech are also used in situations analogous to those evoking acoustically similar, but inarticulate, sounds in the monkeys.

The nonspeech portion of the human vocal repertoire not only has an acoustic homology in macaques, but is also used in a fashion that parallels the use of the related vocal patterns of the Japanese monkeys. It is, therefore, tempting to examine one portion of the Japanese monkeys' repertoire, the articulated sounds, and to speculate on them. Is it possible they are related to man's articulated sounds by use in analogous situations as well as in production and acoustics?

Both girney sounds and human speech sounds combine voiced and articulated components. Girneys are morphologically the most variable sounds in the monkeys' repertoire; they are also utttered with the greatest temporal lability, different pattern variants being given in rapid succession. Although they are uttered within the general framework of establishment or maintenance of affinitive contact, this can occur in a plethora of circumstances. Girneys are evoked chiefly in conjunction with the peculiar attraction to an infant of females without a youngster, between females in clusters, or between female grooming dyads. Their principal use is by subordinate animals in hesitant initiation of contact and while adopting postures and gestures of submission. This social use suggests that they may be employed to signal nonaggressive intent to the dominant animal who is approached.

The girneys are quiet sounds, audible for a few meters at most. Although they are used chiefly at very close range, they are also interchanged between potential grooming partners before contact is established. The subordinate female, uttering girneys, initiates the encounter, and the dominant may begin to reciprocate vocally. As the distance between them is then diminished by the subordinate, the utterances from both increase in rate and degree of temporal overlap. After affinitive or "friendly" proximity is established, the two monkeys may hold and face each other while rocking back and forth, girneying, then settling into huddled drowsiness.

Girneys are the only sounds repeatedly interchanged between two animals and are also the only ones given in a tête-à-tête fashion. They seem to aid in welding a social bond, particularly among unrelated individuals. Their use is not restricted to females. Males use girneys in consortship solicitation and in mounting, whether vigorous copulatory mounts, the brief play mounts of young animals, or slow and deliberate greeting mounts of mature males. In these situations, the sounds may serve as indicators of nonaggressive state just as they do for grooming and huddling relationships of females.

Since the pattern of delivery of these variable sounds is highly labile, the succession of elements of varied morphologies has the potential for being exploited with communicative significance in affinitive situations. Although no evidence was sought in this study as to whether these girneys are used in such a fashion, it has already been demonstrated that the Japanese monkey has the capacity to exploit variability in pattern morphology and temporal lability in delivery of different acoustic features. It is not unlikely that small differences between the acoustic shape or temporal sequence of the girneys may indicate minor differences in the various tête-à-tête circumstances, just as different patterns of *coo* reflect different nuances in contact situations and transitional utterances reflect changes in arousal. The variability of girneys given in one situation with no changes noted in the vocalizing monkey may indicate that this portion of the repertoire is relatively divorced from tight constraints imposed by internal states. Dissociation from emotional determinants would predispose these sounds to be precursors of human language (cf. Itani, 1963; Marler, 1969a).

In summary, it is suggested that girneys are the portion of the repertoire homologous to speech in having an articulated component. They possess the lability in form required for meaningful elaboration into speechlike sequences. They are used in establishing affinitive contact between unrelated individuals; perhaps, by signalling reassurance or subordination, they promote these friendly interactions. This function may be analogous to that served by the prehistoric antecedents of speech.

It may not be too farfetched to entertain the possibility that man's ancestors, most likely living in open-country habitat and with ready visual access to each other, also employed an intergraded signal repertoire.

What may be the remnants of such a repertoire can still be seen in the vocal development of human infants and in prosodic phenomena. One portion of this repertoire may have been employed in facilitating non-agonistic face-to-face social encounters. If there is a shared heritage with the repertoire of the Japanese monkey, then it may be found in the articulated sounds that are used in the analogous circumstances tied to subordination and friendliness.

Such sounds may have played an important role for early hominids in promoting sociality beyond immediate kin groupings. Certain patterns of cooperative behaviors are at times accomplished more efficiently by larger groups, e.g., hunting or foraging parties. Other kinds of interactions optimally cross family lines as well, e.g., mating. Harmonious pairs or clusters of unrelated individuals might necessitate that interactants mutually reassure each other of friendly intent. If proximity allowed by certain sounds helped yield cooperation which was advantageous, then the capacity for employing these patterns might be strengthened by natural selection. Furthermore, the opportunity for socially significant associations to develop between particular temporal sequences or morphological variants of such sounds would be enhanced as greater amounts of time were spent in the affinitive and advantageous interactions. The elaborations of articulated vocal patterns and their sequences might provide a base onto which protohumans could effectively build a communicatory system revealing by vocalizations finely grained aspects of cooperative mood.

Relicts of prelinguistic articulatory reassurances may still be found in human sounds. The babblings and cooings of a human adult female toward the infant of another or the nonsense murmurings of consorting couples are perhaps fading remnants of similar uses of the antecedents of speech. Further suggestions of a shared heritage with monkeys may be evidenced today in both the prosodic aspects of speech of linguistic adults and also by the exploitation of the relationship between vocal patterns and internal states by which pre-linguistic human infants effectively inform parents of their mood.

Although the Japanese monkeys communicate within a system which principally exploits variability in vocal pattern rather than variability in delivery of fixed patterns, they use the system with a demonstrated capacity for both specificity and detail. This capacity and the modifiability of responses in light of contexts differs in degree from that of human beings but is not incomparable. These speculations on the relation of articulated monkey sounds with speech do not require any discontinuity or other uniquely provocative postulate concerning the evolutionary history of language.

Acknowledgments

The field study was conducted under the auspices of the Japan–U.S. Cooperative Science Program, J. T. Emlen and R. Motoyoshi, principal investigators. Other participants engaged in parallel studies were S. Ashizawa, M. Kawabe, S. Kawabe, S. Kawamura, Y. Kurita, K. Norikoshi, A. Orii, G. R. Stephenson; their cooperation is gratefully acknowledged. I thank also S. Azuma, K. Hayashi, K. Imanishi, J. Itani, M. Kawai, S. Kondo, S. Mito, D. Miyadi, K. Murofushi, Y. Sugiyama, A. Toyoshima-Nishimura, and the late K. Yoshiba, for access to unpublished data, translations of works in Japanese, logistic aid, and thought-provoking discussions. The generous cooperation of the Miyajima Ropeway Co., and of S. Iwata, Iwatayama Monkey Park, made possible the study, as did the aid of the Kyoto University Primate Research Institute and the Japan Monkey Center. Other help came from S. Horide, Mutsumido-Honten Co., the staff at the Department of Animal Sociology, Osaka City University, the Kanchi family, and gamekeepers Hioka-san and Kunizawa-san.

Assistance in the field was rendered by J. Lund, K. Minkowski, and G. McC. Stephenson. The collaboration in the field of S. Ashizawa, K. Norikoshi, and A. Orii is particularly appreciated. T. T. Struhsaker visited the study areas and offered many valuable suggestions and insights. The able assistance of K. Minkowski facilitated analysis and completion of this work. Critical comments and suggestions offered by K. Minkowski and P. Marler in response to early drafts of this chapter were invaluable and are gratefully acknowledged. I especially thank P. Marler for sponsorship of the research, for innumerable enlightening discussions, and for his generous cooperation throughout the study and manuscript preparation.

References

Altmann, S. A. (1962). A field study of the sociobiology of the rhesus monkeys (*Macaca mulatta*). *Ann. N.Y. Acad. Sci.* **102,** 338–435.

Altmann, S. A. (1965). Sociobiology of rhesus monkeys. II. Stochastics of social communication. *J. Theor. Biol.* **8,** 490–522.

Altmann, S. A., ed. (1967a). "Social Communication among Primates," Univ. of Chicago Press, Chicago, Illinois.

Altmann, S. A. (1967b). The structure of primate social communication. *In* "Social Communication among Primates" (S. A. Altmann, ed.), pp. 325–362. Univ. of Chicago Press, Chicago, Illinois.

Altmann, S. A. (1968). Primates. *In* "Animal Communication" (T. A. Sebeok, ed.), pp. 466–522. Indiana Univ. Press, Bloomington.

Andrew, R. J. (1962). The situations that evoke vocalizations in primates. *Ann. N.Y. Acad. Sci.* **102,** 296–315.

Apfelbach, R. (1972). Electrically elicited vocalizations in the gibbon *Hylobates lar* (Hylobatidae), and their behavioral significance. *Z. Tierpsychol.* **30,** 420–430.

Bastian, J. R. (1965). Primate signalling systems and human languages. *In* "Primate Behavior: Field Studies of Monkeys and Apes" (I. DeVore, ed.), pp. 585–606. Holt, New York.

Bertrand, M. (1969). "The Behavioral Repertoire of the Stumptail Macaque," Bibliotheca Primatologica, No. 11. Karger, Basel.

Bolwig, N. (1959). A study of the behaviour of the Chacma baboon, *Papio ursinus*. *Behaviour* **14,** 136–163.

Borror, D. J., and Reese, C. R. (1953). The analysis of bird songs by means of a Vibralyzer. *Wilson Bull.* **65**, 271–303.

Carpenter, C. R. (1934). A field study of the behavior and social relations of howling monkeys (*Alouatta palliata*). *Comp. Psychol. Monogr.* **10**(2), 1–168.

Carpenter, C. R. (1942). Societies of monkeys and apes. *Biol. Symp.* **8**, 177–204.

Chalmers, N. R. (1968). The visual and vocal communication of free-living mangabeys in Uganda. *Folia Primatol.* **9**, 258–280.

Darwin, C. (1872). "The Expression of the Emotions in Man and Animals." Appleton, New York. (Reprinted by Univ. of Chicago Press, Chicago, Illinois, 1965.)

de Laguna, G. A. (1927). "Speech: Its Function and Development." (Reprinted by Indiana Univ. Press, Bloomington, 1963.)

DeVore, I., ed. (1965). "Primate Behavior: Field Studies of Monkeys and Apes." Holt, New York.

Fienberg, S. E. (1970). The analysis of multidimensional contingency tables. *Ecology* **51**, 419–433.

Fisher, R. A. (1970). "Statistical Methods for Research Workers," 14th Ed. Hafner, New York.

Fossey, D. (1972). Vocalizations of the mountain gorilla (*Gorilla gorilla beringei*) *Anim. Behav.* **20**, 36–53.

Fretwell, S. (1969). Ecotypic variation in the non-breeding season in migratory populations: a study of tarsal length in some Fringillidae. *Evolution* **23**, 406–420.

Frisch, J. (1959). Research on primate behavior in Japan. *Amer. Anthropol.* **61**, 584–596.

Furuya, Y. (1961–1962). The social life of silvered leaf monkeys, *Trachypithecus cristatus*. *Primates* 3(2), 41–60.

Goodall, J. (1965). Chimpanzees of the Gombe Stream Reserve. *In* "Primate Behavior: Field Studies of Monkeys and Apes" (I. DeVore, ed.), pp. 425–473. Holt, New York.

Green, S. (1972). Communication by a Graded Vocal System in Japanese Monkeys: A Field Study. Doctoral dissertation, Rockefeller Univ., New York.

Grimm, R. J. (1967). Catalogue of sounds of the pig-tailed macaque (*Macaca nemestrina*). *J. Zool.* **152**, 361–373, 7 plates.

Hall, K. R. L., and DeVore, I. (1965). Baboon social behavior. *In* "Primate Behavior: Field Studies of Monkeys and Apes" (I. DeVore, ed.), pp. 53–110. Holt, New York.

Hays, W. L. (1963). "Statistics for Psychologists." Holt, New York.

Hill, W. C. O., and Booth, A. H. (1957). Voice and larynx in African and Asiatic Colobidae. *J. Bombay Natur. Hist. Soc.* **54**, 309–321.

Imanishi, K., and Altmann, S. A., eds. (1965). "Japanese Monkeys." Published by S. A. Altmann.

Itani, J. (1963). Vocal communication of the wild Japanese monkey. *Primates* **4**(2), 11–66.

Jay, P. (1965). The common langur of North India. *In* "Primate Behavior: Field Studies of Monkeys and Apes" (I. DeVore, ed.), pp. 197–249. Holt, New York.

Jolly, A. (1966). "Lemur Behavior: A Madagascar Field Study." Univ. of Chicago Press, Chicago, Illinois.

Joos, M. (1948). Acoustic phonetics. *Language* **24** (2), Monogr. Suppl. No. 23.

Jürgens, U., and Ploog, D. (1970). Cerebral representation of vocalization in the squirrel monkey. *Exp. Brain Res.* (*Berlin*) **10**, 532–554.

Jürgens, U., Maurus, M., Ploog, D., and Winter, P. (1967). Vocalization in the squirrel monkey (*Saimiri sciureus*) elicited by brain stimulation. *Exp. Brain Res.* (*Berlin*) **4**, 114–117.

Kaufman, I. C., and Rosenblum, L. A. (1966). A behavioral taxonomy for *Macaca nemestrina* and *Macaca radiata* based on longitudinal observation of family groups in the laboratory. *Primates* **7**, 205–258.

Kawai, M., Yoshiba, K., Ando, S., and Azuma, S. (1968). Some observations on the solitary male among Japanese monkeys. A pilot report for a socio-telemetrical study. *Primates* **9**, 1–12.

Kendall, M. G. (1970). "Rank Correlation Methods," 4th Ed. Griffin, London.

Klüver, H., and Bucy, P. C. (1937). Preliminary analysis of functions of the temporal lobes in monkeys. *Arch. Neurol. Psychiat.* **42**, 979–1000.

Kozelka, R. M., and Roberts, J. M. (1971). A new approach to non-zero concordance. *In* "Explorations in Mathematical Anthropology" (P. Kay, ed.), pp. 214–225. MIT Press, Cambridge, Massachusetts.

Liberman, A. M., Shankweiler, D. P., and Studdert-Kennedy, M. (1967). Perception of the speech code. *Psychol Rev.* **74**, 431–461.

Lieberman, P. (1967). "Intonation, Perception, and Language," Res. Monogr. No. 38. MIT Press, Cambridge, Massachusetts.

Lieberman, P., Harris, K. S., Wolff, P., and Russell, L. H. (1971). Newborn infant cry and nonhuman primate vocalizations. *J. Speech Hearing Res.* **14**, 718–727.

Leiberman, P., Crelin, E. S., and Klatt, D. H. (1972). Phonetic ability and related anatomy of the newborn and adult human, Neanderthal man, and the chimpanzee. *Amer. Anthropol.* **74**, 287–307.

Marler, P. (1957). Specific distinctiveness in the communication signals of birds. *Behaviour* **11**, 13–39.

Marler, P. (1961). The logical analysis of animal communication. *J. Theor. Biol.* **1**, 295–317.

Marler, P. (1965). Communication in monkeys and apes. *In* "Primate Behavior: Field Studies of Monkeys and Apes" (I. DeVore, ed.), pp. 544–584. Holt, New York.

Marler, P. (1967). Animal communication signals. *Science* **157**, 769–774.

Marler, P. (1968). Aggregation and dispersal: two functions in primate communication. *In* "Primates: Studies in Adaptation and Variability" (P. Jay, ed.), pp. 420–438. Holt, New York.

Marler, P. (1969a). Vocalizations of wild chimpanzees—an introduction. *Recent Advan. Primatol.* **1**, 94–100.

Marler, P. (1969b). *Colobus guereza*: territoriality and group composition. *Science* **163**, 93–95.

Marler, P. (1970). Vocalizations of East African monkeys. I. Red colobus. *Folia Primatol.* **13**, 81–91.

Marler, P. (1972). Vocalizations of East African monkeys. II. Black and white colobus. *Behaviour* **42**, 175–197.

Marler, P. (1974). A comparison of vocalizations of red-tailed monkeys and blue monkeys, *Cercopithecus ascanius* and *C. mitis*, in Uganda. *Z. Tierpsychol.* (in press).

Miller, R. E. (1971). Experimental studies of communication in the monkey. *In* "Primate Behavior: Developments in Field and Laboratory Research" (L. A. Rosenblum, ed.), Vol. 2, pp. 139–175. Academic Press, New York.

Miyadi, D. (1965). Social life of Japanese monkeys. *In* "Science in Japan" (A.H. Livermore, ed.), pp. 315–334. Amer. Ass. Advan. Sci., Washington, D.C.

Moynihan, M. (1964). Some behavior patterns of platyrrhine monkeys. I. The night monkey (*Aotus trivirgatus*). *Smithson. Misc. Collect.* **146**(5), 1–84.

Moynihan, M. (1966). Communication in the titi monkey, *Callicebus*. *J. Zool.* **150**, 77–127.

Papez. J. W. (1937). A proposed mechanism of emotion. *Arch. Neurol. Psychiat.* **38,** 725–744.

Poirier, F. E. (1970). The Nilgiri langur (*Presbytis johnii*) of South India. *In* "Primate Behavior: Developments in Field and Laboratory Research" (L.A. Rosenblum, ed.), Vol. 1, pp. 251–383. Academic Press, New York.

Pribram, K. H. (1962). Interrelations of psychology and the neurological disciplines. *In* "Psychology: A Study of a Science" (S. Koch, ed.), Vol. 4, pp. 119–157. McGraw-Hill, New York.

Reynolds, V., and Reynolds, F. (1965). Chimpanzees of the Budongo Forest. *In* "Primate Behavior: Field Studies of Monkeys and Apes" (I. DeVore, ed.), pp. 368–424. Holt, New York.

Rohlf, F. J., and Sokal, R. R. (1969). "Statistical Tables." Freeman, San Francisco, California.

Rowell, T. E. (1962). Agonistic noises of the rhesus monkey (*Macaca mulatta*). *Symp. Zool. Soc. London* No. 8, pp. 91–96.

Rowell, T. E., and Hinde, R. A. (1962). Vocal communication by the rhesus monkey (*Macaca mulatta*). *Proc. Zool. Soc. London* **138,** 279–294.

Schaller, G. B. (1963). "The Mountain Gorilla: Ecology and Behavior." Univ. of Chicago Press, Chicago, Illinois.

Schneirla, T. C. (1950). The relationship between observation and experimentation in the field study of behavior. *Ann. N.Y. Acad. Sci.* ***51,*** 1022–1044, Art. 6.

Sebeok, T. (1967). Discussion of communication processes. *In* "Social Communication among Primates" (S. A. Altmann, ed.), pp. 363–369. Univ. of Chicago Press, Chicago, Illinois.

Siegel, S. (1956). "Non-parametric statistics." McGraw-Hill, New York.

Smith, W. J. (1963). Vocal communication of information in birds. *Amer. Natur.* **97,** 117–125.

Smith, W. J. (1965). Message, meaning, and context in ethology. *Amer. Natur.* **99,** 405–409.

Smith, W. J. (1968). Message-meaning analyses. *In* "Animal Communication" (T. A. Sebeok, ed.), pp. 44–60. Indiana Univ. Press, Bloomington.

Smith, W. J. (1969). Messages of vertebrate communication. *Science* **165,** 145–150.

Sokal, R. R., and Rohlf, F. J. (1969). "Biometry." Freeman, San Francisco, California.

Stern, T. (1957). Drum and whistle "languages": an analysis of speech surrogates. *Amer. Anthropol.* **59,** 487–506.

Struhsaker, T. T. (1967). Auditory communication among vervet monkeys (*Cercopithecus aethiops*). *In* "Social Communication among Primates" (S. A. Altmann, ed.), pp. 281–324. Univ. of Chicago Press, Chicago, Illinois.

Sugiyama, Y. (1965). Short history of the ecological and sociological studies on non-human primates in Japan. *Primates* **6,** 457–460.

Thompson, R. F. (1967). "Foundations of Physiological Psychology." Harper, New York.

Tokuda, K. (1961–1962). A study on the sexual behavior in the Japanese monkey troop. *Primates* **3**(2), 1–40.

Van Lawick-Goodall, J. (1968). A preliminary report on expressive movements and communication in the Gombe Stream chimpanzees. *In* "Primates: Studies in Adaptation and Variability" (P. Jay, ed.), pp. 313–374. Holt, New York.

Winter, P. (1969). The variability of peep and twit calls in captive squirrel monkeys (*Saimiri sciureus*). *Folia Primatol.* **10,** 204–215.

Winter, P., Ploog, D., and Latta, J. (1966). Vocal repertoire of the squirrel monkey (*Saimiri sciureus*), its analysis and significance. *Exp. Brain Res.* (*Berlin*) **1,** 359–384.

Facial Expressions in Nonhuman Primates*

William K. Redican

Department of Behavioral Biology, and
Department of Psychology,
California Primate Research Center
University of California, Davis, California†

In light of man's characteristic emphases on spoken and written means of communication, it is often easy to overlook the tremendous amount of information that is transmitted nonverbally between two or more individuals. If this is true for humans, it is even more true of many nonhuman primates who rely upon visual displays, such as postures and facial ex-

* Supported by a National Science Foundation traineeship to the author and by National Institutes of Health Grants RR00169, HD04335, and MH22253.

† Present address: Department of Psychobiology and Physiology, Stanford Research Institute, Menlo Park, California.

pressions, as their principal mode of social communication. This review examines one particular and very important means of visual communication in nonhuman primates—facial expressions—by presenting an overview of the research in both laboratory and field in recent years. The general schema frequently relies on J.A.R.A.M. van Hooff's (1962, 1969) and R. J. Andrew's (1963b, 1964) important contributions. A survey of field literature was also undertaken to complement their observations that were made mostly on captive zoo animals.*

I. SENSE MODALITIES

The priority of the visual system over other sense modalities in nonhuman primates has been underscored by a number of field workers. Hall and DeVore (1965, p. 88) concluded that the major system mediating behavior between individual baboons involves visual cues such as facial expressions and intention movements (e.g., head-bobbing), whereas auditory, tactile, and olfactory cues are in descending order of importance. Thus, for the baboon, "many hours of the day are spent in almost complete silence" (p. 88). DeVore estimated that 15–25% of all signals by baboons are vocal (Ploog and Melnechuk, 1969, p. 438), and in the closely related rhesus monkey, the estimate was only 5.1% (Altmann, 1967b, p. 330). Among patas monkeys (*Erythrocebus patas*), another terrestrial primate, social interaction is also quite silent. Patas were never heard to vocalize even in the morning or evening, according to Hall's (1968, pp. 66–67) study. Rather, communication of intention and social status in a group of patas is principally achieved by gestures or movements perceived through the visual channel (Hall, 1968, p. 107). An interesting example of the primacy of vision in much social communication is furnished by Washburn and Hamburg (1968, p. 462), who noted that recorded vocalizations without accompanying visual gestures did not elicit a response in monkeys (no species specified). Cole (1963) similarly stressed the role of vision—at the expense of touch and especially audition—for the pigtailed monkey (*Macaca nemestrina*). However, Moynihan (1969, p. 338) countered that this generalization possibly could not be extended to all Old World (catarrhine) and certainly not to all New World (platyrrhine) monkeys. This is an important distinction, as we shall see below in the discussion of platyrrhine facial displays, since it is generally true that terrestrial or semiterrestrial primates rely most on visual communication, whereas arboreal primates are similarly dependent

* The literature review for this chapter was concluded early in 1972.

on auditory communication. One would do well to keep this in mind only as a general trend, however, since visual systems are by no means neglected in arboreal species such as the Nilgiri langur (*Presbytis johnii*), for which the major system mediating interindividual behavior in dominance situations still involves categories of visual cues quite similar to those noted above for the terrestrial baboon (Poirier, 1970, p. 171).

As might be expected, the primacy of vision for primates is also operative at the expense of olfactory, tactile, and gustatory modalities. Insectivores, rodents, carnivores, and ungulates probably make use of more olfactory and tactile signals and do so more frequently than primates (Moynihan, 1970b, p. 91). Once again this is expressed as a trend; one should avoid a "phylogenetic scale" model in accounting for these phenomena, as gustation or olfaction, for example, are not completely neglected in communication systems of nonhuman primates. Among prosimians, the galago (*Galago crassicaudatus, Galago senegalensis*) mother can discriminate her own infant from others by means of olfactory and auditory cues alone (Klopfer, 1970). Ploog notes that olfaction is certainly used by some primates in sexual behavior, in identifying strangers in a group, and in environmental exploration (Ploog and Melnechuk, 1969, p. 439). R. P. Michael and his associates have clearly demonstrated the role of olfaction in sexual contexts even among Old World monkeys. Bilateral ovariectomy of female rhesus monkeys abolishes ejaculation and regular behavioral variations in males which are normally correlated with the menstrual cycle (Michael *et al.*, 1967). Adult male rhesus monkeys may show a marked decline in frequency of mounting during the female's luteal phase, even though the female continues to present during this period (Michael and Welegalla, 1968). Anosmic rhesus males did not respond to estrogen-treated females until their olfaction was restored, after which they performed an operant response to criterion in order to gain access to females (Michael and Keverne, 1968; Michael, 1969). Recently, Michael, Keverne, and Bonsall (1971) isolated what they believe to be a sexual pheromone from vaginal secretions of female rhesus monkeys. Topical application of short-chain aliphatic acids on the sexual skin of ovariectomized females elicited a high degree of mounting attempts and ejaculations by rhesus males. (Mounting and ejaculation remained at a low level both before and after treatment.) Apparently, pheromonal substances have greater appeal to male monkeys than red sexual skin (Michael and Keverne, 1970).

One should also not overlook olfactory marking among many primates, especially prosimians. The secretions of scent glands are of importance in both spatial and sexual interactions (see, e.g., Lemuridae, Petter, 1965; Jolly, 1966; *Callithrix,* Epple, 1967). The reader is referred to Ralls'

(1971) recent review of mammalian scent marking for more information on the topic.

Considerable theoretical interest also centers on the multiplicity of social and environmental factors that predicate a principal reliance on either the visual or auditory systems. Frank Geldard (1960, p. 1583) has distinguished a number of factors that influence the selection between auditory and visual coding of a message. For example, the auditory system is utilized: (*1*) when rapidly successive data are to be resolved in temporal discrimination (the relatively shorter latency in auditory reaction time is indicative of this function); (*2*) when the recipient is preoccupied or in a state of reduced alertness and unexpected messages such as a warning must "break in"; (*3*) where flexibility of message transmission is important through, for example, intergradation of inflection; and (*4*) where visual reception is less available, such as when the physical environment interferes with visibility or the recipient is oriented in the wrong direction. Conversely, visual coding is called for: (*1*) if the message involves spatial orientation or guidance ["vision is the great spatial sense, just as audition is the great temporal one" (p. 1583)]; (*2*) when fine discrimination is needed; and (*3*) when the recipient must make a relatively prompt selection of data from a large array of information. Geldard's description of audition as a temporal rather than spatial system is readily apparent if one has ever attempted to identify which of many monkeys in a group is the source of a particular vocalization. If several monkeys are vocalizing simultaneously the task is even more challenging. Marler (1965, p. 547) also points to the role of visual cues in locating the signaler in space. Auditory cues, on the other hand, must rely on a relatively imprecise binaural comparison of signal characteristics. He notes that visual signaling is not without its drawbacks, since an animal must be conspicuous both morphologically and behaviorally while doing so. This is counteracted by the fact that visual cues are given more frequently in close range of conspecifics, that is, within an animal's social group. The display of a purely visual signal in this context is not going to make an animal any more conspicuous to predators than he already was before the signal, except when he must stand clear of concealing undergrowth, for example, to execute the display. Within a group of animals, then, auditory cues are inefficient in many circumstances and unnecessarily add to an animal's conspicuousness in his environment. Over greater distances, where there is an increased chance that a potential predator may intercept the signal, auditory cues with their decreased spatial specificity involve less risk to the signaler.

The interplay between audition and vision should certainly not be overlooked. Marler (1965, p. 566) interpreted the field data for genera such as the baboon as indicating that close-range vocalizations attract the gaze of the recipient, who can then receive additional information from visual signals emitted by the sender. Schaller (1965, pp. 344–345) came to basically the same conclusion for gorillas (*Gorilla g. beringei*), which must occasionally call attention to themselves through dense vegetation before visual gestures or facial expressions become functional. This may not be universally applicable, however. For example, Kawabe (1970, p. 291) suggests that the loud vocalizations of siamang gibbons (*Hylobates syndactylus*) may function to *prevent* visual or physical contact between groups. In the case of the langur (*P. johnii*), "a dominant animal need not vocalize to emphasize its state of agitation or call the attention of a subordinate" (Poirier, 1970, p. 172). Presumably an individual is likely to maintain attentive visual contact more reliably with a dominant rather than a subordinate animal. Thus the utilization and interaction of different sense modalities depend not only on ecological factors (e.g., dense vegetation, presence of predators) but on social variables such as the dominance status of sender and recipient.

The importance of visual communicatory systems is illustrated in a number of corresponding morphological features in nonhuman primates. There is an evolutionary trend—evident in the anthropoid more so than prosimian families—toward greatly increased frontality of the eye orbits (Napier and Napier, 1967, p. 18). As the angle between the optical axes is reduced, there is an increased degree of stereoscopic vision since the two retinas can focus concurrently on a single point in space, but this takes place at the expense of a decrease in the total field of vision (Napier and Napier, 1967, p. 18; Andrew, 1964, p. 235; van Hooff, 1962, p. 101). This relatively high degree of visual control over the area immediately in front of the head in primates has led, among other things, to a decay of facial tactile vibrissae (whiskers) which thus freed the muscles formerly attached to them for functioning in facial displays (van Hooff, 1969, pp. 12–13). Andrew (1964, p. 294) notes, for example, that the presence of vibrissae in other animals has probably interfered with the evolution of facial expressions to any great extent. When alert, for example, rodents keep their whiskers in a constant state of motion which could clearly interfere with the development of elaborate facial expressions (van Hooff, 1962, p. 100). Furthermore, in cold-blooded animals the facial musculature is limited primarily to that which opens and closes the mouth, eyes, and external nares. The skin covering the head is also largely immobile (e.g., in the reptiles it may be even plated). Thus, the facial expressions of such animals are limited to intention movements of biting which may have become ritualized. The transition to homiothermic metabolism in early mammals may have been a significant factor in the enhancement of facial mobility (Gregory, 1963, pp. 40–41). Extreme temperature changes called for an insulating layer of tissue and other heat-regulating mechanisms such as sebaceous glands (whose secretions help maintain the skin's pliability), sweat glands, and hair follicles. In general, the facial muscles of primates are much more mobile than in other animals, a feature that is facilitated by the lack of a well-defined muscle fascia (van Hooff, 1962, p. 102). An interesting exception to this, however, is the relative immobility of the ears in primates (Andrew, 1964, p. 235). Within the primate order itself, the general facial musculature becomes progressively differentiated from Prosimii through Pongidae and Hominidae. The exception to this trend is the auricular musculature (Hill, 1957, p. 28). Nocturnal primates (e.g., galagos and tarsiers), as might be expected, have highly mobile ears that aid in sound localization in the night, and, by comparison, in many of the anthropoid families ear movement is largely restricted to retraction or flattening (Napier and Napier, 1967, p. 20). This signal function of these and other morphological features will be discussed in greater detail below.

A detailed explication of neuroanatomical or neurophysiological bases of facial displays is not germane to this review; however, it is interesting to note that for the rhesus monkey, at least, there is a much greater dependence on subcortical structures in the control of facial expressions than is evident in man. Myers (1969) found that only mild contralateral loss of facial expression resulted from temporal, prefrontal, and precentral cortical lesions, in contrast with much more severe results in human subjects (i.e., temporary contralateral paralysis). Myers also found that damage to "centrally placed deeper-lying structures [p. 2]" than the cortex or hemispheral white matter in human subjects interfered with affective but not voluntary responses of contralateral facial musculature. He apparently did not investigate subcortical centers for the rhesus; however, if it is the case that structures in the diencephalon (especially the hypothalamus) are critical for integrated patterns of facial expression in nonhuman primates, this would seem to provide anatomical support for Jarvis Bastian's (1965, p. 598; 1968, p. 33) suggestion that nonhuman primate signals are related to the immediate emotional or arousal states of the signaler and recipient. At the same time, however, it would do well to keep in mind Neal Miller's recent remark that no part in the brain could be comfortably assessed as playing no part in communicative behavior (Ploog and Melnechuk, 1969, p. 473).

In summary, we have seen that the visual modality mediates a large portion of social interactions among nonhuman primates, especially ground-dwelling diurnal species. Visual cues are highly efficient in a spatial sense just as auditory cues are in a temporal sense. The former are used primarily for within-group communication, whereas the latter are used for long-distance communication in which the hazards of auditory conspicuousness are balanced by the difficulties of locating an auditory cue in space. These and other sense modalities overlap and interact. Morphological features of primate faces are well-adapted for elaborate expression through complex musculature, flexible facial skin, lack of vibrissae, and in some cases bright areas of coloration.

II. NONHUMAN PRIMATE FACIAL EXPRESSIONS

A. Threat

The threat display will serve as a good introduction to several other expressions, since virtually the entire repertoire of morphological components are brought into play in the forms of facial threat. There is no unitary form of facial threat characteristic of nonhuman primates for either most genera or most social situations. However, certain patterns that utilize one or more specific components are consistently observed, often in an additive manner (i.e., at higher intensities of threat, elements are often added to more fundamental units). The question of what specifically constitutes a threat is a difficult one and may be approached from several perspectives. Briefly, one might attend to the structural features of the display, the animal's internal state when displaying, or the social antecedents or consequences of the display. It would be helpful to keep these in mind as the following sections are read, but we shall postpone a more detailed discussion of the topic until we have examined the literature.

The most characteristic threat configuration is what van Hooff (1969) refers to as the "staring open-mouth face" (p. 25). One of the most conspicuous components is the slightly to fully open mouth, formed as the mouth corners are brought forward and the lips are tensed into a circular aperture. The teeth are not usually visible—due in large part to the contraction of the orbicularis oris muscle and the relatively pronounced development of the cheeks in primates—except when the mouth is opened very widely. In the rhesus macaque the lower teeth may be partially exposed but usually not the upper (van Hooff, 1969, p. 27; Andrew, 1964, p. 251; Hinde and Rowell, 1962, p. 4). The open-mouth component is fairly widespread throughout the primate order (Figs. 1 and 2). In the common tree shrew (*Tupaia glis*) and several insectivores, however, there is no noticeable movement of the corners of the mouth, and the

FIG. 1. A threat by an adult female rhesus monkey. The ears have been flattened and the eyebrow has been raised. Notice that only the bottom row of teeth is visible.

teeth are conspicuously exposed (Andrew, 1964, p. 251). In the ring-tailed lemur (*Lemur catta*) and sifaka (*Propithecus verreauxi*), the open mouth with teeth covered appears during "mobbing noises" directed toward ground predators and during play-biting (see below) (Jolly, 1966, p. 135). Andrew (1964, p. 252) also noted that the teeth are covered by the black lemur (*Lemur macaco*), but not by *L. catta*. The lesser mouse lemur (*Microcebus murinus*) was observed to at least open its mouth in threat. The situation is less confusing in the Cercopithecidae and Ceboidea, which more uniformly keep the mouth covered in threat (Andrew, 1964, p. 252). Field studies of the black mangabey (*Cercocebus albigena johnstoni*) in Uganda (Chalmers, 1968, p. 259) and laboratory studies of the patas monkey (*Erythrocebus patas*) (Hall *et al.*, 1965, p. 27, 30) and bonnet (*Macaca radiata*) and pigtailed macaques confirm the above descriptions for the open-mouth component.

The evolutionary bases of the open-mouth component of the threat are fairly straightforward. Opening the mouth per se is suggested (Andrew, 1964, pp. 250–251) as an intention movement of biting, whereas the tensing of musculus orbicularis oris to form the "O" configuration of the mouth is viewed as a respiratory reflex during expiration. This is clearly seen in the lemur: during expiration part of the orbicularis oris contracts and constricts the nostril back to its resting position. Andrew (1963b, p. 19) suggests that this contraction may be exaggerated during any violent expiration associated with vocalization in threats, so that the entire muscle contracts and rounds out the mouth aperture.

FIG. 2. Displaying piloerection, raised eyebrows, flattened ears, and open mouth, a mature adult male baboon comes to an abrupt halt in front of an antagonist. The latter is crouching and screaming. Note the difference in the degree of vertical lip retraction. (Photograph courtesy of Tim Ransom.)

There are scattered references to "snarling" in the literature. All major groups of primates can raise the upper lip on one side to bite at food with the premolars (Audrew, 1963b, p. 18), although the communicatory significance of this ability is not clear. It is presented as a threat gesture in the male chacma baboon (*Papio ursinus*) (Bolwig, 1959, p. 142, 153; Hall, 1962, p. 293), and the male Celebes macaque (*Macaca maurus*) (Bernstein, 1970, p. 281), and it occurs rather dramatically in the drill and mandrill (*Mandrillus leucophaeus* and *Mandrillus sphinx*) (van Hooff, 1969, pp. 43–44). Andrew (1964, p. 252; 1963b, p. 75) also claims that it appears in *Pan, Hylobates,* and *Pongo.* For *Mandrillus,* Andrew (1963b, p. 18) interprets it as a threat, whereas van Hooff (1969, p. 44) suggests that it is a type of fear grimace (see below), a position which seems more tenable to this writer. Hill (1970, p. 486) noted that snarling is displayed by mandrills while greeting. It is often accompanied by lateral head movements and alternates with lipsmacking. As we shall see, this lends rather strong support to the association of snarling with grimacing rather than threatening. Indeed, Bernstein (1970, p. 273) notes that drills and mandrills threaten with their mouths *closed,* while the head is jerked forward and downward sharply. In any case, in view of its very rare description in field and laboratory studies, it seems likely that the snarl has not evolved as a threat gesture to the extent that it has in the Canidae, for example (for a review, see Fox, 1970).

A movement that seems to function as a protective response in most forms of threat display is flattening the ears against the head, thus reducing the amount of target area available to a potential aggressor (Figs. 3, 4, and 5). Once again it is R. J. Andrew who has stressed its protective function. He notes (Andrew, 1965, p. 88; 1964, p. 303) that it is a response characteristic of many mammals ("from mice to monkeys") when startled by a noise or by some other sudden or noxious stimulus, when something is thrust at the face, or when the individual advances to investigate or bite an object. In most species of mammals

FIG. 3. An adult male baboon strides toward his opponent during an agonistic interaction. Note the extreme ear-flattening. His expression might best be described as a tense-mouth face. (Photograph courtesy of Tim Ransom.)

FIG. 4. A feral adult female baboon perceiving her own reflection in a window. Ear-flattening and brow-raising are displayed in extreme degrees. (Photograph courtesy of Tim Ransom.)

the ears are erect and oriented directly at an opponent during the first stages of a threat, but many turn them backward and down in later stages (Guthrie and Petocz, 1970, p. 586). It is interesting to note that the patas apparently flattens its ears not only when threatening but when eating as well (Hall *et al.*, 1965, p. 31). This is also true of rhesus monkeys (Redican, personal observation). It seems possible that the frequently vulnerable position of an animal that has fixed its attention on a food object and, perhaps, paused momentarily to eat it may be responsible for the appearance of a potentially protective response in this situation.

As noted above, there is a general reduction in ear mobility in primates which apparently is related to the reduced use of the ears in diurnal animals when determining the direction of sounds (Andrew, 1963b, p. 98). Reduction of postauricular muscles in the tree shrew (a diurnal prosi-

FIG. 5. Adult and juvenile female baboons respond to their reflected image in a window by flattening their ears and raising their eyebrows. (Photograph courtesy of Tim Ransom.)

mian) is greatly advanced, although one ear can be flattened independently of the other when scanning the environment (Andrew, 1964, pp. 236–237). In lemurs and lorises, the pinna is flattened against the side of the head when threatening an equal or more dominant animal, in response to novel stimuli, or especially when leaping forward (Andrew, 1963b, p. 99). Jolly (1966, p. 91) reports that ear-flattening occurs in *Lemur catta* and *Lemur macaco* when staring threateningly, when engaging in all aggressive encounters, and while tail-marking or waving. Nocturnal galagos can move their large ears independently from a position directly forward to flat against the head, and bushbabies (*Galago senegalensis*) have been seen to flatten them when grabbing at prey (Andrew, 1964, p. 237; 1963b, p. 99).

Further reduction of ear mobility can be seen in most Cercopithecoidea and Ceboidea, although the ears are still retracted in both confident and aggressive threat (van Hooff, 1969, p. 26; Andrew, 1963b, p. 99). Quick flattening and erection of the ears are characteristic of most marmosets in defensive and aggressive threats (Epple, 1967, p. 62). The bonnet and pigtailed macaques are particularly likely to flatten their ears in intense threat (Kaufman and Rosenblum, 1966, p. 212), and the rhesus does so when biting (Sade, 1967, p. 100). Hinde and Rowell (1962), moreover, mention ear movement during a threat in the rhesus in two circumstances:

first, if "the head is held momentarily at the end of a forward jerk, [when] the ears may be moved rapidly backwards and forwards" (p. 7), and second, during a "backing threat," in which an individual aggressively approaches an object eliciting fear while in a crouched posture (Hinde and Rowell, 1962, pp. 7–8). The configuration around the mouth is not made clear in the latter case, and it is possible that there are considerable elements of the fear grimace (see below) involved. Hinde and Rowell (1962, p. 4) also mention that ear-flattening takes place in a running attack, if not in threat.

Not all Old World monkeys and apes move their ears when threatening. According to van Hooff (1969, pp. 26–27), ear movements are difficult to detect in several apes, the patas monkey, the Abyssinian colobus (*Colobus guereza*), and most guenons except the vervet (*Cercopithecus aethiops*). Unlike the marmosets described by Epple, at least one species of tamarin (*Saguinus geoffroyi*) does not seem to flatten and erect the ears during agonistic interactions (Moynihan, 1970a, p. 58). Hall *et al.* (1965, p. 27) described slight backward movement of the ears in a threatening patas monkey and suggested that this limited movement is owing to the fact that patas' ears lie very close to the head and are heavily furred. [The threat display per se, however, actually occurs more frequently in the patas than in baboons and macaques (Hall *et al.*, 1965, p. 47).] Dense hair growth behind the ears is also described for the eastern highland gorilla (*G. g. beringei*) (Schaller, 1963, p. 208). Chimpanzees' ears are not flattened during threats (Reynolds, 1970, p. 376).

A final major component of the threat display has to do with the very important areas around the eyes. During a threat the eyes themselves are invariably fixed directly at the opponent and the iris is completely exposed in a generally widely open eye (van Hooff, 1969, p. 25) (Figs. 6 and 7). Behavior associated with a direct stare is one of the most interesting topics in primate social communication. It is no doubt the most universal component of threat for nonhuman primates and is a basis on which further elaboration of the threat display takes place. As Altmann (1968, p. 60) notes, merely looking at a social partner is a common method for directing social signals, and when so used is not a threat. However, when it is prolonged, e.g., 3–5 seconds for vervet monkeys (Struhsaker, 1967c, p. 7), and particularly when it is accompanied by other postures or gestures such as head-bobbing or an aggressive stance, it is quite another matter. A direct stare even by itself is a mild form of threat and is specifically mentioned as such in a large number of studies for a wide range of genera: Lemuridae (Jolly, 1966, p. 56, 103); *Macaca mulatta* (Chance, 1956, p. 2; Altmann, 1962, p. 397; Hinde and Rowell, 1962, p. 7; Andrew, 1964, p. 397; Sade, 1967, p. 100; Lindburg, 1971,

FIG. 6. Two mature adult male baboons simultaneously stare and raise their brows toward a third. This took place during a sequence of rapidly executed parallel movements and displays performed by these individuals. (Photographs courtesy of Tim Ransom.)

p. 40); *Macaca radiata* and *Macaca nemestrina* (Kaufman and Rosenblum, 1966, p. 213); *Macaca arctoides* (Bertrand, 1969, p. 69); *Cercocebus albigena johnstoni* (Chalmers, 1968, p. 261); *Papio ursinus* and *Papio anubis* (Hall and DeVore, 1965, p. 92); *Papio hamadryas* (Kummer, 1967, p. 69; 1968, pp. 34–36); *Cercopithecus aethiops* (Struhsaker, 1967a, p. 284; 1967c, p. 7, 51); *Erythrocebus patas* (Hall and Mayer, 1967, p. 215); *Presbytis entellus* (Jay, 1965a, p. 236); *Pan troglodytes schweinfurthii* (Goodall, 1965, p. 466; van Lawick-Goodall, 1968a, pp. 318–319; Nishida, 1970, p. 49); and *Gorilla g. beringei* (Schaller, 1963, p. 290; 1965, p. 341, 347).

The social contexts and consequences of a direct stare, as might be expected, show some degree of variability. A stare by a bonnet or pigtailed macaque may lead to physical withdrawal of the percipient, a grimace, or such submissive postures as crouching (Kaufman and Rosenblum,

FIG. 7. Eyebrow-raising and staring often serve as low-intensity threats, as in this adult female baboon. She is presenting to the adult male on the right and simultaneously threatening the juvenile on the left. Notice that all three individuals have flattened their ears. (Photograph courtesy of Tim Ransom.)

1966, p. 213). A stumptailed macaque who is being "scrutinized" or stared at may redirect the attention by staring at another monkey (which may signal the beginning of overt redirected aggression) or even at an empty point in space (Bertrand, 1969, p. 72). A stare often prevents a stumptailed macaque from engaging in some activity or may stop an activity in progress. Thelma Rowell (1963, p. 39) reports that when a rhesus "aunt" was being too possessive of an infant, a simple glance from the latter's mother at a distance of 12 feet was sufficient to bring about the infant's immediate release. Bertrant suggests that the efficiency of the stare and other expressions depends on the "ranks and personalities of the performer and the addressee" (p. 70). For example, a stare from the alpha male directed toward a low-ranking subordinate is sufficient to send the latter running away screaming, whereas a similar stare directed toward the beta male might only succeed in preventing him from stealing food. (This contextual variable is an important one and will be returned to often.) In the case of the hamadryas baboon, Kummer (1967, p. 66) reported that a stare by an adult male is sufficient to halt an aggressive attack and lead to withdrawal by the aggressor. The efficacy of a stare is brought out clearly in the following interaction among hamadryas in an enclosure:

> Subadult male 4a saw that adult male α looked at him. Male 4a started lip-smacking and approached α, finally grimacing and screeching intermittently with knees bent. Male α merely stared at him. Two feet from them a 1-year-old female explored the wire net, turning her back to the males. Suddenly, 4a looked at her, pulled her down, and embraced her. Gradually, his screeching diminished; the female jumped off and he ran away (Kummer, 1967, p. 69).*

In most Old World monkeys the response to a threat–stare depends not only on the dominance status of sender and recipient but particularly on whether or not the former is standing or running toward the latter. If the sender is running, of course, the recipient is more likely to flee (Struhsaker, 1967b, p. 1197). Struhsaker suggests that the stare can function as an aggressive or defensive threat gesture, depending once again on the context, for vervet monkeys of both sexes and all ages (Struhsaker, 1967c, p. 7, 51). For the chimpanzee, a "glare" with lips compressed sometimes occurs prior to chase or attack *or* prior to copulation. If given by an adult male, it often leads to presenting by a female (van Lawick-Goodall, 1968a, pp. 318–319). In the gorilla the stare occurs during intragroup dominance interactions (in which case it is usually of short duration) and between males of different groups after charging displays. Moreover, a stare by a dominant gorilla often leads to physical displacement (e.g., from a resting place) of the percipient, and as expected, it is most commonly given by adult males, although juveniles and adult females were sometimes observed to do so (Schaller, 1963, p. 290; 1965, p. 341, 347). An interesting exception to the above forms of interaction was reported by K. R. L. Hall for the patas monkey. Female patas express their dominance over another female in part by persistent harrying of the subordinate:

> Harrying consists of walking fairly slowly after the subordinate and causing her over and over again to move away. It consists particularly of walking-toward, without looking the subordinate in the face, that is, with the action of turning the head from left to right and *looking past* the subordinate to either side (Hall, 1967, p. 270).*

So here we have a clear instance of a mild form of threat involving not a direct stare but its converse. By not "recognizing" the other animal, in effect, the aggressor is escaping most of the responsibility for responding defensively. That is, if the dominant individual does not look at the subordinate, it need not respond to facial threats from the latter. The burden of response is thus on the subordinate and, unless it actually attacks, it will have to give way and be harried from place to place. It would be interesting to see how frequently this tactic is used by individuals of varying dominance statuses. For example, if the two individuals are of nearly equivalent rank, the dynamics may be quite different from the interaction between a highly dominant and a very subordinate individual.

Data on the physiological accompaniment of visual gestures such as a stare are conspicuous in their absence. There is, however, a very interesting study by Wada (1961) in which the efficacy of direct eye contact is clearly demonstrated in macaques' brainstem responses. Electrical stimuli were administered to the cortex of macaques and the subsequent brainstem responses were recorded. Wada (1961, p. 41) found:

> When the animal discovered that it was being watched, the response was depressed as long as the animal could see the experimenter. . . . Such [brainwave] flattening regularly occurred whenever the animal realized that the experimenter's gaze was fixed on it. This specific nature of the most powerful suppression of the response—that is, the direct meeting of the experimenter's gaze and that of the monkey—suggests concentrated focusing of discriminatory attention of a quality necessary for self-preservation in the monkey.*

Although this variable was not carefully controlled for, it appears that it was the experimenter's gaze and not merely physical proximity that elicited the response.

Just as we have found the direct stare to be a virtually universal component of threat, gaze aversion is one of the most frequent *responses* to a direct threat among nonhuman primates. Fundamental to the response, as Altmann (1967b, p. 332) suggests, is the fact that avoiding visual contact is basically a means of avoiding social interaction. As Kaufman and Rosenblum (1966) expressed it for bonnet and pigtailed macaques, "the failure to receive a communication often obviates the need to respond to it appropriately" (p. 215). In an important article, Chance (1962, pp. 72–76) similarly suggested that turning away one's eyes in a social encounter literally cuts off any threatening social stimuli from further perceptual processing. In so doing, the individual's tendency to flee and his level of arousal are reduced without having to withdraw from the interaction physically. At the same time, one might add, the reorientation of one's gaze can have vital effects on the motivation of the other individuals in an agonistic interaction. By shifting one's visual orientation away from an opponent, one is eliminating a set of stimuli often associated with a direct stare (i.e., a potential threat), and the chances of eliciting overtly aggressive behavior in the opponent are thereby diminished. It should be understood that gaze aversion is not restricted to responding to a direct threat from another animal, as monkeys will behave similarly when confronted with a frightening object; e.g., Hinde and Rowell (1962, p. 11) note that a rhesus will look attentively in practically every direction other than that of a frightening object. However, our principal interest will remain in the context of interanimal social communication.

Studies of free-ranging and captive primates have established the occurrence of gaze aversion in a wide range of species. *Lemur catta* lowers its eyelids when stared at and looks away from the source of the stare (Jolly, 1966, p. 91, 99). Bertrand (1969, p. 70) noted that stumptailed macaques (*Macaca arctoides*) usually avoid looking in the eyes of a dominant animal and avert their eyes when looked at or stared at. Bonnet macaques avoid the direct stare of a more dominant individual, and looking away is a common subordination gesture for males and females (Rahaman and Parthasarathy, 1968, p. 262; Simonds, 1965, p. 182). When it occurs alone, Altmann (1962, p. 397; 1968, p. 55) ranks avoiding either visual contact or stares as a gesture ranking just below a grimace in intensity of submissiveness, as does Jay (1965a, p. 237) for the hanuman langur (*Presbytis entellus*). In closely related baboons (*Papio ursinus* and *P. anubis*), looking away was interpreted by Hall and DeVore (1965, p. 93) as an indication of fear, uncertainty, or escape tendencies. Once again the patas monkey proves to be somewhat atypical, as no forms of submissive or appeasement gestures were observed in the captive (Hall and Mayer, 1967, p. 218) or feral (Hall *et al.*, 1965, p. 32) individuals. Recall that this is in contrast to the frequent threats by patas. Nevertheless, following an actual or imminent attack the recipient generally avoids eye contact with the aggressor (Hall *et al.*, 1965, p. 30), and ear-flattening has been observed in a fearful patas monkey (Bolwig, 1963, p. 320). Several interesting instances of gaze aversion are reported for gorillas. They characteristically refrain from directly staring at human observers during low-intensity "alarm" situations, and submissiveness and/or appeasement is generally indicated by averting their eyes and turning their head from side to side (Schaller, 1963, p. 209, 292; 1965, p. 355). In one related instance a gorilla approached within 30 feet of Schaller: "When I began to shake my head, he immediately averted his face, perhaps thinking that I had mistaken his steady gaze for threat. Then, when I in turn stared at him, he shook his head" (Schaller, 1964, p. 136). Peter Marler (1965, p. 571) interprets gorillas' head-shaking as an extreme or elaborated form of gaze aversion, a position that fits well with Schaller's observations. Van Hooff (1969, p. 73) also views head-shaking in macaques as ritualized looking-away movements. Drills and mandrills also rotate their head from side to side when displaying their characteristic figure-eight or "snarling" expression (van Hooff, 1969, p. 44; Bernstein, 1970, p. 281), which seems to be further evidence that this is not a threat expression in these species. In contrast, when threatening the head is jerked forward and up and down in these species (van Hooff, 1969, p. 29; Bernstein, 1970, p. 273).

FIG. 8. Eyelid coloration is an important means of emphasizing the area around the eyes. Shown above is an adult female sooty mangabey.

Two further areas important in the facial communication of threat are the eyelids and eyebrows. The eyelids are surrounded by conspicuously colored areas of skin in many primate species, including mangabeys, guenons, and guerezas, and they seem to be among the few morphological features in nonhuman primates that have evolved exclusively for communication (Jay, 1965b, p. 559; Ploog and Melnechuk, 1969, p. 444) (see Figs. 8 and 9). There is some degree of structural and functional variability of eyelid displays in many species. In most Cercopithecinae, a threat exposes the bright skin of the upper eyelid and the region immediately above it [e.g., *Theropithecus gelada* (Hill, 1970, p. 596)], and in the chacma baboon, for example, the effect may be accentuated by a rapid up-and-down blinking of the eyelids (van Hooff, 1969, p. 25; Hall, 1962, p. 312). A threat display in pigtailed macaques may expose a light-blue eyelid coloration which is particularly prominent in animals whose face has been exposed to sunlight (L. A. Rosenblum, personal communication). Van Hooff (1969, p. 26) was "struck by the relative immobility and lack of expression" in the faces of groups such as *Erythrocebus patas*, *Cercopithecus*, and *Colobus abyssinicus*. In contrast, Hall *et al.*, (1965, p. 39) did find that patas monkeys employ their darkened eyebrows and contrasting light eyelids in threatening. Struhsaker (1967a, p. 284; 1967c, pp. 7–8) also noted that feral vervets raise their eyebrows and expose their eyelids in threat in a relatively frequent and conspicuous pattern. In vervets the eyelids and the area immediately above them are particularly brightly colored, and, in conjunction with a stare, eyelid

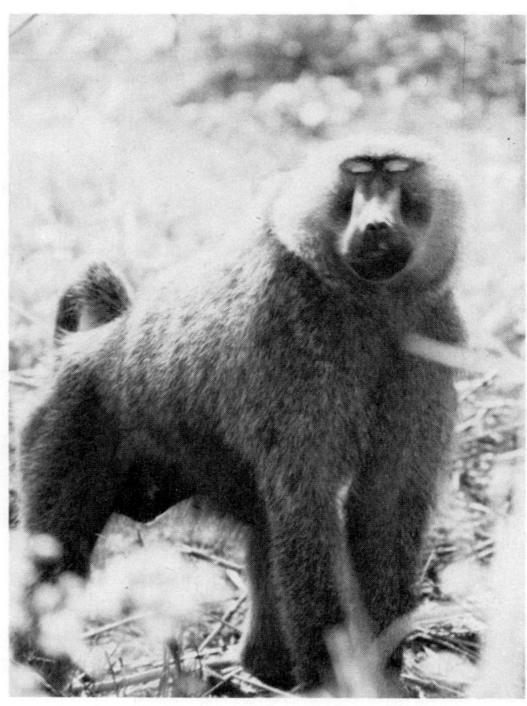

FIG. 9. Rapid eyelid blinking may enhance a stare–threat, as in the case of this adult male baboon threatening another. Notice the slight piloerection. (Photograph courtesy of Tim Ransom.)

exposure is associated with intense defensive and aggressive agonism in all ages and both sexes (Struhsaker, 1967c, pp. 7–8, 51). Somewhat surprisingly, Struhsaker (1967b, p. 1198) suggested that there is little structural variation in eyelid coloration among Cercopithecinae, while noting that the frequency with which eyelid displays occur can vary between species (e.g., Kenya baboons expose eyelids in threat more often than vervets in the same area) and perhaps even populations. However, a perusal of *Cercopithecus* morphology uncovers a great deal of variability in the very genus that Struhsaker studied. Moreover, this variability is quite instructive in demonstrating the intimate relationship between an organism's ecology, morphology, and communicatory patterns. *Cercopithecus aethiops*, the only consistently terrestrial species, has white or yellow patches of fur above the eyes, whereas the arboreal *Cercopithecus mitis* shows little differences in eyelid pigmentation in its predominantly black face and has a relatively immobile facial musculature. There are also intergradations in pigmentation in species which are both arboreal and terrestrial (e.g., *Cercopithecus lhoesti* and *Cercopithecus neglectus*) (Napier and Napier, 1967, pp. 109–110; Gartlan and Brain, 1968, pp. 280–281). Arboreal and terrestrial habitats clearly exert different selection pressures on communication systems. One of many intervening variables involved may be the intensity and characteristics of dominance and aggressive interactions involved in the two habitats.

Closely associated with eyelid movements are those of the eyebrows. Brow movements often emphasize the set of stimuli associated with the eyes themselves. Coss (1968, p. 277) suggested that "the brow structure complements the discoid conspicuousness of the two eyes because the shape of the brow follows a disk-like arc." In many of the Cercopithecoidea, eyebrow lowering may function to emphasize a direct agonistic stare, but, in intense threat, scalp retraction readily supplants it (Andrew, 1963a, p. 1035). In the Hominoidea, a threat is accompanied by lowering and converging the brows, responses that Andrew (1963b, p. 18, 100; 1964, pp. 253–254) suggested aid in optical convergence and focusing as well as exaggerating the fixed stare in threat. Van Hooff (1969, p. 73) noted that frowning may help to steady the eye in visual fixation. Finally, Charles Darwin, in the much-cited "The Expressions of the Emotions in Man and Animals" (1872, pp. 227–228), proposed that the lowering of the brows has to do with the exclusion of excess light in fixating or scanning distant objects. It is curious, however, that he found apes never to frown "under any emotion of the mind" (p. 143), whereas van Hooff (1969, p. 26) reports frowning for *Pan* in threat and Schaller (1963, p. 209) observed gorillas "scowling" in mildly disturbing situations.

In most New and Old World monkeys the more typical movement of the eyebrows during threat is upward, as might be anticipated by the previous discussion of eyelid displays. Upward brow movement has been recorded for *Macaca nemestrina* and *M. radiata* (Kaufman and Rosenblum, 1966, p. 212), *M. mulatta* (personal observation), *Erythrocebus patas* (Hall *et al.*, 1965, p. 27, 30; Hall and Mayer, 1967, p. 215), and *Theropithecus gelada* (Bramblett, 1970, p. 329). It is also an important gesture for the hamadryas baboon. Kummer (1968, p. 35) reported a sequence in which a young male harem leader simply stared and raised his brows at his straying female, which subsequently returned and presented to him. Andrew (1963b, p. 18) emphasizes that the response in the Cercopithecidae represents a by-product of scalp withdrawal following ear-flattening, although Sade (1967, p. 100) suggested that the two responses may be independent. In man and the chimpanzee, in whom ear movement is more restricted, Darwin (1872, pp. 278–282) suggested that brow-raising is a response that facilitates the rapid scanning of the environment after perception of a startling or frightening stimulus. Scalp withdrawal in threat has apparently evolved independently in New and Old World monkeys, and in both cases it occurs with other facial protective responses (e.g., during play) (Andrew, 1964, p. 254).

To restate a point made earlier, the various components noted above—mouth-opening, ear-flattening, brow and eyelid movements, and direct staring—may all occur in a facial threat display or they may be seen more-or-less independently of one another. It is thus misleading to focus on a "representative" threat expression, as a raised eyebrow may at times be as efficient a threat (in terms of eliciting a response in conspecifics) as a montage of components.

Social situations in which the threat display takes place have probably emerged haphazardly (if not intuitively) in the above discussions, but a few general remarks are in order. In most cases it is extremely difficult to extrapolate causal mechanisms leading to the display other than in instances such as territorial infringement or competition for food or sex partners. If the circumstances were to be subsumed under the most frequent single category, it would no doubt be dominance–subordinance interactions. Indeed, the dominant animal in an interaction more frequently displays a threat expression (van Hooff, 1969, p. 30). In stumptailed macaques, Bertrand (1969, p. 75) noted its occurrence in dominant animals to prevent subordinates from taking food or from coming closer, and it also occurred during the introduction of new members into a group (cf. lipsmack, below). [It may be worth noting, however, that Kaufmann (1967, p. 79) found displacement to be a more reliable index of dominance than threat displays for rhesus on Cayo Santiago.]

Sex differences in frequency of threats have been reported but not always in the same direction. For example, Mason, Green, and Posepanko (1960, pp. 75, 80) suggested that female rhesus monkeys are more likely than males to threaten a set of controlled stimuli, but Møller *et al.* (1968, p. 346) found a higher frequency of social and redirected threats in male rhesus monkeys. Rowell (1966, p. 434) observed facial threats more frequently among females in a captive group of olive baboons (*P. anubis*), especially middle-ranking individuals. The adult male in the group was threatened more than twice as often as he himself threatened (unlike hamadryas males), but these frequencies are difficult to interpret since females greatly outnumbered males to begin with. Bernstein (1971, pp. 41–42) recently commented on sex differences of aggressive displays and his remarks bear careful attention:

> In a sexually dimorphic species such as the mangabey, adult male aggression has a greater potential for resulting in severe injuries than does equivalent adult female aggression. The consequences of a single bite or slash involving male

canine teeth may be far greater than multiple bite injuries produced by females. The behavior of adult males, at least in part, reflects this potential. Adult males went through long periods of elaborate maneuvering and signaling prior to contact, and fights tended to be brief explosive episodes as contrasted with the often prolonged and persistent attacks of females who would repeatedly bite and chew on tails or limbs.

It would appear, then, that male sooty mangabeys (*Cercocebus atys*) would be expected to threaten more frequently than females of the species. Across several species, then, there have been an equal number of reports for males and females with regard to a higher frequency of threatening.

Struhsaker (1967c, pp. 10, 52) has repeatedly emphasized that a threat may occur in either an aggressive or defensive context, the distinction between them often being made on the basis of the animal's posture. In the case of the aggressive threat, it is not always followed by physical attack, especially if the recipient displays signs of submission [e.g., *Macaca arctoides* (Bertrand, 1969, p. 75); *M. radiata* and *M. nemestrina* (Kaufman and Rosenblum, 1966, p. 211)]. Chalmers (1968, pp. 261–263) is one of the few researchers who has reported frequency data concerning the context and responses to threat and other displays. His data for the black mangabey are presented in Tables I and II. The threat–stare led to a far greater frequency of flight by the recipient of the threat than either attack or approach–remain behaviors. Moreover, there was some tendency for a threatening animal to attack shortly afterward, but in twice as many instances he did not.*

The threat display often takes place during copulation, but it may be directed at either of the individuals, or even toward an inanimate object. In the patas monkey, for example, when an adult male begins to be sexually interested in a female he is much more likely to threaten other animals in the group who approach. Even infants, who are normally immune from such displays, are threatened on such occasions (Hall, 1968, p. 84). After the male dismounts, the female vervet may lunge toward the male and repeatedly threaten him by staring, gaping, and exposing her eyelids (Struhsaker, 1967c, p. 25). In the rhesus, Altmann (1962, p. 390) found that when a female threatens other males from a distance, a nearby male will often mount her, a gesture interpreted as an indicator of sexual receptivity.

Zumpe and Michael (1970, pp. 12–13, 16, 17) focused their attention on redirected threats (i.e., those not directed at the partner) in copulating pairs of rhesus monkeys. Dominant males were especially likely to threaten away from their partner. While threatening in this manner, males

* No statistical tests were reported for these data.

TABLE I

FREQUENCY OF RESPONSES BY THE SENDER AFTER VARIOUS GESTURES [a]

Gesture	Behavior of displaying monkey after gesture		
	Attacks	Remains with [b]	Flees
Stare	6	12	0
Yawn	9	16	19
Lipsmack	0	21	9

[a] Adapted from Chalmers (1968, p. 262).
[b] Includes grooming, copulating, and sitting by the other monkey.

TABLE II

FREQUENCY OF RESPONSES BY THE PERCIPIENT AFTER VARIOUS GESTURES [a]

Gesture	Reaction of percipient		
	Attacks	Approaches or remains with [b]	Flees
Stare	0	2	12
Yawn	1	7	6
Lipsmack	0	14	0

[a] Adapted from Chalmers (1968, p. 262).
[b] Includes grooming, copulating, and sitting by the other monkey.

repeatedly glanced over their shoulder toward the females, who often responded by threatening in the same direction as the male. Redirected threats were most frequent when the male's sexual interest was high, and these threats were often seen just before the first mount in a series; they increased in intensity between mounts and ceased after the male had ejaculated. Females also exhibited redirected threats, especially when they were sexually receptive. Subcutaneous or intracerebral implants of estrogen elevated the frequency of this type of threatening in females. This association between hormonal state and frequency of redirected threatening may provide a clue as to why reports of sex differences in frequency of facial expressions have so often disagreed. It is at least possible that the females in the several studies cited above were observed during different hormonal states.

Now that we have examined the literature on this expression, we shall be better equipped to deal with the criteria by which this material has been assembled under the heading of threat. Persuing this first from a structural perspective, we have seen that a basic facial cue exhibited by a threatening primate involves a fixed and direct stare. Brow-lowering and subsequently brow-raising and ear-flattening, exposure of eyelid coloration, and occasionally eyelid blinking and some form of vocalization accompany the stare in increasing levels of arousal. This is perhaps the most facile method of analyzing facial threat displays. That is, one might say that an animal that arranges its face in the above configurations is by definition displaying a threat. Clearly, these are necessary but not sufficient criteria. Not only do identical components appear in several categories of expression, but responses such as ear-flattening may appear when an animal is doing nothing more than eating. By necessity one must interpret a single component in terms of which others are displayed (e.g., ear-flattening with or without a stare) in describing a threat.

A second major approach to threat expression might focus on group social behavior prior to and following the display. This applies to both senders and percipients. A threat may be preceded by an animal approaching too closely to the displayer of the threat or to the displayer's offspring, siblings, or consort partner. Threats may also be elicited by possession of or interest in food or space resources. Consequences of a threat frequently include increased or maintained social distance, flight, presenting, grimacing, or crouching, but far less frequently overt attack. On the basis of these criteria, one might deduce the function of a threat to be a means of establishing, maintaining, or modifying social relationships in an agonistic context without overt physical combat. An idealized interaction, then, would be one in which one animal approached another too closely, the latter threatens the former, and the former flees without having been attacked. It should be apparent, however, that such clear-cut exchanges are not always the case. Recall Kummer's observation of a hamadryas female that returned to the vicinity of the male (i.e., reduced social distance) that had just threatened her. Neither is there uniformity in the situations or stimuli that elicit threat displays. A particular stimulus, such as the proximity of another individual, may or may not elicit a threat, depending, among other things, on the relationship between the potential threatener and other members of the social group. For example, rhesus females being groomed by a male are far less likely to threaten a nearby individual than when they are alone (Redican, unpublished data). Since both social antecedents and consequences of threat displays may vary to such a remarkable degree, the lesson is clear that behavioral observations must include more precise quantitative descriptions of these events. The need is even more urgent in more ambiguous expressions, as we shall shortly see.

An additional perspective from which the criteria for threat expressions may be approached is one that stresses the displayer's motivational state. Briefly, an individual exhibiting a threat is said to be in both an aggressive and fearful internal state. At one extreme, a predominantly aggressive animal might be expected simply to chase, attack, or display what has been described as a "tense-mouth face" (van Hooff, 1969, p. 22). In this case the eyes are fixed and staring, the mouth is tightly closed with lips compressed and perhaps drawn slightly inward, eyebrows are usually lowered, and ears flattened (Hinde and Rowell, 1962, p. 4; Bolwig, 1963, p. 320; 1964, p. 186; van Lawick-Goodall, 1968a, pp. 318–319; van Hooff, 1969, pp. 22–25). The expression is seen by the initiator in a chase, attack, or stealthy approach. It seems reasonable to interpret the components of this expression to be principally protective responses in an animal that intends to engage in an imminent physical altercation. The elaborate configurations of a threat display are absent, however. On the other hand, a predominantly fearful animal would be expected to flee, grimace (see following), avert his gaze, crouch, or present. The open-mouth threat display can thus be viewed as the outcome of both fearful and aggressive internal states. It is thus not entirely accurate to describe a threatening animal simply as aggressive—a not uncommon practice in the literature.

B. Grimace

With the discussion of many of the structural components of facial expressions accomplished, we can proceed through this and the remaining expressions with fewer diversions. The features of the grimace appear to be fairly uniform throughout the wide range of genera studied, including the ape [e.g., the gorilla (Schaller, 1963, pp. 209–210, 219) and orang-utan (MacKinnon, 1971, p. 174)]. The mouth is opened in varying degrees and the corners of the mouth and lips are retracted, thus exposing the teeth quite clearly (van Hooff, 1969, p. 40). In more extreme forms of the expressions in the rhesus, the ears are usually flattened and the head drawn back on the shoulders (Hinde and Rowell, 1962, p. 11) (Figs. 10 and 11). The eyes may be oriented toward the opponent or, more commonly, may rapidly alternate toward and away from the other individual (van Hooff, 1969, p. 40; Møller *et al.*, 1968, p. 344). Robert Miller (1967, p. 133) also noted this rapid glancing directed toward a frightening object and referred to it as "repetitive peeking." In the context of agonistic interactions, G. Mitchell (personal communication) suggests that it is highly advantageous for a fearful subordinate to be aware of the location and intention movements of a potentially aggressive animal, but at the same time he must avoid staring at the latter since this is a form of threat. The resulting compromise is a series of quick glances. In a fearful animal, then, direct eye contact with other individuals is typically of high frequency but low duration; the converse generally holds for a threatening animal.

Just as the threat expression is most frequently given by the more dominant animal in an interaction, the grimace is given by the subordinate in most cases. Rowell (1966, p. 437) found that 97% of grimaces in a captive group of olive baboons were given by a subordinate to a dominant animal. Van Hooff (1969, p. 11) depicts the tendency to flee in most grimacing monkeys as being moderately present, although inhibited (Hinde and Rowell, 1962, p. 11), whereas that of attacking is not evident at all. He thus interprets the gesture as having an appeasement function, for not only does it not elicit an attack in the recipient but it also interrupts aggressive behavior in the dominant monkey. We shall take a closer look at this type of behavioral analysis at a later point.

The grimace seems to take place in three major contexts. The first has to do with overtly aggressive behaviors in which an animal that is the recipient of an actual or impending attack will display the grimace [e.g., *Macaca assamensis* (Osman Hill and Bernstein, 1969, p. 11), *Papio ursinus* and *P. anubis* (Hall and DeVore, 1965, p. 101), *Pan troglodytes schweinfurthii* (van Lawick-Goodall, 1968a, pp.320–321, 328, 343)]. For instance, in the vervet monkey the grimace is elicited in situations of intense agonism and as a response to a genital display (Struhsaker, 1967c, p. 52). Since the gesture occurs in agonistic contexts, one might expect sex differences in the frequency with which it occurs. Reports of sex differences again do not always agree. Mason *et al.* (1960, p. 81), Møller *et al.* (1968, p. 346), and Cross and Harlow (1965, p. 43) found that captive rhesus females grimace more often than males, whereas Altmann (1968, p. 59) observed feral rhesus males to screech and grimace more frequently than females: "Females tend more than males toward a strategy of 'come-closer-or-scream.'" Such marked differences between observers is puzzling but by no means uncommon, as we saw above for the threat expression.

The second major context of the grimace has to do with copulation, in which case it is more often displayed by the male. Male chacma baboons were observed to grimace on approximately 40% of copulatory sequences (Hall, 1962, p. 289), but Saayman's (1970, p. 99) figure (85%) is much higher for the same species. The latter author also found that it occurred infrequently in the female, especially after she had been chased by the male prior to mounting (cf. van Lawick-Goodall, 1968a, p. 362, for *Pan*). Hall and DeVore (1965, p. 95) "occasionally" observed grimacing in copulating adult male chacma but not olive baboons. Maxim and Buettner-Janusch (1963, p. 176), however, observed olive baboons "baring [their]

FIG. 10. A grimace in an adult female rhesus monkey.

teeth while emitting low grunts, which increased in frequency and intensity" during mounting. (The sex of the animal thus displaying was not indicated.)

Grimacing during copulation has been frequently reported for macaques [e.g., pigtailed macaque males (Tokuda *et al.*, 1968, p. 289)]. Copulating male rhesus (Southwick *et al.*, 1965, p. 152) and pigtailed (Tokuda *et al.*, 1968, p. 289) macaques may not only grimace but emit a sharp, high-pitched screech vocalization as well. The vocalization and possibly the grimace seem to occur more frequently during mounts leading to ejaculation, but this is not invariable. Tokuda *et al.* (1968, p. 289) observed the screech and grimace only at the moment of ejaculation in pigtailed males. Among 7 rhesus males observed in our laboratory, only 1 displayed both patterns, and they appeared either at the ejaculatory mount or during several consecutive nonejaculatory mounts (Redican, unpublished data). Altmann (1962, p. 385) also records this type of vocalization (described as "eee") but not the grimace during copulation. It seems probable that, since the two are so highly associated, he may have listed them as one category. In addition to this close relationship with vocalization, the grimace may function as a type of appeasement gesture for the female that looks back toward the male from time to time and that, in view of the circumstances, may have a high tendency to flee. The grimace and vocalization do not appear to accompany invariably sexual arousal per se, however, since, at least in captive adult rhesus males, masturbation to ejaculation may be accomplished in solemn silence with no discernible facial expression. (Occasionally the lower lip may be drawn

FIG. 11. A Subadult male baboon grimaces and leans away at the close approach of a mature adult male. (Photograph courtesy of Tim Ransom.)

downward and the teeth slightly exposed.) A more precise understanding of these behaviors must await further research into questions such as the influence of nearby conspecifics on sexual behavior patterns. There is a possibility, for example, that a mounting male's screech may serve as a spacing mechanism by inducing other adult males to maintain an appropriate social distance.

The third and final social context in which grimacing occurs has to do with forms of greeting. Greeting will be viewed here as behaviors occurring upon first meeting after separation, regardless of the dominance status of the animals concerned (Andrew, 1964, p. 228). Greeting responses may also occur when a subordinate is trying to establish contact with a dominant individual (Andrew, 1964, p. 228), or, one might add, at the approach of the latter to a subordinate animal. Thus, as Marler (1968, p. 436) suggests, the grimace is often used to evade another animal with a minimum of aggressive arousal—as in the first context mentioned above—but it can also be used in approaching another animal, which once again points to the importance of context in assessing the communicative significance of facial expressions (Fig. 12). Several species of lemur grin when approached by a dominant animal (Jolly, 1966, p. 56), and Struhsaker (1967a, p. 289) observed a juvenile female vervet grimacing as she approached the alpha female. Similarly, in the rhesus monkey the approach of a dominant animal will elicit a grimace, and any assertion of dominance will result in

FIG. 12. A young adult male baboon displays a wide grimace as he approaches a mother and young infant to handle the latter. (Photograph courtesy of Tim Ransom.)

an increased intensity of grinning (Andrew, 1964, p. 58). The close association of grimacing and greeting is again made for the rhesus by Hinde and Rowell (1962, p. 12): "Since the frightened grin is often given on the approach of a dominant animal, it may appear to be (and may perhaps evolve into) an expression of greeting. Greeting, in fact, may usually involve an element of fear."

Bernstein and Draper (1964, p. 86) introduced an adult male rhesus into a captive group of juveniles (2½ to 3½ years old) and found that grimacing was the most common behavior exhibited by the juveniles in the first hour. Juveniles occasionally approached the male but usually grimaced and withdrew before coming too close. "Lipgrinning," yawning (see following), and fleeing were typical responses to an introduction of an adult male into a captive group of sooty mangabeys (Bernstein, 1971, p. 36). Bertrand (1969, p. 79) reported that stumptailed macaques also grimace when approached or looked at (stared at) by a dominant monkey or human observer. As indicated above, greeting behavior is not restricted to the subordinate animal. As two chimpanzees approach each other, both dominant and subordinate may grin, the latter also vocalizing (van Lawick-Goodall, 1968a, p. 365). Carpenter (1940) reported that when an adult female gibbon (*Hylobates lar*) perceived a newly introduced juvenile in her cage, she "approached it repeatedly, slightly opening her mouth, showing her teeth and protruding her tongue" (p. 101). Carpenter (1940) also described a very instructive instance of greeting in two gibbons: "The facial expression involves a muscular pattern which may best be described as being similar to the human smile. The corners of the mouth are drawn back, baring the teeth, and the tongue is often protruded" (p. 167). This observation is pertinent to much of the theoretical speculation surrounding the evolution of the human smile to be discussed shortly.

Once again, R. J. Andrew stresses the protective characteristics in the evolution of the grimace by pointing to several converging lines of selective pressure. In primitive mammals (e.g., the opossum) which greatly rely on gustation and olfaction in exploring the environment, the need for an efficient protective response to expel noxious objects from the mouth is clear. Withdrawal of the mouth corners and lips, together with occasional protrusion of the tongue, facilitate such an expulsion. Andrew is careful to point out, however, that these responses may not have the same function in the insectivores and primates (Andrew, 1963b, p. 8; 1964, p. 281). *Lemur catta* grins not only when smelling scent marks of other lemurs but when threatened by superiors and during all "spat" calls. Jolly (1966, pp. 135–136) thus concluded that the connection between protective responses and defensive ones (e.g., grimace) seems to be well-established in the lemur.

An important factor in the grimace concerns the high-pitched vocalizations that accompany fearful situations in many primates. In Old World monkeys, apes, and many prosimians, screams are accompanied by withdrawal of the mouth corners. Andrew (1963b, p. 17; 1965, p. 91) suggested that the contraction of neck muscles (e.g., the platysma) may provide functional support for such intense vocalizations and at the same time aid in withdrawing the mouth corners. He also suggested that the grimace has undergone considerable evolution in this respect. In *L. catta* the grimace appears with vocalizations of lesser intensity than shrieking, which Andrew views as a drop in the threshold of grinning toward the level of most catarrhines; the latter often remain silent when grimacing (e.g., *Macaca*) (Andrew, 1963b, p. 17, 34; van Hooff, 1969, p. 75). Grimacing in chimpanzees may or may not be accompanied by "squeak" vocalizations, but both variants appear in similar social contexts (van Lawick-Goodall, 1968a, pp. 320–321).

Finally, Andrew allows for the possibility that teeth exposure as a preliminary to biting may be a variable in the evolution of the grimace. But, as we have noted above, the grimacing animal rarely follows up a grimace with an overt attack (although this is not to say that this was the case in earlier evolutionary history). Furthermore, primates make greater use of canines and incisors in biting—teeth that can be readily exposed without going so far as grimacing (Andrew, 1964, p. 282). In view of these considerations, intention to bite seems of lesser significance in the evolution of the grimace than the other factors discussed. Nevertheless, Andrew (1964) importantly notes that "[mouth corner] withdrawal could subsequently be used either in expelling objects from the mouth or in biting with the molars. . . . The fact that withdrawal may have two functions does not mean that two entirely different causations of the response are involved" (p. 282).

A detailed discussion of the causation of human facial expressions cannot be undertaken in this paper, but a few comments are in order stemming from many of the observations discussed above. Particular attention has been focused on smiling and laughter in humans, and both Andrew (1965, p. 91) and van Hooff (1969, p. 83) stress the homology between the human smile and the fear grimace. Andrew (1965, p. 91) writes "When a man maintains a fixed smile in response to the verbal attacks of a superior, or when he smiles on meeting a stranger, he is behaving defensively in the same way as his fellow primates." Similarly, van Lawick-Goodall (1971, p. 275) suggested "If the human nervous or social smile has its equivalent expression in the chimpanzee, it is, without doubt, the closed grin." Andrew further suggests that the grimace, which originally was a response to startling stimuli, has evolved in man to be associated also with much smaller (and sometimes pleasurable) changes in stimulation; that is, the smile has come to occur in pleasurable situations as well as fearful or startling ones (Andrew, 1965, p. 92). The ethologist, Eibl-Eibesfeldt (1970, p. 132), associates the origin of the human smile with a "friendly showing of teeth" which is said to occur during social grooming and play:

> A number of primates expose their teeth with a friendly gesture, derived from grooming (e.g., macaques and some lemurs expose the teeth with a gesture that is reminiscent of grinning, when performing grooming movements in the air). The possibility of our smiling response having a similar origin must be considered (Eibl-Eibesfeldt, 1971, p. 244).

The present writer finds it difficult to accept Eibl-Eibesfeldt's analysis for several reasons. First, grimacing—if indeed this is the expression referred to—is less of a "friendly" expression than a fearful one. Second, nonhuman primates certainly do not prominently display their teeth during grooming but rather lipsmack. Third, exposure of teeth is a feature of the play face, but the pattern differs markedly from smiling. Specifically, in the play face the corners of the mouth are not generally drawn back and the teeth tend to be rather widely separated. The converse, of course, applies to both nonhuman grimacing and human smiling. Several reservations may also be made with respect to Bolwig's (1964) assertion that the smile has evolved from either an actual or intended play–bite. He therefore feels that the smile is an expression of "joy and anticipation of play" in nonhuman primates. Not only are there difficulties in dealing with motivational states such as joy, but again the morphological configuration of the play face does not usually involve withdrawal of the mouth corners to expose the teeth, as we shall see below.

There is a certain amount of disagreement concerning the suggested evolution of human laughter. We shall defer a discussion of this topic until we have considered the play face.

In this and other sections, concepts such as appeasement, submission, and dominance have or will be utilized. The reader should be aware that such intervening variables undergo a process of continual reorganization and reinterpretation by behavioral scientists. These terms have been used rather casually in the present paper since this is not the proper time to examine thoroughly the rather complex issues involved. Briefly, it is this writer's impression that a term such as appeasement has been widely established in the literature in the absence of a rigorous body of quantitative evidence. If one understands an appeasement display to be one that terminates or averts the expression of overt hostility toward the displayer, this may be a reasonable analysis for some agonistic interactions. But one has no reason to assume that the paradigm can be generalized to every or even to most occasions on which the grimace, for example, is displayed. Once again the need is apparent for more quantitative data on the events preceding and following facial and other displays in the sender and percipient.

In summary, the grimace involves the prominent exposure of the teeth while the mouth typically remains closed and the corners of the mouth are retracted. Other protective responses may accompany it, and the animal usually avoids fixed eye contact with the source of alarm. Subordinate animals grimace more frequently than dominant animals. The expression occurs in three major social contexts: (1) as a response to actual or imminent aggression; (2) during copulation; and (3) during greeting. It has generally been interpreted as an appeasement or submission display. Some implications for human smiling are made.

C. Open-Mouth Grimace

In a few instances, one or more variations of the grimace have been described, principally by van Hooff. Basically they involve components of the threat and grimace, so that the animal's mouth is relatively widely open but at the same time the teeth are visible. The tendency to flee is said to be higher than in the case of threat. This type of expression often occurs in response to an open-mouth threat by a dominant animal (van Hooff, 1969, pp. 32–40). Further discussion of this expression would not be particularly profitable, since the threat and grimace have been examined in detail. At the same time, the intermediate nature of this expression illustrates that two distinct facial expressions can very gradually merge into one another and that the change or merger itself can be of communicative significance. In passing, one notes that the open-mouth grimace has been observed in the vervet (Gartlan and Brain, 1968, p. 279), the stumptailed macaque (Bertrand, 1969, p. 78), and the chimpanzee (van Lawick-Goodall, 1968a, p. 320). In the latter species, at least, open-mouth grimacing is specifically interpreted as being an an affective continuum with closed-mouth grimacing (van Lawick-Goodall, 1971, pp. 274–275) (Figs. 13 and 14).

Gelada monkeys (*Theropithecus gelada*) display an extraordinary expression which is apparently unique to the species and probably falls in the present category of expression or that of grimace. The long upper lip is raised and flipped so that the pink inner mucosal surface becomes visi-

FIG. 13. A young subadult male baboon screams and displays an open-mouth grimace toward another subadult male shortly after a fight. (Photograph courtesy of Tim Ransom.)

FIG. 14. A subadult male baboon looks back and gives an open-mouth grimace toward the mature male mounting him. (Photograph courtesy of Tim Ransom.)

ble and stands in marked contrast to the dark face. In this "lipflip" expression, the lips are lifted to cover the external nares completely and the eversion may be rapidly repeated (van Hooff, 1969, p. 32; Hill, 1970, p. 524, 596). The social circumstances of this expression in geladas are not particularly clear at this point, but John Crook (1965) described geladas exhibiting the lipflip (or "face lift") immediately before the animal's withdrawal from a dispute, and he associated the expression with a state of anxiety. Hill (1970, p. 596) notes that it occurs in greeting and may be an appeasement gesture. Van Hooff (1969, p. 45) included one form or another of the lipflip as either the "staring bared-teeth scream face" or the "silent bared-teeth face," and in the latter case observed that it frequently alternates with lipsmacking. One also notes that geladas display a distinct lipsmack expression as well (van Hooff, 1969, p. 49; Hill, 1970, p. 596).

D. Lipsmack

This expression is characterized by a relatively unambiguous set of facial components but occurs in perhaps the widest variety of social circumstances. The lower jaw moves up and down but the teeth do not meet. At the same time the lips open and close slightly and the tongue is brought forward and back between the teeth, so that the movements are usually quite audible. Zuckerman (1932, p. 230) noted several decades ago that lipsmacking actually involves more facial companents than the lips themselves. To the best of the present writer's observational capacity, the actual smacking noise appears to be made by the tongue breaking contact with the roof of the mouth and/or upper lip or row of teeth, rather than by the lips themselves parting. The tongue movements are often difficult to see, as the tongue rarely protrudes far beyond the lips. The ears are usually flattened and the brows pulled back. The head may be tilted upward or bobbed repeatedly up and down during the display (van Hooff, 1969, pp. 47–48; Hall, 1962, p. 309; Anthoney, 1968, pp. 359–360). As in the case of other facial expressions, one or more of the above components may appear independently.

Van Hooff (1969, pp. 48–49) distinguished at least two forms of the "lip-smacking face": the "tongue-smacking face" and the "true lip-smacking face." The former is described as being displayed with a lower rate of facial movement than the latter and with a greater degree of tongue protrusion. The present writer feels it wiser at this point to postulate a unitary basic form of lipsmacking with, however, varying rates of facial movements. Perhaps, as a consequence of rapid jaw and lip movements, the element of tongue protrusion is at a minimum in rapid lipsmacking. Few if any field or laboratory observations on lipsmacking familiar to this writer have been reported in the form of the above discrete categories.

One can distinguish several circumstances in which this expression occurs. It is more frequently observed in the animal taking the initiative during grooming, especially in *Macaca* and *Papio: M. radiata* (Rahaman and Parthasarathy, 1968, p. 263); *M. mulatta* (Hinde and Rowell, 1962, p. 15); *M. arctoides* (Blurton Jones and Trollope, 1968, p. 388); and *Papio ursinus* and *P. anubis* (Hall, 1962, pp. 303–304; Hall and DeVore, 1965, p. 105). It also ocurs in *Cercopithecus aethiops* (Struhsaker, 1967c, p. 5). Why behavior such as lipsmacking should occur in grooming is not difficult to understand. When one animal grooms another it often chews and ingests any debris or particles of skin that are uncovered. These may be picked off by hand or removed directly with the mouth [e.g., chacma and olive baboons (Hall, 1962, p. 304; Hall and DeVore, 1965, p. 105)]. As van Hooff (1969) expressed it, "the whole [tongue-smacking] process gives the impression that the animal is tasting the particle" (p. 48). The lip movements are very often not continuous throughout grooming. There may be no lipsmacking at all for rather long periods, even though the animal is grooming quite intently. The resumption of lip movements often apparently coincides with the discovery of a particle on the groommee's fur, since it often immediately precedes removal of debris with hand or mouth.

The motor patterns typical of lipsmacking have much more widespread communicatory significance. Andrew (1964, p. 302), for example, stressed their extension from routine grooming during the care of an infant to other situations such as courtship. The close relationship between lipsmacking and grooming is illustrated in *Erythrocebus patas*, which moves the mouth and lips very slowly during grooming. The more rapid lipsmacking pattern seen during grooming in *Macaca* and *Papio* is thus not evident in patas monkeys (Hall *et al.*, 1965, p. 40). These same authors also reported (Hall *et al.*, 1965, p. 47) that the groomer rarely picks off the particles of debris with its mouth in this species (a pattern common in *Papio*, for example), so this particular component of the grooming pattern may be of some importance in shaping the form of lipsmacking.

In perhaps the majority of social interactions in which it occurs, the lipsmack has been interpreted as an appeasing or submissive display. When it is given at the start of grooming bouts in the rhesus, for example, or when given by a frightened subordinate to a dominant animal, it seems to reduce fear and/or aggression in the party concerned (Hinde and Rowell, 1962, p. 16). Three similar interpretations were made independently for the bonnet macaque: Simonds (1965, p. 182) viewed it as a clear indication of subordination; Kaufman and Rosenblum (1966, p. 215), referring also to the pigtailed macaque, found that it was frequently placating or submissive and was generally associated with presenting, crouching, or physical withdrawal; and Rahaman and Parthasarathy (1968, p. 260) noted that if a subordinate accidentally makes eye contact with a dominant animal the former often lipsmacks. A subordinate crabeating macaque (*Macaca fascicularis*) often may approach and begin lipsmacking after a dominant animal raises his brows toward him (Mitchell *et al.*, no date, p. 9). Geladas also frequently lipsmack in response to an eyelid display (Bramblett, 1970, p. 329). Thelma Rowell (1966, p. 438), however, found that *dominant* baboons (*P. anubis*) lipsmacked to subordinates in 85% of such interactions. Furthermore, baboon males lipsmacked more frequently than females (Rowell, 1966, p. 438; 1967, p. 506), although the converse was found for rhesus monkeys in an experimentally produced stressful situation (Rowell and Hinde, 1963, p. 241). Immature chacma baboons occasionally lipsmack when presenting to adult males, which may also lipsmack. On rare occasions, adult males simultaneously lipsmack and present to each other (Saayman, 1971, pp. 48–49). Ransom (1971) observed lipsmacking and tongue protrusion in adult, male olive baboons being mounted by another male. The male being mounted might also reach back toward the other animal in much the same way as a female might. As in the case of baboons, the most dominant individuals in a group of mangabeys (*Cercocebus albigena*) were observed to lipsmack most frequently (Chalmers and Rowell, 1971, p. 5). However, the dominant mangabeys lipsmacked to each other rather than to subordinates. In addition, dominant mangabeys were the objects of more lipsmacking displays than were subordinates, but this relationship was not seen in baboons by these authors. In the vervet, lipsmacking "seemed to express nonaggression, thus permitting the close proximity of subordinate and dominant individuals" (Struhsaker, 1967a, p. 291), and was observed in all ages except infants (Struhsaker, 1967a, p. 290). In this species, lipsmacking and teeth-chattering sometimes occurred when an animal was supplanted from an area by a more dominant animal (Struhsaker, 1967c, p. 15). Langurs move their tongue in and out of their mouth as a submissive gesture, which Jay (1965a, p. 237) evaluated as intermediate between gaze aversion and grimacing.

In most of the above situations, lipsmacking serves as a pacificatory gesture. According to van Hooff (1969, p. 50), it gives rise to aggression in neither the sender nor recipient. This hypothesis is quantitatively supported by Chalmers' data for the black mangabey (see Tables I and II) which show that in no instance following a lipsmack did either the sender or recipient attack. Chalmers (1968, p. 263), therefore, interprets it as a conciliatory gesture that "suppresses" any tendency to flee or attack on the part of the recipient of the expression. In this species the lipsmack was often accompanied by a pronounced lateral shaking of the head, a response that Chalmers (1968, p. 261) viewed as peculiar to mangabeys. We have seen above in several species, however, that this motor pattern is frequently associated with situations evoking gaze aversion. An analogous process may be operative in the mangabey, for when they were seen to lipsmack alone the sender was never seen to flee, whereas if headshaking was also performed he was just as likely to flee as to remain (see Table III) (Chalmers, 1968, p. 263).

Lipsmacking is often observed during copulation, when it seems to serve a pacificatory or reassuring function similar to the above instances. Its mode of occurrence may vary widely across species in this context, however. Thus, both male and female bonnet macaques lipsmack when copulating (Rahaman and Parthasarathy, 1968, p. 263), whereas the female rhesus does so much more frequently than the male. When female rhesus monkeys are being mounted they may look back toward the male, clasp his hair, and/or lipsmack. Zumpe and Michael (1968, p. 119) suggested that this set of behaviors almost invariably occurs during the ejaculatory mount, such that the female begins to turn her head and lipsmack just prior to the male's ejaculatory contraction. Lipsmacking alone, however, certainly occurs during nonejaculatory mounts, although it appears to be less prominent in the early sequences of mounts (Redican, unpublished data). On the other hand, lipsmacking was not observed at all in copulating stumptailed macaques by Bertrand (1969, p. 82) or only in the female during the early stages of copulation (Blurton Jones and Trollope, 1968, p. 337). In the chacma baboon, Bolwig (1959, p. 150) observed lipsmacking prior to and during copulation, although he did not state whether one or both sexes lipsmacked. In the same species, Hall (1962, p. 289) did not notice lipsmacking during mounting, although he noted that it may have occurred beforehand. In the chimpanzee, lipsmacking has been suggested to occur rarely, being largely replaced by the grimace (van Hooff, 1969, p. 67; Reynolds, 1970, p. 377). Van Lawick-Goodall (1968a, p. 362; 1968b, p. 202, 218, 241, 264, 286, 308), on the other hand, reported both slow and rapid lipsmacking in several circumstances in chimpanzees, including social and asocial grooming and copulation (chiefly in the case of males). Nishida (1970, p. 58) also observed lipsmacking in an adult male chimp that was approaching a presenting estrous female.

A further situation in which lipsmacking occurs is in forms of greeting. For example, an established rhesus male may lipsmack to a female just introduced to an enclosure (Hinde and Rowell, 1962, pp. 15–16). Lipsmacking in greeting seems especially pronounced in baboons. Bolwig (1959, p. 155) found that the chacma baboon lipsmacks when making a "friendly" approach to another, especially when approaching an infant held by its mother. When Hall (1962, p. 308) introduced a strange male into a troop of chacma baboons, by far the largest proportion of responses to and from him involved lipsmacking and ear-flattening. Apparently the lipsmack is also displayed much more readily than the grimace in greeting (Andrew, 1965, p. 92). In stumptailed macaques, lipsmacking functions not only as a greeting gesture but as an appeasement or reassuring gesture as well (Bertrand, 1969, p. 82).

Lipsmacking is often directed toward newborn or young infants. Adult male and female hamadryas often lipsmack and reach out to touch infants as they pass by (Zuckerman, 1932, p. 261). Hall and DeVore (1965, p. 83, 107) also found that chacma and olive baboons lipsmack when approaching a neonate and its mother. Rowell, Din, and Omar (1968, pp. 471–472) made similar observations of olive baboons in Uganda: "Adult females [and to a lesser extent adult males] take the keenest interest in the new-

TABLE III
LIPSMACKING WITH AND WITHOUT HEADSHAKING
RELATED TO RESPONSE OF DISPLAYING MONKEY [a]

Gesture	Behavior of displaying monkey after gesture	
	Flees	Does not flee
Lipsmack only	0	10
Lipsmack and headshaking [b]	9	11

[a] Adapted from Chalmers (1968, p. 263).
[b] Lipsmacking and headshaking together are associated with a significantly greater ($p = 0.024$) tendency to flee.

born infant. They peer at its face, lip-smack and present to it if they 'catch its eye,' touch or try to touch it and groom it." Adult female yellow baboons (*Papio cynocephalus*) frequently lipsmack both to black and to brown infants, although the pattern changes somewhat after the infant changes color (Anthoney, 1968, p. 362). Female rhesus monkeys—not necessarily the mothers—also lipsmack to newborns (Rowell, 1963, p. 381; Rowell *et al.*, 1964, p. 221; Kaufmann, 1966, p. 22). Hall and DeVore interpreted the lipsmack in these contexts to be a prelude to or an accompaniment of greeting.

Surprisingly, lipsmacking or a pattern similar to it also occurs in many agonistic encounters. On some occasions the lipsmack is a mild form of threat for the rhesus (milder than a direct stare), and, although it was infrequently observed in Altmann's (1968, p. 54, 58) field study, it was displayed more often by adult males than females in affective interactions. Altmann's (1962) earlier description of a lipsmacking rhesus seems to allow for this particular context: "In the extreme form, as given by adult males, the monkey walks briskly and directly toward the other individual, smacks its lips with head extended, ears and scalp pulled back, then quickly lifts the head (still lipsmacking) and simultaneously turns around, lowers the head to its normal position, and walks briskly away" (p. 378). Hinde and Rowell (1962, p. 15) also assign a degree of slight aggression to lipsmacking in rhesus monkeys, although their position is not entirely clear on the subject. The black mangabey lipsmacks slowly when grooming but very rapidly (approximately 5 per second) during agonistic encounters (Chalmers, 1968, p. 261). Similarly, the Celebes black ape (*Cynopithecus niger*/*Macaca nigra*) displays an accelerated lipsmack during aggressive interactions with slightly greater jaw movements and less tongue protrusion than its normal lipsmack pattern (Bernstein, 1970, p. 279). The increased rapidity of lipsmacking during agonistic interactions may, of course, have to do with an aggressive animal's heightened level of arousal. The patas monkey similarly has two types of lipsmacking (again the slower during grooming) and the agonistic significance of the faster variation seems clear. An adult male may lipsmack loudly during a high-intensity threat sequence while he is moving toward the other animal, and in one reported case an adult male chased a younger adult male and lipsmacked while so doing (Hall *et al.*, 1965, p. 29; Hall, 1967, p. 265; Hall and Mayer, 1967, p. 216). This type of lipsmack was observed in male–female aggressive interactions only when the male was the aggressor and then only by the male, but it occurred with identical frequency during female–female altercations (Hall and Mayer, 1967, p. 216). Hall (1967, pp. 277–278) described a typical agonistic episode as follows:

> In an episode involving several animals, we find that the aggressor will lip smack noisily while looking toward others in the group who may then join with her, also lip smacking, in attacking, for example, the adult female scapegoat of the group. . . . This lip smacking seems to differ from that which occurs in grooming interactions only in its great rapidity and noisiness, but the expressions which accompany it—the threat face and so forth—are entirely different. Possibly this lip smacking is an exaggerated form of that which signifies a relaxed, friendly relationship, and is thus used, in the aggressive context, as a signal which, combined with head turning, may enlist aid from the aggressor's associates. If this is so, then it is only secondarily a threat-intention signal.*

Hall's suggestion is an interesting one and deserves careful attention in future research. Enlisted aggression is certainly not unusual in primates (e.g., A threatens B, and C attacks B), and lipsmacking may be one variant of this complex type of interaction. This appears to be the context for a form of agonistic lipsmacking among rhesus monkeys observed by Lindburg (1971, p. 61):

> Subadult and large juvenile males most often used this expression and did so in tense interactions with high-ranking adults. The intent appeared to be to divert attention from themselves to other individuals. . . . Although these "diversionary" patterns rarely led to attack on a third animal, they apparently minimized the likelihood of attack on the threatening individual.

It is not an easy task to disentangle the functional significance of agonistic lipsmacking, and it is clear that more data are needed. Perhaps lipsmacking in this context may still function as a sort of appeasement gesture if the threatening animal anticipates an attack in return.

Another possible explanation as to why lipsmacking or expressions similar to it occur in aggressive interactions may have to do with the rapid opening and closing of the mouth during threat in a small number of reported species. For example, van Hooff (1969, p. 28) cites an "aberrant" threat pattern in *Cercopithecus diana* in which the mouth is quickly opened and closed in succession. Threatening male geladas "chomp the jaws in a chewing movement with mouth partially open and lips drawn over and covering upper and lower teeth" (Bernstein, 1970, p. 281). Carpenter's (1940, pp. 98–99) description of an aggressive gibbon is also quite instructive: "The excited, angered gibbon. . .will often open and close its mouth with tensed lips as if smacking them together and this emotional expression is shown in many situations which provoke aggressive excitement." (One notes, however, that the lips are tense and tongue-protrusion is not evident.) In a gesture that may be related, the orang-utan smacks its lips when threatening by "pressing the lips together, drawing air inward and quickly separating the lips to make a sharp smacking or kissing sound" (Davenport, 1967, p. 256). MacKinnon (1971, p. 172) described a quite similar threat pattern in orang-utans and noted that it may be followed by grunting vocalizations and branch-shaking or -dropping. It seems possible that in some groups of primates, at least, rapid lip movement may be a remnant of a rapid opening and closing of the mouth associated with the more typical threat display. This is not to say that other factors have not been important as well (e.g., bruxism), for if anything is clear about the lipsmack it is that it occurs in a tremendous variety of circumstances and it would be less than wise to suggest a unitary evolutionary basis.

We have not yet, in fact, exhausted the possible bases of lipsmacking, as a very important factor has to do with infantile behavior patterns. Andrew (1963b, p. 99) suggested that movements of the mouth during nursing may generalize to the form of lipsmack given in greeting, and this interpretation is also offered by Chevaliér-Skolnikoff (1968, p. 4) in the case of *M. arctoides*, whose infants lipsmack by 2 or 3 weeks of age. Thus, lipsmacking may avert aggression in many situations because its infantile components elicit behaviors from conspecifics that are characteristically directed toward infants. Anthoney (1968) in particular puts great stress on the importance of nursing movements in the ontogeny of this expression, and his hypotheses will be examined in greater detail below. In this writer's view, the external similarity between nursing and lipsmacking movements is still not sufficient to establish a firm association between the two patterns. The process by which nursing movements generalize to serve as a greeting expression, for example, remains to be more carefully explicated. There are some brief observations that lend support to the hypothesis that the earliest lipsmack patterns are *anticipatory* nursing or nonnutritive sucking movements. On two occasions in our laboratory, the close approach of a human observer preceded slow repetitive lip movements in a 1-month-old female rhesus infant. She did so while maintaining visual orientation toward the observer and physical (nonventral) contact with her mother. After a few seconds of looking and lipsmacking, she quickly turned her head and established nipple contact with her mother. It seemed as if the infant wanted to catch a last look before contacting the nipple but began anticipatory motor movements before turning. Thus, in a mildly threatening situation (such as the approach of another monkey), one of the first responses a young infant might make would be to regain the mother's teat. As the infant matures one would expect it to explore visually approaching strangers for lengthier intervals before returning to the nipple. As we have said, the motor pattern may come into play before the infant turns away. What the approaching monkey sees, therefore, is an infant standing, looking, and perhaps smacking his lips. It should also be recalled that the monkey approaching the infant if and when the latter is making these mouthing movements is likely to be lipsmacking as well (e.g., a female approaching a mother–infant pair). The infant may then come to associate this pattern with the approach of another monkey who, in most cases, is likely to initiate an affiliative interaction with the infant. A rudimentary basis might thereby be established for the lipsmack's occurrence in later affectionate or greeting interactions. Some sort of convergence may be operative with grooming patterns as well, since not only does an infant actively make nursing movements with its mouth but it may observe lipsmacking by its mother as it is being groomed. Much of the above is unadulterated speculation, but, at the present state of our behavioral understanding, it seems a reasonable hypothesis that the multiplicity of ontogenetic factors underlying the development of lipsmacking might contribute to the great number of social circumstances in which lipsmacking appears in adult life.

In summary, lipsmacking occurs in a very wide variety of social circumstances, including grooming, copulation, greeting, and antagonism. It is most commonly but not exclusively a pacificatory, affiliative, or appeasement gesture. Lipsmacking patterns may have their origins in neonatal nursing or nonnutritive sucking movements which have persevered and have been reinforced and modified in subsequent grooming interactions. Slow-to-moderately fast lipsmacking is usually seen in grooming, copulation, or greeting, whereas very rapid patterns are seen in antagonistic interactions as a mild threat display.

E. Grin–Lipsmack

An expression that shows several characteristics of both grimacing and lipsmacking has been reported for several species of macaques. Van Hooff (1969, p. 54) referred to a "teeth-chattering face," which he noted often alternates rapidly with lipsmacking, typically in subordinate individuals.

The lips are retracted vertically and the lower jaw is moved rapidly up and down so that the teeth usually meet in a chattering pattern. In rhesus macaques, at least, very brief episodes of lipsmacking may be frequently interspersed with vertical lip retraction so that it appears as if the animal were lipsmacking and grimacing simultaneously. Lip closure may not be as prominent in other macaques displaying this expression, however, especially in stumptailed macaques.

Stumptailed macaques may retract their lips, exposing and often chattering the teeth, in an expression similar to a slow lipsmack and a grimace occurring together (Mitchell *et al.*, no date, p. 5). Two stumptails may exhibit this expression when embracing in ventral–ventral contact, for example, or in response to the close proximity of a human observer. Blurton Jones and Trollope (1968, p. 367) described a grin–lipsmack in the same species which was said to originate in either a lipsmack or grimace. It occurred after hiting and threatening by two individuals and before they settled down to grooming (Blurton Jones and Trollope, 1968, p. 375). It also occurred in males immediately before and during copulation, occasionally by females during copulation, and by infants whose genitalia were being manipulated (Blurton Jones and Trollope, 1968, pp. 388–389). Also in stumptails, Bertrand (1969, p. 80) observed teeth-chattering which often alternated with lipsmacking or followed or preceded a grimace. In the bonnet macaque, Kaufman and Rosenblum (1966, p. 222) described similar "clonic jaw movements" which involves rhythmic contractions of the brow, retraction of the lips, and exposure of the teeth accompanied by chattering. It occurred most often in adult males as a prelude to copulation. [Bertrand (1969, p. 80) similarly noted for stumptails that teeth-chattering often alternates with lipsmacking or follows or precedes a grimace.] In the little-studied Assamese macaque (*Macaca assamensis*), a gesture described as a "lipgrin" with characteristics similar to those above was seen to precede and accompany positive social interactions, especially grooming, during which typically lipsmacking eventually replaces the lipgrin (Osman Hill and Bernstein, 1969, p. 11). Bobbitt, Jensen, and Gordon (1964, p. 73) described a "silly grin," with features similar to those described above, for the pigtailed macaque, and Hansen (1966, pp. 118–119) used the identical term for an expression given rarely by rhesus mothers to retrieve their infants. Baysinger, Brandt, and Mitchell (1972) frequently observed this expression in mothers who had been recently separated from their infants. Hinde and Spencer-Booth (1967, pp. 173–175) take issue with Hansen by asserting that "grinning" does not seem to be specifically adapted for retrieving rhesus infants, although they suggest that a correlation between a mother's grinning and an infant's return to the mother may be the common results of a fearful set of stimuli.

It seems clear, however, that Hansen was not referring to a grimace by itself, but an expression with several characteristics of a lipsmack as well.*

F. Yawn

Although the yawn is not considered a communicatory facial expression by several investigators, van Hooff (by omission) apparently among them, its role as such has been clearly established in several species. To be sure, yawning may result from a surplus of CO_2 or an insufficient supply of O_2 in the brain's arterial system. The external features of a fatigue yawn and a display yawn are quite similar. The entire set of teeth, including the canines, may be exposed, and as the yawn reaches a climax the head may be thrown back, the eyes tightly closed, and the ears flattened against the head (Møller *et al.*, 1968, p. 344) (Fig. 15). In geladas, the long upper lip is often completely everted as the mouth is opened (Zuckerman, 1932, p. 249).

Both human and perhaps nonhuman observers are faced with a difficult

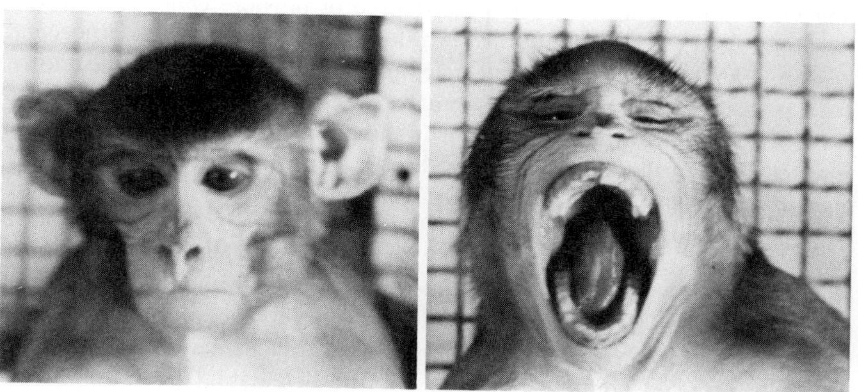

Fig. 15. Yawning juvenile rhesus male. These two photographs were taken within seconds of each other and illustrate how radical ear movements can be in nonhuman primates. This male's canines have not yet erupted.

* There is a certain amount of disagreement concerning the expression(s) described here. Van Hooff (1969, p. 54) and Bernstein (1970, p. 279) do not include the silly grin of *Macaca nemestrina* or *Macaca mulatta* as examples of grin–lipsmacking. In these cases I choose to "lump" rather than "split." Furthermore, Bernstein (1970, p. 279) claims that the expression is not found in *Macaca fascicularis*, in direct contrast to van Hooff (1969, p. 42), while the latter (p. 43) claims its absence in *M. nemestrina!*

task in distinguishing the different forms of yawning in all but extraordinary circumstances. In several births of rhesus monkeys observed in our laboratory, a number of neonates yawned within moments after birth, and another did so within 4 hours of birth. Clearly these instances of yawning were indicative of a metabolic and not a social need. In the case of adults the situation is understandably more complex. At least one observer, however, described a discernible difference in the two forms of yawning in a home-reared hamadryas baboon (Macdonald, 1965, p. 150). In circumstances less intimate than one's home, however, an observer faces an almost impossible task in distinguishing the two types of expression. If possible, one might correlate yawning with phenomena such as heightened muscular tonus, scratching, muzzle-wiping, rapid visual scanning, or other subtle behavioral cues. Our approach in this section will be to describe the social contexts in which yawning occurs and sidestep the issue of metabolic need versus communicative intent. We are more interested here in the role of this expression in an animal's ongoing social system than in the etiology of the expression per se.

The yawn occurs in two closely related circumstances. The first is typically described, with a certain amount of necessary ambiguity, as occurring when an animal is under some degree of tension, conflict, or stress. Yawning often ocurs in situations of mild stress in the rhesus (Hinde and Rowell, 1962, p. 20) and bonnet macaque (Rahaman and Parthasarathy, 1968, p. 264). Pigtailed and bonnet macaques, especially males, are said to yawn when the group or individual is threatened but when the physical manifestation of dominance relationships is inhibited, such as when a human observer enters a group area (Kaufman and Rosenblum, 1966, p. 216). Particularly in pigtails, the male may yawn once after copulation (Kaufman and Rosenblum, 1966, p. 217). Adult male rhesus in laboratory environments also frequently yawn between mountings in a copulatory sequence but very rarely during pelvic thrusts or grooming bouts (even though the latter may last longer than 30 minutes) which are interspersed between series of mounts (Redican, unpublished data). Bertrand (1969, p. 97) reports that the higher-ranking males of a group of stumptails yawned when observed steadily by a human, but two subordinate males of the same age instead engaged in what were described (Bertrand, 1969, p. 97) as "other displacement activities." Bertrand also observed tension yawns in *Macaca sylvanus* and *Macaca silenus*. Adult male patas may yawn when tense and alert in the presence of a human observer, and Hall *et al.* (1965, p. 34) suggested that a baboon would have barked in situations where a patas yawned, an hypothesis amenable to testing in a laboratory environment. When two captive groups of sooty mangabeys were introduced to each other, tail arching and yawning were the first displays observed. Attack and defense often followed (Bernstein, 1971, p. 38). Struhsaker and Gartlan (1970, p. 56) recently reported that yawns in agonistic situations are given only by adult and subadult males among patas. Yawning as an aggressive gesture is also limited to males among Lowe's guenons (*Cercopithecus campbelli lowei*) (Bourlière *et al.*, 1970, p. 313). In a captive group of vervets, only males were observed to exhibit "canine displays" (Durham, 1969, p. 92). Stress-induced yawning thus consistently emerges as an expression more characteristic of males than females.

The vervet once again proves to be of interest. Stress-induced yawns were observed in Uganda (principally in adult and subadult males in response to the presence of human observers) by Hall, Gartlan, and Struhsaker, but not at all in Kenya by Struhsaker (Hall and Gartlan, 1965, p. 53; Struhsaker, 1967c, p. 55). This difference is puzzling, and any explanation must await further field observations. Hall and Gartlan made an important point in suggesting that although yawning can be expected to be available to all ages and sexes, the "differential occurrences are probably related to the age and social status of the individual, which determines the distance of its approach to the [human] observer" (p. 54). Similarly, the frequency of yawning is conceivably a function of spatial arrangements among the individuals themselves and the frequency and types of agonistic interactions in which they engage.

FIG. 16. A young adult male directs a yawn away from a nearby consort pair during a harassment episode. (Photograph courtesy of Tim Ransom.)

A number of observations are available for macaques and baboons. In captive rhesus, yawning was displayed more frequently by older male animals (Møller *et al.*, 1968, p. 346). In an interesting study, Rowell and Hinde (1963, p. 237) found that captive rhesus displayed lipsmacking in mildly stressful situations (in order of increasing stress: offering of food by a human observer and having a human standing and looking close by the cage). However, there was a significant increase in yawning only in the latter circumstances and in an even more stressful one (i.e., a human in a grotesque mask and cape, no less, jumping up and down). Thus, yawns appear to supplant lipsmacks in more highly stressful circumstances. Tension yawns have been frequently reported for baboons, including *Papio hamadryas* (Kummer, 1968, p. 54). Bolwig (1959, p. 146) noted that yawning and scratching often occurred in feral chacma males when "sitting on guard" or watching other baboons; however, yawning was never seen in two captive baboons from the same area. Maxim and Buettner-Janusch (1963, p. 175) observed yawning in feral olive baboons when observed from a great distance. Chacma and olive baboons may also yawn when in conflict, in which case the gaze may or may not be directed toward the source of disturbance (Hall and DeVore, 1965, p. 100) (Fig. 16). Subsequently, Hall (1967) distinguished between directed and un-

FIG. 17. A yawning animal may be oriented directly toward or away from an adversary. This photograph illustrates the former in an altercation between 2 adult male baboons. The male on the left is crouching and screaming. (Photograph courtesy of Tim Ransom.)

directed yawns, the latter of which occur frequently and "would appear to be simply a tension or arousal indicator" (p. 278). A final note concerning baboons in this context comes once again from Niels Bolwig (1959, p. 147), who reports that when surrounded by chacma baboons in the field, he often yawned to assess its effects, which often resulted in a general centrifugal emigration of nearby baboons.

Yawning is also relatively common in the apes; e.g., white-handed gibbons (*Hylobates lar*) are reported to yawn when frustrated, "annoyed," "bored," or when in conflict (Carpenter, 1940, p. 168). Chimpanzees commonly yawn in stressful situations, particularly when a human observer is nearby (van Lawick-Goodall, 1968a, p. 329; Nishida, 1970, p. 60). Finally, gorillas occasionally yawn when in close proximity to human observers; e.g., one silverback male yawned more than 20 times in 15 minutes as he sat within 60 feet of Schaller (1963, p. 205).

In the second major set of circumstances, a yawn can apparently function as a type of overt threat display when the animal is oriented directly toward its opponent (Fig. 17). It functions as such in the chacma and olive baboon, in which case it is always accompanied by ear-flattening, eyebrow-raising, and other threat gestures (Hall and DeVore, 1965, p. 100; Hall, 1967, p. 278). *Macaca arctoides* is also reported to yawn directly in threat (Bertrand, 1969, p. 95). *Lemur catta* yawns during agonistic scent-marking displays, but Jolly (1966, p. 91) does not specify the orientation of the animals. Chalmers (1968, pp. 261–262) rates a yawn as just below a direct stare in terms of aggressiveness for black mangabeys and reported some interesting data concerning the effects of direction of yawning (Table IV). An animal is significantly more likely to flee after yawning away from his opponent than yawning toward him. These data would seem to support Hall and DeVore's distinction between directed yawns as threats and nondirected yawns as signs of stress.

TABLE IV
Direction of Yawning Related to Subsequent Behavior of Displaying Monkey [a]

Gesture	Behavior of displaying monkey after gesture	
	Flees	Does not flee
Yawn toward	4	14
Yawn away [b]	15	11

[a] Adapted from Chalmers (1968, p. 262).

[b] Yawning away is associated with a significantly greater ($p < 0.05$) tendency to flee.

It seems reasonable that much of the communicatory significance of yawning has to do with canine display, as this is the only facial expression considered thus far in which they are consistently and prominently exposed. [Even during snarling the teeth are exposed only at the premolar level (Andrew, 1963b, p. 65).] Coss (1968, p. 277) hypothesized that the canine profile superimposed—but only on occasion, one might note—against a red tongue or cheek has the same visual properties as tusks, horns, and claws displayed by many animals (i.e., tapering contours ending in a sharp point). To test the efficacy of this pattern, he presented a zig-zag pattern of pointed lines and a similar but rounded pattern to a group of human subjects and found that the former pattern elicited a significantly greater degree of pupil dilation. The importance of canine display in yawning nonhuman primates is underscored by John Crook's suggestion (in Vine, 1970, p. 293) that the large canines of the herbivorous gelada monkey are of functional significance only in yawning and "face lift" (lipflip) displays (an apparent exception to the above generalization). As Napier and Napier (1967, p. 22) note, canines certainly have nonalimentary functions among baboons since they have no need for carnivorous adaptations. The pronounced sexual dimorphism in canine size among Cercopithecidae and Pongidae (Napier and Napier, 1967, p. 22) also fits well with the observation that males yawn more frequently than females. However, it would be unwise to restrict the function of canines exclusively to display (even for herbivorous species), since canines are clearly useful in defense against predators and in intraspecific hostilities. For example, Saayman (1971, p. 41) occasionally observed severe slashes inflicted by canines in aggressive encounters among chacma baboons. One must also account for the obvious degree of wear in the canines of old adults in noncarnivorous species.

The yawning pattern among patas monkeys warrants a certain amount of scrutiny. In this species the head is lowered before yawning to an angle 45° or less from the horizontal. The yawn is then executed and the head quickly raised up. To paraphrase Struhsaker and Gartlan (1970, p. 56), the mouth is opened widest when the head is lowered most, so that a yawning patas' teeth are probably not visible to an observer situated in front of the displayer. This would not appear to be an obstacle for more lateral observers, however, which would more often be the case if patas' yawns are directed away from another animal. The role of canine display in yawning thus still seems to be tenable.

The evidence seems clear that yawning can function as a social cue. Altmann (1967b, p. 338) for one, agrees that a *directed* yawn should be interpreted as a semantic message. However, an *averted* tension yawn may only be secondarily communicative. A parsimonious view of an averted yawn would not allow to conclude that it is performed with the intention of eliciting a response from another animal by the very fact that it is not so directed. At the same time, however, it may elicit social responses in a conspecific (e.g., increased distance), so its communicative status is not easily disputable.

In summary, yawning often occurs in monkeys and apes experiencing some degree of conflict or stress. When oriented directly at another animal, it is a form of threat slightly less intense than an open-mouth staring threat. It is displayed more frequently by males than females. Canine display is apparently one of its most important communicative functions.

G. Play Face

Monkeys and apes often display a particular facial configuration when playing. The mouth is rather widely open and the mouth corners usually retracted only slightly; the upper lip may be tensed and curled over the upper incisors in some species (van Hooff, 1969, p. 63; Bolwig, 1964, p. 180), but patterns of teeth exposure vary (Fig. 18). This basic description holds for such species as the stumptailed macaque (Bertrand, 1969, p. 100), the patas monkey (Hall *et al.*, 1965, pp. 40–41), and the vervet (Struhsaker, 1967c, p. 32). The expression may be displayed by adults as well as by younger animals. For example, on frequent occasions adult, male rhesus monkeys exhibit the expression when playing with rhesus infants with whom they are caged in our laboratory (Redican and Mitchell, 1974). In their play episodes, a great deal of play-biting takes place on the part of both animals. The infant's mouth is held widely open with the lips usually concealing only the upper row of teeth, but the male's canines are usually prominently exposed. Degree of teeth exposure thus seems to be a highly variable characteristic in the play face. The infant's ears are often markedly flattened, but the adult's are rarely so. This could conceivably reflect different levels of arousal or need for protective responses during play.

The play face has been described for all major groups of apes. In the chimpanzee, at low-intensity play the lower lip is retracted to show the lower teeth, and at higher intensities all of the anterior teeth may be exposed (van Lawick-Goodall, 1968a, p. 318; 1968b, p. 258). Loizos (1969, p. 263) observed the play face in *Pan* most frequently in social play, but also during acrobatics and exploratory forms of solitary play. The play face appeared in virtually all of the play sessions observed by van Lawick-Goodall (1968b, p. 258). Once displayed the play face recurred throughout most of the episodes, but contrary to Loizos' (1969, p. 235)

FIG. 18. Two 3-month-old infant baboons in play, one showing the open mouth, raised brows, wide-open eyes, and flattened ears of a play face. (Photograph courtesy of Tim Ransom.)

observations, play was usually initiated without the expression being given at the beginning of a play episode. This issue is not quite clear at the present time, for at another point van Lawick-Goodall (1968b, p. 244) described instances in which play was clearly initiated with a play face: infants slightly older than 5 months "frequently try to initiate play, pulling their mothers' hands towards them with play faces." (Two mothers typically responded with play; two did not.) As Bolwig (1964, pp. 179–182) suggested, a major variable underlying the expression seems to be readying the mouth for a play-bite. Moreover, Loizos (1969, pp. 263–264) made the interesting suggestion that a pronounced exposure of the teeth during play may tend toward a form of grimace, since there may be a need for appeasement gestures in many socially ambiguous play situations. One notes, however, that a screech would seem to be less ambiguous and far more efficient in this circumstance. Schaller (1963, p. 209) also recorded that playing gorillas partially open their mouths with the corners drawn back into a "smile," but the gums and teeth are not exposed. Clarence Ray Carpenter (1971) recently described an interesting sequence of behaviors in the gibbon colony at Hall's Island, Bermuda. Individual gibbons may stagger, fall, or somersault energetically while the mouth may be held open, the head thrown back, and the eyes closed. Carpenter hypothesized that this pattern functions as an invitation to other gibbons to play. By "cutting up" (Carpenter's ingenious term) like this, one animal may generate affective excitement in others which may eventually induce social play. Carpenter has observed a play face in gibbons during both solitary and social play. (In the latter case it is given by both partners.) The teeth were usually visible when the gibbons were viewed directly from the front. MacKinnon (1971, pp. 181–183) provided photographs and a description of the play face in orang-utans: the mouth is open with lips curled back tightly; the teeth may be covered but at least one set is visible and utilized in play-biting; facial wrinkles are exaggerated; and the eyes are normally open. He noted that the play face and "a particular jaunty approach communicate a desire to play" (p. 174) and that the play face subsequently becomes greatly exaggerated during actual play.

The play face is particularly interesting in view of its possible association with human laughter. Bolwig (1964, p. 182) hypothesized that both smiling and laughter have derived from this expression, but we have taken exception to the former assertion above. Van Hooff (1969, p. 83) and Blurton Jones (1969, pp. 450–451), among others, associate the play face more specifically with laughter. There is a reasonable amount of support for this position. Loizos (1969, p. 235) noted "soft guttural exhalations" in playing chimpanzees. Van Lawick-Goodall (1968b, p. 244) presented particularly strong evidence supporting the association between play and laughter: "During play chimpanzees frequently make a series of low panting noises . . . which sound roughly like human laughter." Play with laughter is more frequent in younger than in older chimpanzees (van Lawick *et al.*, 1971). What is particularly interesting is that 3 infants first emitted this vocalization approximately 1 week after they first exhibited the play face. A note of caution is in order, however, as the data available thus far more specifically associate play than the play *face* with laughter-like vocalizations. The soft guttural exhalations or low panting noises are frequently noted in playing chimpanzees and so also is the play face, but researchers have not specifically described the vocalization and facial expression as concurrent. This observation is not intended to be a major criticism of the theoretical link between laughter and the play face—especially in view of van Lawick-Goodall's developmental association between the two—but is rather a call for more detailed observations of ongoing play interactions.

Other theories remain to be briefly discussed. Andrew generally sees the development of laughter as being synonymous with the smile, since childhood laughter can be evoked by threats of aggressive behavior from adults or older children and in such situations can avert attack. Second, he believes that the association of vocal laughter with the grimace is a result of the connection between grimacing and vocalization (Andrew, 1963b, p. 82; 1965, p. 91). Although it may be the case that a child's laugh is not often followed by the execution of an aggressive act, this does not necessarily imply an appeasing or submissive function (and, hence, an association with the grimace). The child's laugh may more simply be interpreted as some form of play invitation, for example, and not a particularly submissive display. Moreover, one suspects that the tonal characteristics of vocalizations during grimacing differ markedly from those during play. If this is the case, Andrew's second suggestion does not appear tenable. Konrad Lorenz (1966, p. 172, 284) interprets smiling and laughter as a ritualized form of redirected *threat*, "a view made plausible by the exposure of the teeth" (Eibl-Eibesfeldt, 1970, p. 132). The evidence does not seem to support Lorenz's views on this issue if we have been correct in viewing teeth exposure as being largely absent in threat. In the future, attention might profitably be focused on situations evoking human smiling and/or laughter in an effort to assess whether there are two quantitative gradients of a unitary response [as Lorenz (1966, p. 171) believes] or whether they are responses differing in both origin and form.

H. Protruded-Lips Face

A male pigtailed monkey who has access to an estrous female may show a "most peculiar response" (van Hooff, 1969, p. 56) which takes place as he smells the genital area of the female. The lips are markedly protruded with the upper lip curled slightly in that direction and the lower lip pressed tightly against it. The mouth itself remains closed and the ears and brows are retracted (van Hooff, 1969, p. 57). [Cole (1963, p. 112) tersely describes the expression as being "smattered."] Tokuda *et al.* (1968, p. 288) also report its display in a male pigtail when approaching an estrous female, and it appears to induce the female to present (van Hooff, 1969, p. 58).

This expression occurs most frequently (van Hooff, 1969, p. 58) if not exclusively (Jensen and Gordon, 1970, p. 267; Kaufman and Rosenblum, 1966, p. 222) in the pigtailed macaque, but it is not restricted to sexual interaction. For example, researchers at the University of Washington have found that pigtailed mothers occasionally give a similar expression to retrieve their infants. The expression is referred to as a LEN, since it involves pursing the *lips*, pulling back the scalp and *ears*, and extending the *neck* (Bobbitt *et al.*, 1964, p. 73). The pigtailed infant is usually looking at its mother from a distance when the latter gives the expression. Following the display the mother and infant are likely to engage in more proximate interaction and contact behavior, such as cradling and nursing. The infant characteristically takes the initiative in reducing social distance, and, if it does not, the mother more frequently repeats the LEN than she approaches the infant (Bobbitt *et al.*, 1969, p. 120; Jensen and Gordon, 1970, p. 271). In a recent study, Castell and Wilson (1971, p. 209) noted that pigtailed mothers LENned most frequently when their infants began to spend considerable amounts of time away from them. However, there was no specific mention of a retrieval function for the LEN by these authors. They also observed infants giving a LEN toward their mothers, but the expression is still clearly more of a maternal expression since mothers LENned toward their infants 12 times more frequently than infants LENned toward their mothers (Castell and Wilson, 1971, p. 207).

A pigtailed mother may also give this expression if her infant makes direct eye-to-eye contact with her (Jensen and Bobbitt, 1965, p. 64). Similarly, Castell and Wilson (1971, p. 207) reported that mothers initially LENned to their infants while cradling them. This position would seem to allow eye-to-eye contact, but Castell and Wilson do not indicate the direction of the mother's or infant's visual orientation during the LEN.

Referring to the sexual context, Andrew (1964, p. 283) suggests that the expression has to do with "savoring" an olfactory sensation, but he also notes that protective components may be present as well since the female may respond aggressively to the male's advances. Van Hooff (1969, p. 72), on the other hand, hypothesized that the expression may be the closed phase of a lipsmack, an interpretation that seems more applicable to the several contexts in which the expression occurs. For example, it may lead to the return of an infant because of an association with grooming by the mother. Moreover, the appearance of the LEN in infants as young as 3 weeks (Castell and Wilson, 1971, p. 207) suggests a developmental onset similar to that of the lipsmack, as we shall see below. Tokuda *et al.*'s (1968, p. 292) hypothesis that pigtailed macaques LEN when Japanese macaques would lipsmack lends support to the functional similarity between the two expressions.

I. New World Monkeys

Up to this point we have generally avoided any detailed discussion of platyrrhine facial expressions because they present some unique characteristics. The development of facial-musculature mobility took place independently in the ancestors of Cercopithecoidea and Ceboidea. In general, most Ceboidea have a very limited repertoire of facial expressions (Andrew, 1964, p. 236; Moynihan, 1969, p. 323). For example, referring to the night monkey (*Aotus trivirgatus*), Moynihan (1964) simply remarks that "they do not have any facial expressions" (p. 80), and most of their nonfacial visual displays are not highly ritualized (Moynihan, 1969, p. 322). The titi monkey (*Callicebus moloch*) has a more expanded set of expressions, although it is said not to communicate a great deal with its face (Moynihan, 1969, p. 323). Both *C. moloch* and *Callicebus torquatus* may bare their teeth by lowering the bottom lip while the mouth remains closed or nearly closed. This is their most conspicuous facial expression, Moynihan suggests, largely because the white teeth stand out against the black face (in *C. moloch*, at least). Nevertheless, it is rarely given: one juvenile displayed it when picked up by hand and another did so during an intense agonistic interaction with an adult while attempting a number of escape movements. Moynihan concludes that the tendency to escape in these circumstances was much stronger than the tendency to attack (cf. van Hooff); however, the expression differs from the grimaces given by *Cebus* and *Lagothrix* insofar as the corners of the mouth are not drawn back (Moynihan, 1966, pp. 88–90). This species also has a prominent expression in which rapidly executed jaw, lip, and, perhaps, tongue movements take place together with a direct stare and a number of body postures (e.g., lowered head) and movements (e.g., tail-lashing). At a reasonably close range the display is distinctly audible. W. A. Mason (personal communication) interprets the display as a threat. Interestingly enough, a titi monkey disturbed by the presence of a human may suddenly turn its head sideways, without moving the rest of its body appreciably. Highly alarmed titis which cannot escape may lower their head so that the face is oriented downward, much as night monkeys do (Moynihan, 1966, p. 85). The similarity to head-shaking associated with gaze aversion in Old World monkeys and apes seems to be a tenable hypothesis.

Behavior patterns that may be analogous to lipsmacking are often seen in New World monkeys. For example, rhythmic in-and-out protrusion of the tongue occurs in many marmosets during sexual, affiliative, and aggressive social encounters (Epple, 1967, pp. 50–51). Hampton, Hampton, and Landwehr (1966, p. 278) record its occurrence in copulating male Pinché tamarins (*Saguinus oedipus*). Lipsmacking per se is described for similar social circumstances among marmosets but only in males (Epple, 1967, p. 51). Epple suggested that, since lipsmacking and tongue protrusion occur in both aggressive and affiliative contexts, they may indicate a willingness for social contact rather than an appeasement motive. Similarly, Hampton *et al.* (1966, pp. 281–282) interpreted the repetitive tongue movements as part of a recognition or greeting system. Pinchés executed the display not only during copulation but when approached by humans shortly after being captured in the wild. The response habituated after several days. It was also noted in subjects seeing themselves in a mirror for the first time. The animal taking the initiative in allogrooming (either male or female) among rufous-naped tamarins (*Saguinus geoffroyi*) may flick the tongue in and out slightly. Similar movements may occur before or after eating or even when an animal is looking at accustomed feeding areas. In many of the latter cases the animals appear to be "tasting" something or anticipating doing so (Moynihan, 1970a, p. 56). Moynihan also observed a much more exaggerated tongue-protrusion pattern in the same species during which the tongue is extended almost completely and the tip usually curled upward. The tongue is either held motionless or vibrated rapidly, and the expression may alternate with the less extreme form of tongue-flicking. The exaggerated form seems to be mainly characteristic of males during copulation or when confronting unfamiliar conspecifics, especially adult females (Moynihan, 1970a, pp. 56–57).

A type of tongue-protrusion expression similar to the patterns in marmosets and tamarins is seen in the howler monkey (*Alouatta villosa*). In this species, copulation is almost invariably preceded by rhythmic in-and-out and up-and-down tongue movements by the male *and* female (Carpenter, 1934, p. 89; Andrew, 1963b, p. 99; Bernstein, 1964, p. 95). Bernstein found that in most instances the female initiated them, but the male responded with similar movements. This gesture appears to be specific to sexual interactions for the howler and is directed toward the partner only when he or she is in a correct position to see it (Carpenter, 1934, p. 89), a pattern that seems to point to its communicatory significance. Moreover, this gesture appears in howlers in spite of the fact that they have only rarely been observed to allogroom, even though skin parasite infestation may be a serious problem (Carpenter, 1960; Richard, 1970, p. 253). Allogrooming is also relatively rare in *Saguinus geoffroyi* (Moynihan, 1970a, p. 15). It would be somewhat reckless to evaluate the tongue-protrusion displays in platyrrhines as directly comparable to lipsmacking in catarrhines, since the forms of the expressions leave room for several interpretations. The occurrence of tongue-protrusion during copulation and greeting seems to associate it with lipsmacking; however, the role of gustation in the evolution of the former expression may be of far greater importance than in the latter. Normative longitudinal data would clearly be of value here to assess whether tongue-protrusion is an extension of nursing movements, for example, as is suggested for lipsmacking. Moreover, the functional relationship of grooming to tongue-protrusion is not quite as clear for the New World species as for most Old World species. (For example, whether or not the gesture is a remnant of motor patterns associated with ingestion of particles from the groomee's fur is an open question. The presence of tongue-protrusion displays in species that only rarely groom also raises interesting questions.)

Some form of facial threat display takes place in a number of platyrrhines. Thus, eyebrow-raising occurs during aggressive threatening in several marmosets (*Callithrix jacchus, Callithrix leucocephala, Callithrix a. argentatus*). Brows are lowered during threats in *Callimico goeldii,* and the lips are also protruded in two tamarins (*Saguinus oedipus* and *Saguinus spixi*) (Epple, 1967, p. 49). Hampton *et al.* (1966, p. 282) described brow-lowering in *S. oedipus* to such a degree that the eyes were virtually obscured. The social circumstances of the display were not described. Since the animals maintain a fixed stare during the display, one assumes it occurred in the context of a threat. Moynihan (1970a, p. 58) noted that rufous-naped tamarins may occasionally display a very marked frown, perhaps also accompanied by partial eye-closure, apparently as a "purely hostile" expression. Moreover, unlike *Callithrix* species (and many Old World monkeys, for that fact), *S. geoffroyi* apparently does not flatten or erect the ears during agonistic interactions. As we found earlier in the case of eyelid coloration, the areas of white on this tamarin's face may be emphasized by piloerection and a number of movements or postures during displays.

The only New World genera that seem to have a large number of facial expressions are spider monkeys (*Ateles*), woolly monkeys (*Lagothrix*), and especially capuchins (*Cebus*) (Moynihan, 1969, p. 323). This seems to be particularly true of the grimace; e.g., *Cebus* and *Lagothrix* give a wide grimace during greeting (Andrew, 1964, p. 282). In fact, according to Andrew, in several genera of platyrrhines, there are more visual displays accompanying the grimace than in the catarrhines (e.g., the greeting display for *Cebus* begins with a grimace and may include teeth-chattering, head-shaking, blinking, scalp movements, and vocalization) (Andrew, 1963b, p. 94; 1964, p. 282). Returning once again to marmosets, one finds that the facial skin of many species becomes markedly pale during grimaces. Most marmosets increasingly flatten their ears against the head as a function of increasing fearfulness. Incisors and often canines may become visible as the corners of the mouth are withdrawn (Epple, 1967, p. 50). According to Moynihan (1970a, p. 56), retraction of the mouth corners in grimacing may be only very slight in tamarins (*S. geoffroyi*).

Perhaps the most important ecological variable that has contributed to a diminished use of facial expressions in New World monkeys is their largely arboreal habitat. It is clear that visual signals are not an efficient means of communication in dense foliage. In addition to this, Chalmers (1968, p. 277) points out that the quality of forest illumination is limited. Monkeys often have to look upward against a bright background of sky to see another individual, and details of expression and subtle features of coloration are difficult to distinguish in a silhouetted monkey.

These factors also influence communication in arboreal Old World monkeys, and a brief digression to some of these species may help to clarify the role of arboreal ecology in New World monkeys' visual communication. The communication system of the Old World arboreal black mangabey, for example, is much more vocal than visual. Only 172 visual gestures were observed in a field study versus 645 vocalizations, and the latter figure may be an underestimation (Chalmers, 1968, p. 278; Chalmers and Rowell, 1971, p. 5). The largely arboreal Nilgiri langur (*Presbytis johnii*) lacks expressions and components such as lipsmacking, ear-flattening, eyebrow-raising, yawning, and eyelid-blinking (Poirier, 1970, p. 175). Poirier suggested that the absence of these gestures may be due to a lack of facial-musculature specialization. Ayer's (1948, p. 79) early monograph on *Presbytis entellus,* however, suggested no such deficiency in the facial musculature, and if *P. johnii* were to be thus characterized it would stand in contrast not only to *P. entellus* but to most other Old World monkeys as well. The silvered leaf monkey (*Presbytis cristatus*) also has an impoverished repertoire of visual displays, lacking, e.g., the lipsmack, and Furuya (1961–1962, p. 51) attributed this to the relatively infrequent social contact between individuals of this species.

Not only are ecological variables of importance in influencing modes of facial expression, but a species' social structure may certainly be critical. For example, a very important reason for differences in communication systems between Old and New World monkeys may have to do with the latter's apparent lack of conspicuous dominance relationships (see Moynihan, 1969, p. 312). Moreover, we have noted earlier that visual communication is most efficient at close distances, so species that have a relatively dispersed spatial distribution, such as many New World monkeys, may not rely on visual cues as much as auditory cues. Gartlan and Brain's (1968, p. 281) comments are well-taken and are here quoted at length:

> It seems likely that with increasing territoriality and tightening of group structure there may be an increased reliance on facial rather than on vocal communication. Reliance on facial communication . . . both necessitates and tends to maintain certain aspects of the social system—primarily its integrity as a unit and its physical closeness, since facial expressions are efficient signals only over short distances. . . . These factors [i.e., morphological features such as whitened eyelids] ensure that the social organization of savanna animals will be both qualitatively and quantitatively different from that of arboreal animals.

Thus, we are cautioned against postulating a one-directional relationship between social or ecological structure and modes of communication. That an animal is arboreal may cause it to neglect facial communication, but at the same time a lack of such skills may ensure that it remains primarily arboreal.

In summary, New World monkeys make less frequent use of facial expressions than Old World monkeys and apes, largely because of their more arboreal habitat. A number of similar expressions are seen in catarrhines and platyrrhines, however, and forms of grimacing and possibly lipsmacking may be very complex in the latter.

III. DEVELOPMENTAL TRENDS

Observations on the ontogeny of facial expressions seem particularly difficult to obtain. A major reason for this is that neonates seem to display remarkably few facial expressions. Rowell (1963, p. 35) comments that, in a group of captive infant rhesus monkeys, their faces were "curiously expressionless." Similarly, Hinde, Rowell, and Spencer-Booth (1964, p. 642) and Baysinger *et al.* (1972) observed facial expressions only very rarely in longitudinal studies of captive rhesus monkeys. Rowell (1963, p. 37) offers an interesting interpretation of this trend:

> The blank expression of the first two months can be regarded as having communicative value in its own right. This would be comparable to the juvenile plumage of birds or the first coat of mammals, which lack the provocative patterns of the adult and so, along with their other functions, proclaim that the individuals wearing them are not to be brought into the scheme of adult competition.

One can always raise the possibility that a neonate's neuromuscular systems are not well-developed enough to allow for elaborate facial expressions. However, these systems are clearly capable of coordinating the complex set of movements associated with nursing, so this obstacle does not seem to be a major one.

An infant monkey's impoverished repertoire of expressions is in contrast to the more frequent number of vocalizations. Infants coo, screech, girn, gecker, bark, click, chirp, and emit any number of uncategorizable vocalizations (depending on the species of the primate and the acuity of the human observer). A neonate might vocalize, for example, when dislodged from the teat or when it has lost contact with the mother altogether (Rowell, 1963, p. 35). It will be recalled that we took great pains earlier in this paper to establish the point that an animal that vocalizes frequently may be highly conspicuous not only to conspecifics but to predators as well. And now we find that highly defenseless neonates may vocalize a great deal. (Sooner or later most generalizations meet their demise.) Upon reflection, however, some sense can be made of this.

An infant's vulnerable position demands an efficient, unambiguous, and virtually foolproof set of communicatory signals if it is to survive. Vocalizations clearly serve these ends. It is well-known from field studies, for example, that an infant's screech may mobilize not only its mother but a large part of a troop to its defense. By contrast, a purely visual cue in critical circumstances may fail to communicate the infant's state reliably and may thereby put its survival in jeopardy (e.g., adults may be too far away or oriented in the wrong direction to perceive a purely visual distress cue). A vocal system may, indeed, make an infant more conspicuous to predators, but it would seem to be even more hazardous for the infant not to have a highly reliable communication system available very shortly after birth (cf. a human infant's cry). As we have seen, however, many or most nonhuman primates eventually shift to predominantly visual modalities by adulthood. For example, the frequency of coo vocalizations in rhesus decreases markedly with age (Møller *et al.*, 1968, p. 346). This trend was illustrated in a recent study by Brandt *et al.* (1972) of mother–

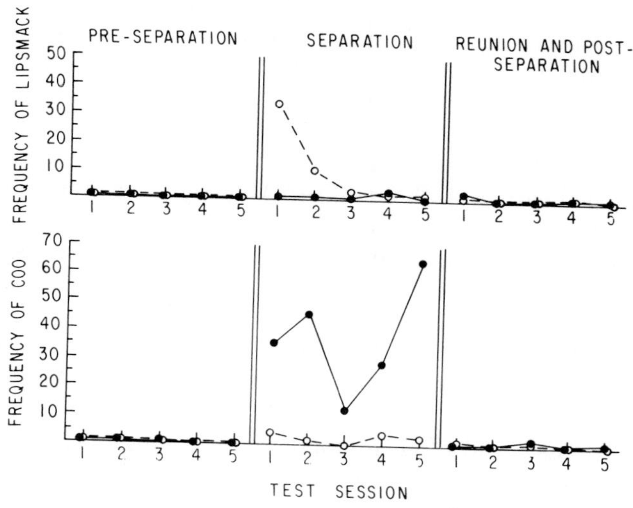

FIG. 19. Frequencies of a representative facial expression and vocalization during pre-separation, separation, and reunion and post-separation phases of mother–infant separation in rhesus monkeys. The three phases shown above each lasted for 2 days, during which time data were recorded on five sessions. Adapted from Brandt *et al.* (1972).

infant separaion in rhesus monkeys. As shown in Fig. 19, an infant's communicative response to such a stressful event as separation is overwhelmingly vocal, whereas the mother's response consists largely in facial expressions. Note that frequency of the mother's expressions declines rapidly across test sessions, whereas the infant's vocalizations do not, which may be one indication that infants find separation to be much more stressful than do mothers.

Data on the subject are somewhat sparse, but the first social expressions to appear in a developing nonhuman primate seem to be the lipsmack and play face. Infant rhesus monkeys make slow lipsmack movements with the lips slightly pursed as early as 3 days after birth (when an adult monkey or human had looked closely into the infant's face). By 6 weeks, adult-type lipsmacking in response to a slightly alarming individual appeared (Hinde *et al.*, 1964, p. 642). Similarly, when very young stumptailed monkeys make eye-to-eye contact with another monkey, they often make "a particular pucker-faced expression which is accompanied by slow mouthing movements" (Chevaliér-Skolnikoff, 1968, p. 3). This expression and movements are very similar to the ones characterizing a nursing infant, and by 2 or 3 weeks they appear to elaborate into lipsmacking in social situations.* On the other hand, grooming per se does not appear in rhesus infants until ten or twelve weeks (Rowell, 1963, p. 37, 48). It is difficult to interpret how invariable the latency is between onset of lipsmacking and grooming for various species, largely because the data are not available, but, if the former does appear consistently sooner, this may point toward the greater (but not exclusive) role of nursing than of grooming movements in the ontogeny of lipsmacking. As in the case of many other mammalian and especially primate behaviors, an early infantile behavior persists into adulthood and serves a social function for the mature animal.

As mentioned earlier, Anthoney (1968) puts great emphasis on the importance of nursing in his discussion of the development of lipsmacking in yellow baboons. On the premise that neonatal sucking patterns are identical to those of lipsmacking, he suggested that they subsequently generalize to involve sexual and greeting behaviors. He noted that a black infant's pink face is frequently the object of lipsmacking by nearby adult females which may groom the mother at the same time. When permitted to leave the mother, the infant is occasionally the object of lipsmacking by an adult female and often responds by following her, lipsmacking in the process. The female often nuzzles the infant in the genital area while lipsmacking, and if the latter is male she will often lipsmack directly on

* The only exception to the early occurrence of lipsmacking was recorded by Struhsaker (1967a, p. 290), who observed the expression in all ages *except* infant vervets.

his penis (which shortly becomes erect if it is not already so), and lipsmacking may thereby become associated with sexual arousal as well as nursing. Lipsmacking toward the genitalia gives way to embracing as the infant's coat turns to brown, and during an embrace an infant may lipsmack to one of the female's nipples. An estrous female may also present to a brown male while lipsmacking over her shoulder. Baboon infants seem to be attracted by lipsmacking movements and the nipples of adult females. Other pink objects such as male penises (particularly for females) and perhaps the tongue of the lipsmacking animal are also said to eventually elicit lipsmacking. As the infant matures, the pink face and perineum of black infants, in turn, elicit lipsmacking as does the pink sexual skin of a mature female. Thus, Anthoney suggests that lipsmacking motor patterns become associated with nursing and sexual arousal in early infancy and gradually generalize to a number of stimuli and social contexts. The troop Anthoney studied did not contain any adult males, so it is not possible to assess the role that they may play in the development of this expression. One also notes that a thorough analysis of social systems must take into account not only stimulus–response associations but especially the complex of emotional bonds and motivational states that characterizes nonhuman primates. Anthoney's schema might profit by an increased emphasis on the latter phenomena.

The play face is another expression that appears early. It has been observed at 5–6 weeks in rhesus infants (Hinde *et al.*, 1964, p. 642) and 4–8 weeks in stumptailed infants (Chevaliér-Skolnikoff, 1968, p. 5). These are also approximately the ages at which "rough-and-tumble" play was seen to be clearly established among rhesus infants (Harlow, 1969, p. 341). The play face was first seen in chimpanzees at 11–24 weeks (van Lawick-Goodall, 1968b, p. 244), a trend that illustrates the advanced developmental state of the monkey relative to the ape at birth.

Some form of "pouting" is sometimes observed in infants of several species, although the age of onset is not clear (Fig. 20). Infants may pout, for example, when away from the mother, and the pouting appears to function as a call for contact comfort or the nipple (van Hooff, 1969, pp. 59–60, 70). Van Hooff's description of pouting includes raised eyebrows and a slightly opened mouth with mouth corners brought forward and lips protruded. The frequency with which this expression is accompanied by a vocalization (i.e., coo) is not clear at present. Chimpanzees may show the expression even in adulthood. Van Hooff (1969, p. 60) noticed it frequently when chimps were shown attractive objects such as food (see also van Lawick-Goodall, 1968a, pp. 323–325; 1971, p. 375). One factor underlying the persistence of this expression into adulthood in chimps may have to do with its marked similarity to the facial configura-

FIG. 20. A form of pout face in an 8-month-old infant baboon. Ears are slightly flattened and the area around the mouth is drawn together in a pucker. (Photograph courtesy of Tim Ransom.)

tion during "hooting," a common vocalization in this species. In another ape, the orang-utan, mild disturbance may be accompanied by a pouting of the lips (MacKinnon, 1971, p. 174).

The next expression to appear developmentally seems to be the grimace (Figs. 21 and 22). In a study by Bernstein and Mason (1962, p. 288), the grimace became a frequent expression in response to a stressful stimulus at a much later developmental stage than did lipsmacking. The two expressions were separated by approximately 18 months. In rhesus infants the grimace is first seen around 9 or 10 weeks or as late as 22 weeks (Rowell, 1963, p. 49; Hinde *et al.*, 1964, p. 643). Kaufmann (1966, p. 22) reported a slight grimace in a 58-day-old, feral rhesus infant after being aggressed by an adult male. Harlow (1969, p. 380) reported "responses indicative of social fear" in rhesus at 90 days. In contrast to yawns and social threats, the grimace declines in frequency as the animal matures. Frequency of grimacing declined markedly in rhesus monkeys from the first through the seventh year in stressful experimental situations (Cross and Harlow, 1965, p. 43). Møller *et al.* (1968, p. 345) found that 39-month-old rhesus monkeys grimaced only half as frequently as 19-month-old animals. The pattern emerging from these observations seems to point to 2–3 months as the typical age of onset for grimacing (for the rhesus,

FIG. 21. A partially formed grimace in a male, 6-month-old, isolate-reared rhesus monkey. Shortly after this photograph was taken, the infant showed a narrow row of teeth. There is considerable evidence associating grimacing with at least one form of human smiling.

that is). It may be fruitless to try to pinpoint an exact age for his and other expressions, however, since the age at which a grimace is first displayed certainly depends greatly on the social situations that the individual experiences. For example, a rhesus infant in our laboratory was seen to grimace as early as 1 or 2 months of age, but it so happens that the animal was witnessing the spectacle of the photographer virtually doing somersaults in order to elicit the expression.

Threat expressions characteristically emerge later in development than any of the above expressions (Figs. 23 and 24). Rowell (1963, p. 37) first observed low-intensity threat expressions in captive rhesus at 5 months of age. In langur infants, "subtle" threats appeared in late subadulthood or early adulthood, whereas grimaces were present since infancy (Jay, 1965a, p. 220). (Interestingly enough, agonistic staring appeared at the intermediate period between grimaces and threats.) Kaufmann (1966, p. 22) also noted a late onset of threatening in rhesus on Cayo Santiago: "The infants seemed to learn slowly the meaning of aggressive signals. Males threatened infants with direct, open-mouthed stares and head-bobbing, and occasionally a male hit an infant and briefly pinned it to the ground. The infants completely ignored this hostile behavior. . . . " The one exception Kaufmann reported was the 58-day-old infant mentioned earlier

Fig. 22. A yearling male baboon displaying a grimace during a weaning interaction. (Photograph courtesy of Tim Ransom.)

which crouched and displayed a slight grimace after being hit twice by an adult male. Møller *et al.* (1968, p. 345, 347) report a much higher frequency of not only threats but yawns in 39- vs. 19-month-old juvenile rhesus, and yawns, social threats, redirected threats, and especially experimenter-directed threats were all more frequent in males than females (see also Cross and Harlow, 1965, p. 48; Harlow, 1969, pp. 343–344). Feral juvenile and infant bonnet macaques are said never to yawn (Rahaman and Parthasarathy, 1968, p. 264), but captive juvenile rhesus are known to do so when under stress (see Fig. 15). Research at the Wisconsin laboratory suggests that "externally directed threat responses" (Harlow, 1969, p. 380) appear early in the first year and increase steadily to the sixth year for rhesus monkeys (see also Cross and Harlow, 1965, p. 48).

The overall developmental pattern thus emerging is a sequence beginning with gregarious or socially cohesive expressions (lipsmacking and play face) followed by fearful (grimace) and then aggressive (threat and

FIG. 23. Raising her eyebrows, a 5-month-old infant baboon gives a stare–threat from the relatively secure position of her mother's back. Spatial proximity to other individuals is an important variable affecting visual displays. (Photograph courtesy of Tim Ransom.)

yawn) expressions (see also Rowell, 1963, p. 37). Threats and yawns increase in frequency as a function of age; grimacing, once established, decreases with age. These trends fit well with the schema for emotional development described in Bronson's (1968) excellent review. Briefly, it was suggested that fear of novel stimuli is delayed for a period after birth to enable the organism to either: (*1*) encode the familiar before it is able to discriminate and respond to novelty (D. O. Hebb's model), or (*2*) complete the maturational processes underlying the capacity for fear and establish primary attachments (Hess's and King's models). Bronson (1968, p. 357) concludes that "The pattern of fear responses develops from an initial tendency to immobility (and vocalization), to retreat, to the appearance of aggressive reactions toward the feared object." In terms of one or both of these models, then, we might conceive of lip-

FIG. 24. A 9-month-old male rhesus infant displaying a threat. This expression is characteristically one of the last to develop in nonhuman primates.

smacking and the play face as helping to establish early primary social attachments (e.g., with adult females or peers) which may be of considerable importance for an individual's integration into a social group or ultimately for his survival. To reemphasize a point made earlier, the communicative expressions used by an individual at a given developmental period depend in large part on the social situations in which he finds himself. A very young infant has no real need for appeasement or especially threat gestures as long as the protection of a watchful mother is available. For the most part, the infant is also exempt from agonistic interaction with adults by virtue of a set of infantile sign stimuli (e.g., the baboon's neonatal black coat).

As a growing infant begins to interact more and more with other animals, it becomes progressively less immune from agonistic interactions. It thus seems reasonably that a useful resource to be developed would be some sort of response to avert direct aggression, a function that the fear grimace appears to serve nicely. So by the same time that an animal has encoded the familiar well enough to be fearful of novelty, he is also developing mechanisms of *reducing* that fear to a reasonable level, suggesting that an optimal level of arousal is sought. The onset of threat expressions, on the other hand, awaits an animal's serious competition for desirable goals—status, food, consort partners—largely independently of its mother or other protective adults. A neonate's blank expression may not only be a nonprovocative pattern, as Rowell suggested, but may simply indicate that the protected animal has no great need of expressions until it enters into independent social interaction with other animals.

IV. EXPERIMENTAL DATA

One of the reasons why references to facial expressions are scattered so widely throughout the field literature is that it is difficult to observe this type of phenomenon with any degree of accuracy in the field. It takes a great deal of skill to arrive at the point where the animals under study do not flee altogether at the approach of the observer, and in many instances it thus proves to be impossible to get close enough to the animals to record details of facial expression. This feat accomplished, moreover, the human observer may find that his presence itself is eliciting a great number of expressions (notably yawning, as we have seen). The study of facial expressions lends itself readily to the laboratory environment, where facial configurations not only can be observed at close range but their stimulus and response characteristics can be more carefully controlled and quantified as well. Moreover, cues provided by facial expression and visual orientation can provide the researcher with invaluable tools in assessing nonhuman primates' motivational and affective states—surely one of the most difficult tasks facing an observer. As we shall see in this section, the above two indices have been particularly useful in studying motivational and emotional disturbances associated with maternal deprivation in monkeys.

Robert Miller, I. Arthur Mirsky, and their associates at the University of Pittsburgh have been among the most active in this area of research. In two early studies, Miller *et al.* (1959a,b) demonstrated that slides of facial expressions of other monkeys could serve as efficient cues in establishing an avoidance response to shock. During extinction, slides of fearful and calm monkeys were introduced and significantly more avoidance responses were made to fearful than nonfearful slides. The authors concluded that rhesus are able to discriminate efficiently and instrumentally between the two types of facial expressions. One wishes, however, that they had taken greater care in specifying the characteristics of a fearful expression, as the two pictures furnished as examples of fear are far from unambiguous (e.g., they might also be interpreted as threat expressions). In a later study a cooperative–avoidance paradigm was established in which one monkey had exclusive access to a lever that, when pressed, would avoid a shock, whereas a second subject had exclusive access to the discriminative cue that signaled the onset of shock after several seconds. The respondent subjects were able to see the face of the other monkeys by closed-circuit television and learned to avoid shock by attending to their facial cues (Miller *et al.*, 1963). Miller (1967) subsequently reported that respondent monkeys not only reacted instrumentally but autonomically as well (i.e., cardiac rate greatly increased during avoidance trials), and the communication process was much more effective in avoidance trials than in reward trials. The crucial set of stimuli displayed by the monkeys viewing the warning cue was head and eye movements away from the stimulus together with brief glances toward it (Miller, 1967, p. 133)—a pattern quite similar to the visual orientation of a grimacing or fearful monkey. Miller *et al.* (1967) also found that monkeys raised in social isolation were incapable of using facial expressions as cues in an avoidance response, as they were neither efficient senders nor responders in the cooperative–avoidance paradigm. For a recent review of the research from this laboratory, the reader is referred to Miller (1971). In passing, one wonders how New World monkeys would fare in the Pittsburgh cooperative–avoidance situation. It might be interesting to see if platyrrhines are capable of developing a set of facial cues efficient enough to function in the above instrumental circumstance. Perhaps some acquired or elaborated expressions would subsequently be employed in group social interactions in a free-ranging environment.

Gene Sackett (1965) projected a set of slides to three groups of feral, 12-month isolate, and maternally raised rhesus monkeys on a wall of their home cages. The categories of slides included monkeys during various activities (e.g., fighting, exploring, copulating) and affective states (e.g., threatening, fearful). Sackett recorded duration of visual and tactile exploration of the projected life-sized images and found that feral subjects explored images with sexual content longer than the other groups, while social isolates explored the nonmonkey pictures and pictures without social communication content for longer durations than those evaluated by Sackett as socially significant.

In a later study by Sackett (1966), slides were projected in the home cage of rhesus otherwise raised in isolation from birth to 9 months. He found that threat and infant pictures produced the greatest frequency of vocalizing, disturbance behaviors, playing, and exploration in the monkeys, and there was more vocalization and disturbance with threat than with infant pictures, especially at around 3 months of age. He also allowed the subjects to control the frequency with which the slides were shown and similarly found that the rate of lever-pressing for threat stimuli decreased markedly at 3 months. Sackett (1966) concluded that threat and infant pictures appear to have "unlearned, prepotent, activating properties for socially naive infant monkeys" (p. 1472), and that visual stimulation involved in threat behavior functions as innate releasing stimuli for fearful behavior. It is not clear in what sense Sackett is using the concept of innate releaser. If he is suggesting that a threat expression releases a particular fixed action pattern, this may not be a rigorous interpretation. One is struck with the tremendous fluidity of stimulus and response contexts involved in primate agonistic interactions. A threat expression may elicit flight, avoidance, grimacing, lipsmacking, yawning, or a number of other responses which seem highly dependent on learning in a social group. Threat expressions thus do not appear to function in as rigid a manner as when a spherical stimulus, for example, elicits a pattern of egg-retrieval in several species of birds.

A recent study in our laboratory (Redican *et al.*, 1971) used a methodology similar to Sackett's. Female juvenile rhesus monkeys pressed a lever to project slides of facial expressions onto a screen in a large operant-conditioning box. The subjects showed no differential response rate to adult, juvenile, or infant age categories but did distinguish between expressions within the juvenile (i.e., age-mate) group. This tentatively suggests that Sackett's infants may also have exhibited an age-mate preference rather than a prepotent preference for infant stimuli per se. We also found a diminished response rate to not only threat stimuli but fear grimaces as well. Since both of these expressions are associated with agonistic social interactions, their aversiveness may be due more to their social context than to some form of innate releasing mechanism characteristic—at least for our socially experienced animals.

William A. Mason has contributed a great deal to an understanding of animal communication in his series of studies on the effects of social restriction on the behavior of rhesus monkeys. In an early study (Mason, 1960), he found that pairs of restricted monkeys (separated from their mothers at birth and raised in individual cages thereafter) participated in more serious aggression and less grooming episodes than feral monkeys and that their sexual behavior was often inappropriate, especially in the males. He concluded that social restriction interferes with the establishment of the cue function of many basic social behaviors (e.g., the threat pattern and the presenting and grooming postures) so that appropriate reciprocal responses are not elicited. In tests of gregariousness (Mason, 1961a), restricted rhesus males were generally avoided by socially experienced females. Furthermore, restricted monkeys of both sexes fought more frequently than feral monkeys. Mason suggested that the effective development of elementary forms of social communication is highly dependent on learning, and that their absence in restricted monkeys results in turbulent relationships with conspecifics. This was also underscored in his study of dominance interactions in rhesus monkeys (Mason, 1961b). Stable dominance relationships with few reversals were established among socially experienced subjects, and fighting was infrequent and initiated by the dominant animal when it did take place. On the other hand, among restricted monkeys, there was no stable dominance hierarchy (with many dominance reversals taking place), and fighting was frequent and engaged in by both dominant and subordinate monkeys. Although Mason's studies do not focus specifically on facial expressions, they are included here to demonstrate the failure to acquire effective communication skills in socially restricted individuals. Included in these skills are appropriate facial expressions that function to coordinate patterns of harmonious social interactions.

Other studies of social isolates have further delineated the disturbed patterns of facial communication in social isolates. Harlow (1964, p. 166) noted that "mother raised infants had a greater range of facial expressions than surrogate [raised] babies, and they had some facial expressions that babies raised with surrogates did not develop." We are not informed of the nature of these expressions, however. Mother-reared rhesus monkeys exhibited "social threats" approximately 5 times as frequently as 6-month isolates in a playroom situation as long as 32 weeks after the latter were removed from isolation (Rowland, 1964; Harlow *et al.*, 1964, p. 126). Young social isolates tend to grimace more frequently than normally socialized monkeys during social interactions, but this relationship is reversed by adulthood. Grimacing was observed only in a group of 7-month-old, isolation-reared pigtailed infants and not in a similar group of mother-reared infants (Evans, 1967, p. 264). Rhesus monkeys raised in partial social isolation grimaced more frequently at 1 and 3 years of age than mother-reared monkeys but threatened less frequently (Cross and Harlow, 1965, p. 47). Surrogate-reared rhesus monkeys between the ages of 15 and 19 months of age displayed more grimaces than mother-reared monkeys when paired with unfamiliar male stimulus animals. They also exhibited more threats (Mitchell and Clark, 1968, pp. 121–122). Similar findings were also reported for 3 to 4½-year-old rhesus monkeys raised in isolation for 6 and especially 12 months (Mitchell *et al.*, 1966, p. 572; Mitchell, 1968, p. 143). The 4½-year-old isolates displayed less grimaces and more threats than 3½-year-old isolates. By the time an isolate has matured, it is far less likely to grimace than a normally socialized adult (Brandt *et al.*, 1971, p. 110). Mitchell suggested that as a social isolate matures, it gradually changes from a predominantly fearful animal on emergence from isolation to an extremely hostile one. We have seen that the trend from fearfulness to hostility is characteristic of the emotional development of most primates, although isolates certainly become much more aggressive than normally socialized individuals. In maternally deprived animals, however, the initial period of attachment is disrupted. To what extent this is manifested in disturbance of affiliative facial expressions is not completely clear at this point. Specifically, one would expect affiliative expressions to be particularly infrequent or atypical in isolates. Mother-reared rhesus monkeys were, in fact, observed to lipsmack with a significantly greater frequency than isolates in Cross and Harlow's (1965, p. 48) study. Brandt *et al.* (1971, p. 108) found that adult rhesus isolates very rarely lipsmacked to either juvenile or adult stimulus animals. Control adults, on the other hand, were especially likely to lipsmack toward adult stimulus animals. It is also of interest to note that Evans (1967, p. 264) found that mother-reared pigtail infants LENned more often than isolates, a trend that parallels the data for lipsmacking in rhesus monkeys. Missakian (1969, pp. 53–54) found a tendency for socially deprived adult rhesus males to engage in actual physical aggression toward females, whereas feral adult males displayed threat responses instead. Similarly, feral mothers normally employed at least one of three signals (silly grin, affectional present, and lateral head-shake) in retrieving their young, but none of the socially deprived mothers did so.

Not only are facial expressions assigned inappropriate social cue functions in isolates, but some of the actual patterns of expressions may be atypical. In the case of threat, for example, it may be directed toward a part of the isolate's own body rather than another animal (Cross and Harlow, 1965, p. 44; Brandt *et al.*, 1971, p. 108) (Fig. 25). A mature male iso-

FIG. 25. An adult male rhesus isolate threatening his own hand and clasping himself. This monkey was raised in total social isolation since birth. He generally avoids eye contact with human observers and when threatening he may keep his mouth open for a prolonged period. (Photographs courtesy of J. Erwin.)

late in our laboratory often displays a conventional facial threat; however, he may keep his mouth wide open for exceptionally long intervals (Mitchell, 1970, pp. 4–5) (see Fig. 25). Patterns of visual orientation may also be disturbed, such as patterns of staring and gaze aversion. For example, when adult isolate and feral rhesus males were released in the same cage, the isolates were typically decisively beaten. Even after suffering superficial wounding, however, isolates often slowly approached the ferals and stared directly in their eyes, at times at a distance of 2 or 3 inches. The controls apparently found this sort of behavior to be very disturbing, and even though they physically outweighed the isolates they repeatedly attempted to face away from the isolate's bizarre advance. In several cases, the controls finally withdrew to a corner of the cage and grimaced (Mitchell, 1970, p. 14; 1972; Brandt *et al.,* 1971, p. 111; Fittinghoff *et al.,* 1974).

V. CONCLUSION

We have made liberal use of the term "display" in many instances in this paper, and it seems appropriate to briefly elaborate on this concept since it is of special significance in a study of facial expressions. As Andrew (1964, p. 299) uses the term, it refers to a "pattern of effector activity which, in many or all individuals or a species, serves to convey information, the passage of which is advantageous to the individual or to the social group to which he belongs, to others of the same or other species." Moynihan (1970b, p. 86) similarly stresses its communicative significance and appends the criterion of ritualization, viewed as a specialization in form and/or frequency specifically to permit or facilitate the process of communication. Ritualized displays clearly serve adaptive functions for species since communication is fundamental to the survival of any aggregate of individuals inhabiting the habitats of most nonhuman primates. Andrew (1964, p. 301) stresses that strong selection pressure opposes the evolution of facial expression in part because functions of other facial structures would be disturbed in the process. Moreover, display patterns, such as alarm and threat, tend to be more conservative in evolution than more ambivalent ones, since both occur at critical moments when it is vital for the percipients of the display to react immediately (Moynihan, 1969, p. 322). Moynihan also notes that it may be to the sender's advantage to elicit an immediate response from conspecifics (e.g., efficient threats should enable an individual to avoid physical combat). We might add that submissive displays are often particularly important to an individual's or group's survival since they may also prevent the occurrence of overt aggression between individuals. Moynihan (1969, p. 322) suggests that:

> Any sudden evolutionary change in a pattern of this type [alarm or threat], no matter how minor, must increase the chances that the signal meaning of the pattern will be misinterpreted, or reacted to slowly. Thus, it probably will be selected against, unless it has some really appreciable compensatory advantage.

He further elaborates that clarity, precision, and strength of transmission are useful characteristics of a signal. These can be achieved most efficiently by making the form of the display exaggerated or emphatic, stereotyped, and distinctly different from any other behavior pattern (Moynihan, 1970b, p. 86).

One is reminded of Darwin's (1872) principle of antithesis:

> Certain states of mind lead . . . to certain habitual movements which were primarily, or may still be, of service; and we shall find that when a directly opposite state of mind is induced, there is a strong and involuntary tendency to the performance of movements of a directly opposite nature, though these have never been of any service (p. 90).

Thus it may be that the fear grimace, for example, is antithetical to the threat expression (i.e., teeth-exposure vs. teeth-covering; mouth closed vs. mouth open), just as gaze aversion is antithetical to a direct stare. Van Hooff (1969, p. 29) also suggests that vertical head-bobbing in threat is antithetical to horizontal head-shaking during appeasement. It seems reasonable in any case that it is crucial for a percipient of a facial expression such as a grimace to be able to distinguish it quickly and reliably from another one such as a threat, and there would seem to be marked selection pressure against their convergence into a more similar configuration—if not for their divergence, as Darwin suggested. It is also remarkable that the considerable number of facial expressions that we have discussed in this paper can be discriminated from one another quite readily by a human observer, and there is good reason to assume that monkeys and apes are even more skilled at the task. One must note, however, that facial expressions are also supplemented by a variety of postures and movements which certainly aid in the discriminability of facial displays. Moreover, if primate facial expressions are characterized by a continuous intergradation among expressions—as Bastian (1965, p. 588) suggests for human visual signals—there seems to be ample opportunity for semantic confusion, which again calls for supplementation by other communicatory channels.

One of the more important considerations which thus emerges is the highly influential effect of context on a particular signal. An animal's posture, movements, vocalizations, distance, size, age, or dominance status all interact in affecting the meaning of a particular facial expression. As Lancaster (1967, p. 30) observed, a threat by a juvenile may be ignored in one context but may lead to a dramatic response on the next occasion if he happens to be standing next to his mother while threatening. Kummer stressed that the meanings of signals must be acquired through experience in appropriate contexts. He notes (in Ploog and Melnechuk, 1969, p. 444):

> A raised eyebrow threat means no more than "Stop what you are doing." Experience will teach the receiver whether the threat objects to his eating a particular food or to his sitting close to a particular group member, if he is doing both at the time of the threat. In hamadryas, the raised eyebrow, though a threat signal, usually producing escape, nevertheless comes to make a female approach the male threatener rather than withdraw, for the more she withdraws the more he threatens until she learns it means "Stop being so *far away!*"

One might even describe a context of the individual. Some chimpanzees, for example, show facial expressions that are unique to those individuals (van Lawick-Goodall, 1968a, p. 323). The importance of context once again militates against viewing visual displays such as facial expressions as manifestations of unitary drives or as releasing stimuli for fixed action patterns.

Efforts to interpret many of the phenomena we have reviewed in terms of a drive model have met with little success. As Mason observed, many different subjective states can be expressed by the same signal, and many different signals can elicit the same response (in Ploog and Melnechuk, 1969, p. 443). Thus, a grimace may be given by a submissive individual in an agonistic interaction or by a copulating adult male. A lipsmack or fear grimace can both avert a hostile interaction and can both serve in greeting. Further, monkey mothers in closely related species can retrieve their infants by grin–lipsmacking or by protruding their lips. These observations do not enable one to easily ascribe a particular expression as a manifestation of a corresponding drive. R. J. Andrew has been one of the most vigorous critics of drive theory as it related to displays. He writes (Andrew, 1964, p. 322):

> The drive model has been unusually misleading in the case of displays, since there has been a tendency to ignore the evocation of particular components by particular stimuli, and to explain the components of all social displays as depending on particular combinations of particular intensities of three drives: aggression, fear, and sex. The mechanism whereby different combinations of different drives are supposed to produce different components, rather than different intensities of components, from those produced by other combinations of drives has never been set out clearly.

Stuart Altmann emphasizes the stimulus–response parameters of behavior rather than intensities of drive and has suggested a stochastic model of behavior. This posits individual behavior as dependent on preceding socially significant actions within an individual's interacting group. Sequential dependencies between the action of group members indicate that social communication is taking place, and these sequences can be represented by their mathematical probabilities of occurrence (Altmann, 1965, pp. 490–491). There is thus more to communicative interactions than a signal and the recipient's response, since the relation between the two and past and present mediating influences are also important (Altmann, in Ploog and Melnechuk, 1969, p. 478). Altmann illustrates this by citing the example of a female monkey presenting to a male. If social signals completely determined the behavior of the percipient, the male would always mount and this would be wasteful from an evolutionary point of view; if the signal had no effect, however, the male's response would always occur by chance or would occur independently of the female's behavior. In the latter instance the signal would be maladaptive and would not constitute a social system since there would be an absence of interanimal communication (Altmann, in Ploog and Melnechuk, 1969, p. 478). Altmann's models seem clearly more viable than drive models in understanding sequences of behavior; however, they often demand considerable mathematical sophistication in their application. It is also possible that the assumptions and manipulations of the mathematics themselves may frequently bury behavioral understanding. We have seen how informative Chalmer's (1968) approach has often been by simply reporting the frequencies of behaviors following any particular display pattern in the sender and recipient, and future field research might profitably be reported with this type of analysis as an essential initial step.

One of the most important aspects of facial communication in which research is clearly needed concerns the specific facial areas that are most carefully attended during a facial display. We have seen that in a monkey's threat, for example, the mouth, eyes, eyebrows, and eyelids may all take part. It is not clear whether the percipient attends to all of these briefly or fixates predominantly on one. In view of the frequent emphasis on threat-staring in this paper, the reader may have already surmised that the writer feels the area around the eyes to be of major importance. From subjective experience it seems clear that this is the case in human facial communication, and this feeling is underscored by the observation that although humans lack many of the structural components available to monkeys and apes in facial expressions (e.g., we certainly cannot flatten our ears and few can wiggle them to any degree, and we have no brightly colored eyelids or prominent canines), we possess a striking configuration around the eyes apparently unique among primates, in that our iris is surrounded by a perceptible amount of white matter or sclera. The gorilla's eyes, for example, are totally dark brown except when the eyeball is rotated far to one side, in which case an area of white matter is visible on one corner (Schaller, 1963, p. 71). Indeed, van Lawick-Goodall (1967, p. 197) presents a picture of a chimpanzee with a mutation so that the sclera is as visible as a human's, and it is a more than slightly unsettling sight. Thus, in spite of Andrew's (1964, p. 298) suggestion that the only specifically human component of facial expression is platysma contraction during crying, humans must certainly utilize the white matter in the eyes to perceive cues associated with direction of orientation or states of arousal (e.g., a "glaring" stare vs. a relaxed one, a steady gaze vs. a rapidly shifting one). This configuration in humans is also supplemented, one suspects, by movement of the eyebrows. Although we do not have bright patches of colored skin above our eyes, we have its converse in patches of fur on a hairless background. The functional significance of eyebrows in diverting perspiration away from the eyes has been recognized for some time, but what we are calling attention to here is their communicative function.

If this review has stimulated the reader's interest in human facial expressions, a wealth of literature awaits him. The following sources are by no means intended to be an exhaustive bibliography but are, rather, selected to give a general overview of the field. For an extensive review of the human (and some nonhuman) facial-communication literature up to 1968–1969, a particularly good source is Vine (1970). Much of the psychological literature is also discussed in the Davitz (1964) volume. A brief review may be found in Grant's (1971) article. Paul Ekman and his associates at the Langley Porter Neuropsychiatric Institute in San Francisco have been active in human facial-expression research for several years. Ekman, Friesen, and Ellsworth (1971a) have recently published a critical integration of human facial-expression research conducted since 1914 which also contains a methodological section on experimental procedures in this area. Ekman (1973) has also edited a volume entitled "Darwin and Facial Expression" which contains a number of chapters relevant to Darwin's 1872 opus, including a chapter by Ekman on cross-cultural studies of human facial expression [in addition, see Ekman (1971); Eibl-Eibesfeldt (1970, pp. 408–427); and Cüceloglu (1970)], a chapter by Charlesworth and Kreutzer on facial expressions in children, and a chapter by Chevaliér-Skolnikoff on facial expressions in nonhuman primates.

The literature concerned with human smiling is particularly replete. The following reviews and sources of original data may be helpful: Spitz and Wolf (1946); Ambrose (1961, 1963); Freedman (1961, 1964, 1965); Salzen (1963, 1967); Polak, Emde, and Spitz (1964); Brackbill (1967); Lewis (1969); Marcus (1969); Blurton Jones (1969); Eibl-Eibesfeldt

(1970, pp. 402–407); and Bugental, Love, and Gianetto (1971). Douglas (1971) and Fry (1971) report some interesting data on laughter.

In a large portion of the research on human expressions the emphasis has been on the judgment and classification of categories of affect. The interested reader is referred to Schlosberg (1941, 1952, 1954); Woodworth and Schlosberg (1954, pp. 111–132); Thompson and Meltzer (1964); Leventhal and Sharp (1965); Drag and Shaw (1967); Mordkoff (1967); Sjöberg (1968); Zaidel and Mehrabian (1969); Boucher (1969); Ekman and Friesen (1969a); Ekman, Friesen, and Tomkins (1971b); Ekman, Liebert, Friesen, Harrison, Zlatchin, Malmstrom, and Baron (1971c). Buck, Savin, Miller, and Caul (1969) applied some of the techniques used in their nonhuman primate studies discussed above to human subjects.

Facial expressions and associated phenomena have been of interest to clinicians involved with psychological assessment and therapy. Pertinent research can be found in papers by Dittmann, Parloff, and Boomer (1965); Ekman (1964, 1965); and Ekman and Friesen (1968, 1969b).

In the present review we have frequently been concerned with visual orientation phenomena associated with facial expressions in nonhuman primates. Data and reviews on this topic among humans are available in Gibson and Pick (1963); Exline and Winters (1965); Argyle (1967, 1970); Diebold (1968); Ellsworth and Carlsmith (1968); Gitter and Guichard (1968); Exline, Gottheil, Paredes, and Winklemeier (1968); Ellgring (1970); Heron (1970); Schmidt and Hore (1970); Stephenson and Rutter (1970); Exline (1971); LeCompte and Rosenfeld (1971); Modigliani (1971); and von Cranach (1971). Of particular interest is Robson's (1967) review of the role of mutual eye contact between human mothers and infants. The clinical implications of atypical visual orientation and especially gaze aversion are discussed by Riemer (1949, 1955); Bateson, Jackson, Haley, and Weakland (1956); Goldfarb (1956); Hutt and Ounsted (1966); Altman, Swartz, and Cleland (1970); and Hinchliffe, Lancashire, and Roberts (1970).

Several researchers have emphasized the experimental manipulation of facial stimuli in assessing a person's responses. Fantz's research (e.g., Fantz, 1963, 1965; Fantz and Nevis, 1967) on facial form perception is well-known. Coss (1970) studied the effects of several eyespot configurations on pupillary dilation and brow movements and discusses his findings in relation to phenomena associated with a direct stare in human and nonhuman primates. I highly recommend his delightful review (Coss, 1968) of facial components and patterns of threat and other expressions which may underlie many of man's artifacts and modes of artistic expression.

Eibl-Eibesfeldt (1970, pp. 420–422; 1971) offers interesting data on a greeting pattern in humans which consists of a smile followed by a rapid raising of the eyebrows and nodding of the head. We have discussed similar components in the context of grimacing (e.g., during greeting) in the present review.

A study of the importance of specific facial areas attended to by nonhuman primates would go far in putting human nonverbal communication in greater perspective. For example, lack of direct visual eye contact between a human mother and infant has been suggested as a contributing factor in infantile autism (Hutt and Ounsted, 1966). However, the parameters of direct eye contact between nonhuman-primate mothers and infants are not entirely clear at this point, although it is relatively certain that infants look at their mother's face often. Jensen and Bobbitt (1965, p. 64), for example, maintain that pigtailed macaque infants "look soulfully, directly and, perhaps, naïvely into their mother's eyes . . ." whereas Chevaliér-Skolnikoff (1968, p. 4) suggested that direct eye contact is largely absent between a normal mother and neonatal stumptailed macaque. If, indeed, direct eye contact does take place with any frequency in nonhuman-primate mother-infant relationships, it remains to be clarified at what developmental point it appears and what social functions it may serve. The time seems opportune to look at these and other variables in the ontogeny of facial communication in many species of nonhuman primates, an approach that for the most part has been overlooked. As we have seen, communication by facial expression constitutes a vital means of interaction in ongoing social systems, and its development during maturation of an organism ought to be explicated carefully.

Acknowledgments

This chapter is the product of many individuals' efforts. I am greatly indebted to Gary Mitchell for his thorough and scholarly readings of the manuscript and for his untiring encouragement, critical comments, and suggestions. I also thank Leonard A. Rosenblum for the great care and insight that characterized his suggestions on early drafts of the manuscript. Tim Ransom's extraordinary photographs are without peer as documents of visual communication among primates, and I express my deep appreciation to him for allowing me to include several of them in this chapter. Interested readers can find information on his field study of anubis baboons of the Gombe in his forthcoming book (Bucknell University Press, 1975). Discussions with William A. Mason on a number of issues were of great assistance. Finally, I thank Gabriele Gurski for being kind enough to translate several passages.

References

Altman, R., Swartz, J. D., and Cleland, C. C. (1970). Differential sensitivity of profound retardates to adults' steady gaze. *Psychol. Rep.* **27**, 30.

Altmann S. A. (1962). A field study of the sociobiology of rhesus monkeys, *Macaca mulatta. Ann. N.Y. Acad. Sci.* **102**, 338–435.

Altmann, S. A. (1965). Sociobiology of rhesus monkeys. II. Stochastics of social communication. *J. Theor. Biol.* **8**, 490–522.

Altmann, S. A., ed. (1967a). "Social Communication among Primates." Univ. of Chicago Press, Chicago, Illinois.

Altmann, S. A. (1967b). The structure of primate social communication. *In* "Social Communication among Primates" (S. A. Altmann, ed.), pp. 325–362. Univ. of Chicago Press, Chicago, Illinois.

Altmann, S. A. (1968). Sociobiology of rhesus monkeys. IV. Testing Mason's hypothesis of sex differences in affective behavior. *Behaviour* **32**, 49–69.

Ambrose, J. A. (1961). The development of the smiling response in early infancy. *In* "Determinants of Infant Behaviour" (B. M. Foss, ed.) Vol. 1, pp. 179–201. Methuen, London.

Ambrose, J. A. (1963). The concept of a critical period for the development of social responsiveness in early human infancy. *In* "Determinants of Infant Behaviour" (B. M. Foss, ed.), Vol. 2, pp. 201–205. Metheun, London.

Andrew, R. J. (1963a). Evolution of facial expression. *Science* **142**, 1034–1041.

Andrew, R. J. (1963b). The origin and evolution of the calls and facial expressions of the primates. *Behaviour* **20**, 1–109.

Andrew, R. J. (1964). The displays of the primates. *In* "Evolutionary and Genetic Biology of Primates" (J. Buettner-Janusch, ed.) Vol. 2, pp. 227–309. Academic Press, New York.

Andrew, R. J. (1965). The origins of facial expressions. *Sci. Amer.* **213** (4), 88–94.

Anthoney, T. R. (1968). The ontogeny of greeting, grooming and sexual motor patterns in captive baboons (superspecies *Papio cynocephalus*). *Behaviour* **31**, 358–372.

Argyle, M. (1967). "The Psychology of Interpersonal Behaviour." Penguin Books, Baltimore, Maryland.

Argyle, M. (1970). Eye-contact and distance: A reply to Stephenson and Rutter. *Brit. J. Psychol.* **61**, 395–396.

Ayer, A. A. (1948). "The Anatomy of *Semnopithecus entellus*." Indian Publ. House, Ltd., Madras.

Bastian, J. R. (1965). Primate signalling systems and human languages. *In* "Primate Behavior" (I. DeVore, ed.), pp. 585–606. Holt, New York.

Bastian, J. R. (1968). Animal communication. I. Social and psychological analysis. *Int. Encycl. Social Sci.* **3**, 29–33.

Bateson, G., Jackson, D. D., Haley, J., and Weakland, J. (1956). Toward a theory of schizophrenia. *Behav. Sci.* **1**, 251–264.

Baysinger, C. M., Brandt, E. M., and Mitchell, G. (1972). Development of infant social isolate monkeys (*Macaca mulatta*) in their isolation environments. *Primates* **13**, 257–270.

Bernstein, I. S. (1964). A field study of the activities of howler monkeys. *Anim. Behav.* **12**, 92–97.

Bernstein, I. S. (1970). Some behavioral elements of the Cercopithecoidea. *In* "Old World Monkeys. Evolution, Systematics, and Behavior" (J. R. Napier and P. H. Napier, eds.), pp. 263–295. Academic Press, New York.

Bernstein, I. S. (1971). The influence of introductory techniques on the formation of captive mangabey groups. *Primates* **12**, 33–44.

Bernstein, I. S., and Draper, W. A. (1964). The behaviour of juvenile rhesus monkeys in groups. *Anim. Behav.* **12**, 84–91.

Bernstein, S., and Mason, W. A. (1962). The effects of age and stimulus conditions on the emotional responses of rhesus monkeys: Responses to complex stimuli. *J. Genet. Psychol.* **101**, 279–298.

Bertrand, M. (1969). "The Behavioral Repertoire of the Stumptail Macaque," Bibliotheca Primatologica, No. 11. Karger, Basel.

Blurton Jones, N. G. (1969). An ethological study of some aspects of social behaviour of children in nursery school. *In* "Primate Ethology" (D. Morris, ed.), pp. 437–463. Anchor Doubleday, Garden City, New York.

Blurton Jones, N. G., and Trollope, J. (1968). Social behaviour of stump-tailed macaques in captivity. *Primates* **9**, 365–394.

Bobbitt, R. A., Jensen, G. D., and Gordon, B. N. (1964). Behavioral elements (taxoomy) for observing mother-infant-peer interactions in *Macaca nemestrina*. *Primates* **5**(3/4), 71–80.

Bobbitt, R. A., Gourevitch, V. P., Miller, L. E., and Jensen, G. D. (1969). Dynamics of social interactive behavior: A computerized procedure for analyzing trends, patterns, and sequences. *Psychol. Bull.* **71**, 110–121.

Bolwig, N. (1959). A study of the behaviour of the chacma baboon, *Papio ursinus*. *Behaviour* **14**, 136–163.

Bolwig, N. (1963). Bringing up a young monkey (*Erythrocebus patas*). *Behaviour* **21**, 300–330.

Bolwig, N. (1964). Facial expression in primates with remarks on a parallel development in certain carnivores (A preliminary report on work in progress). *Behaviour* **22**, 167–192.

Boucher, J. D. (1969). Facial displays of fear, sadness and pain. *Percep. Mot. Skills* **28**, 239–242.

Bourlière, F., Hunkeler, C., and Bertrand, M. (1970). Ecology and behavior of Lowe's guenon (*Cercopithecus campbelli lowei*) in the Ivory Coast. *In* "Old World Monkeys. Evolution, Systematics, and Behavior" (J. R. Napier and P. H. Napier, eds.), pp. 297–350. Academic Press, New York.

Brackbill, Y. (1967). The use of social reinforcement in conditioning smiling. *In* "Behavior in Infancy and Early Childhood" (Y. Brackbill and G. G. Thompson, eds.), pp. 616–625. Free Press, New York.

Bramblett, C. A. (1970). Coalitions among gelada baboons. *Primates* **11**, 327–333.

Brandt, E. M., Stevens, C. W., and Mitchell, G. (1971). Visual social communication in adult male isolate-reared monkeys (*Macaca mulatta*). *Primates* **12**, 105–112.

Brandt, E. M., Baysinger, C., and Mitchell, G. (1972). Separation from rearing environment in mother-reared and isolation-reared rhesus monkeys (*Macaca mulatta*). *Int. J. Psychobiol.* **2**, 193–204.

Bronson, G. W. (1968). The fear of novelty. *Psychol. Bull.* **69**, 350–358.

Buck, R., Savin, V. J., Miller, R. E., and Caul, W. F. (1969). Nonverbal communication of affect in humans. *Proc. 77th Annu. Conv. Amer. Psychol. Ass.* **4**, 367–368.

Bugental, D. E., Love, L. R., and Gianetto, R. M. (1971). Perfidious feminine faces. *J. Pers. Social Psychol.* **17**, 314–318.

Carpenter, C. R. (1934). A field study of the behavior and social relations of howling monkeys. *Comp. Psychol. Monogr.* **10**(2), 1–168.

Carpenter, C. R. (1940). A field study in Siam of the behavior and social relations of the gibbon (*Hylobates lar*). *Comp. Psychol. Monogr.* **16**(5), 1–212.

Carpenter, C. R. (1960). "Howler Monkeys of Barro Colorado Island," Film. Pennsylvania State Univ. Libr., University Park.

Carpenter, C. R. (1971). Primate behavior and ecology. AAAS-NSF Fac. Semin. Univ. of California at Berkeley.

Castell, R., and Wilson, C. (1971). Influence of spatial environment on development of mother-infant interaction in pigtail monkeys. *Behaviour* **39**, 202–211.

Chalmers, N. R. (1968). The visual and vocal communication of free living mangabeys in Uganda. *Folia Primatol.* **9**, 258–280.

Chalmers, N. R., and Rowell, T. E. (1971). Behaviour and female reproductive cycles in a captive group of mangabeys. *Folia Primatol.* **14**, 1–14.

Chance, M. R. A. (1956). Social structure of a colony of *Macaca mulatta. Brit. J. Anim. Behav.* **4**, 1–13.

Chance, M. R. A. (1962). An interpretation of some agonistic postures; The role of "cut-off" acts and postures. *Symp. Zool. Soc. London* No. 8, 71–89.

Chevaliér-Skolnikoff, S. (1968). The ontogeny of communication in *Macaca speciosa.* Paper presented at the meeting of the Amer. Anthropol. Ass., Seattle, Wash.

Cole, J. (1963). *Macaca nemestrina* studied in captivity. *Symp. Zool. Soc. London* **10**, 105–114.

Coss, R. G. (1968). The ethological command in art. *Leonardo* **1**, 273–287.

Coss, R. G. (1970). The perceptual aspects of eye-spot patterns and their relevance to gaze behaviour. *In* "Behaviour Studies in Psychiatry" (C. Hutt and S. J. Hutt, eds.), pp. 121–147. Pergamon, Oxford.

Crook, J. H. (1965). "Gelada: The Mountain Baboon of Ethiopia," Film. Univ. of Bristol Psychol. Dep., Bristol, England.

Cross, H. A., and Harlow, H. F. (1965). Prolonged and progressive effects of partial isolation on the behavior of macaque monkeys. *J. Exp. Res. Pers.* **1**, 39–49.

Cüceloglu, D. M. (1970). Perception of facial expressions in three different cultures. *Ergonomics* **13**, 93–100.

Darwin, C. (1872). "The Expressions of the Emotions in Man and Animals." John Murray, London.

Davenport, R. K. (1967). The orang-utan in Sabah. *Folia Primatol.* **5**, 247–263.

Davitz, J. R., ed. (1964). "The Communication of Emotional Meaning." McGraw-Hill, New York.

DeVore, I., ed. (1965). "Primate Behavior." Holt, New York.

Diebold, A. R., Jr. (1968). Anthropological perspectives. Anthropology and the comparative psychology of communicative behavior. *In* "Animal Communication" (T. A. Sebeok, ed.), pp. 525–571. Indiana Univ. Press, Bloomington.

Dittmann, A. T., Parloff, M. B., and Boomer, D. S. (1965). Facial and bodily expression: A study of receptivity of emotional cues. *Psychiatry* **28**, 239–244.

Douglas, M. (1971). Do dogs laugh? A cross-cultural approach to body symbolism. *J. Psychosom. Res.* **15**, 387–390.

Drag, R. M., and Shaw, M. E. (1967). Factors influencing the communication of emotional intent by facial expressions. *Psychon. Sci.* **8**, 137–138.

Durham, N. M. (1969). Sex differences in visual threat displays of West African vervets. *Primates* **10**, 91–95.

Eibl-Eibesfeldt, I. (1970). "Ethology: The Biology of Behavior." Holt, New York.

Eibl-Eibesfeldt, I. (1971). Transcultural patterns of ritualized contact behavior. *In* "Behavior and Environment: The Use of Space by Animals and Men" (A. H. Esser, ed.), pp. 238–246. Plenum, New York.

Ekman, P. (1964). Body position, facial expression, and verbal behavior during interviews. *J. Abnorm. Social Psychol.* **68**, 295–31.

Ekman, P. (1965). Communication through nonverbal behavior: A source of information about an interpersonal relationship. *In* "Affect, Cognition, and Personality:

Empirical Studies" (S. S. Tomkins and C. E. Izard, eds.), pp. 390–442. Springer, New York.

Ekman, P. (1971). Universals and cultural differences in facial expressions of emotion. *In* "Nebraska Symposium on Motivation" (J. K. Cole, ed.), pp. 207–283. Univ. of Nebraska Press, Lincoln.

Ekman, P., ed. (1973). "Darwin and Facial Expression: A Century of Research in Review." Academic Press, New York.

Ekman, P., and Friesen, W. V. (1968). Nonverbal behavior in psychotherapy research. *Res. Psychother.* **3**, 179–216.

Ekman P., and Friesen, W. V. (1969a). The repertoire of nonverbal behavior: Categories, origins, usage, and coding. *Semiotica* **1**, 49–98.

Ekman, P., and Friesen, W. V. (1969b). Nonverbal leakage and clues to deception. *Psychiatry* **32**, 88–106.

Ekman, P., Friesen, W. V., and Ellsworth, P. (1971a). "Emotion in the Human Face: Guidelines for Research and an Integration of Findings." Pergamon, New York.

Ekman, P., Friesen, W. V., and Tomkins, S. S. (1971b). Facial affect scoring technique: A first validity study. *Semiotica* **3**, 37–58.

Ekman, P., Liebert, R. M., Friesen, W. V., Harrison, R., Zlatchin, C., Malmstrom, E. J., and Baron, R. A. (1971c). Facial expressions of emotion while watching televised violence as predictors of subsequent aggression. "Television and Social Behavior," A Report to the Surgeon General's Scientific Advisory Committee, Vol. 5, U. S. Gov. Printing Office, Washington, D.C.

Ellgring, J. H. (1970). Die Beurteilung des Blicks auf Punkte innerhalb des Gesichtes. (Judgment of glances directed at different points in the face.) *Z. Exp. Angew. Psychol.* **17**, 600–607.

Ellsworth, P. C., and Carlsmith, J. M. (1968). Effects of eye contact and verbal content on affective response to a dyadic interaction. *J. Pers. Social Psychol.* **10**, 15–20.

Epple, G. (1967). Vergleichende Untersuchungen über Sexual- und Sozialverhalten der Krallenaffen (Hapalidae). (A comparative investigation of the sexual and social behavior of the marmoset (Hapalidae).) *Folia Primatol.* **7**, 37–65.

Esser, A. H., ed. (1971). "Behavior and Environment: The Use of Space by Animals and Men." Plenum, New York.

Evans, C. S. (1967). Methods of rearing and social interaction in *Macaca nemestrina*. *Anim. Behav.* **15**, 263–266.

Exline, R. V. (1971). Visual interaction: The glances of power and preference. *In* "Nebraska Symposium on Motivation" (J. K. Cole, ed.), pp. 163–206. Univ. of Nebraska Press, Lincoln.

Exline, R. V., and Winters, L. C. (1965). Affective relations and mutual glances in dyads. *In* "Affect, Cognition, and Personality: Empirical Studies" (S. S. Tomkins and C. E. Izard, eds.), pp. 319–350. Springer, New York.

Exline, R. V., Gottheil, E., Paredes, A., and Winklemeier, D. (1968). Gaze direction as a factor in the accurate judgment of nonverbal expressions of affect. *Proc. 76th Annu. Conv. Amer. Psychol. Ass.* **3**, 415–416.

Fantz, R. L. (1963). Pattern vision in new-born infants. *Science* **140**, 296–297.

Fantz, R. L. (1965). Ontogeny of perception. *In* "Behavior of Non-Human Primates: Modern Research Trends" (A. M. Schrier, H. F. Harlow, and F. Stollnitz, eds.), Vol. 2, pp. 365–403. Academic Press, New York.

Fantz, R. L., and Nevis, S. (1967). Pattern preferences and perceptual-cognitive development in early infancy. *Merrill-Palmer Quart.* **13**, 77–108.

Fittinghoff, N. A., Jr., Lindburg, D. G., Gomber, J., and Mitchell, G. (1974). Consistency and variability in the behavior of mature, isolation-reared, male rhesus macaques. *Primates* **15**, 111–139.

Fox, M. W. (1970). A comparative study of the development of facial expressions in canids; Wolf, coyote and foxes. *Behaviour* **36**, 4–73.

Freedman, D. G. (1961). The infant's fear of strangers and the flight response. *J. Child Psychol. Psychiat.* **2**, 242–248.

Freedman, D. G. (1964). Smiling in blind infants and the issue of innate *vs.* acquired. *J. Child Psychol. Psychiat.* **5**, 171–184.

Freedman, D. G. (1965). An ethological approach to the genetical study of human behavior. *In* "Methods and Goals in Human Behavior Genetics" (S. G. Vandenberg, ed.), pp. 141–161. Academic Press, New York.

Fry, W. F., Jr. (1971). Laughter: Is it the best medicine? *Stanford M. D.* **10**(1), 16–20.

Furuya, Y. (1961–1962). The social life of silvered leaf monkeys (*Trachypithecus cristatus*). *Primates* **3**(2), 41–60.

Gartlan, J. S., and Brain, C. K. (1968). Ecology and social variability in *Cercopithecus aethiops* and *C. mitis*. *In* "Primates" (P. Jay, ed.), pp. 253–292. Holt, New York.

Geldard, F. A. (1960). Some neglected possibilities of communication. *Science* **131**, 1583–1588.

Gibson, J. J., and Pick, A. D. (1963). Perception of another person's looking behavior. *Amer. J. Psychol.* **76**, 386–394.

Gitter, A. G., and Guichard, M. (1968). Looking behavior: First looker's direction and focus of gaze. *CRC Rep., Boston Univ.* No. 23, 22 pp.

Goldfarb, W. (1956). Receptor preferences in schizophrenic children. *AMA Arch. Neurol. Psychiat.* **76**, 643–652.

Goodall, J. (1965). Chimpanzees of the Gombe Stream Reserve. *In* "Primate Behavior" (I. DeVore, ed.), pp. 425–473. Holt, New York.

Grant, E. C. (1971). Facial expressions and gesture. *J. Psychosom. Res.* **15**, 391–394.

Gregory, W. K. (1963). "Our Face from Fish to Man." Hafner, New York. (Original printing, 1929.)

Guthrie, R. D., and Petocz, R. G. (1970). Weapon automimicry among mammals. *Amer. Natur.* **104**, 585–588.

Hall, K. R. L. (1962). The sexual, agonistic and derived social behaviour patterns of the wild chacma baboon, *Papio ursinus*. *Proc. Zool. Soc. London* **139**, 283–327.

Hall, K. R. L. (1967). Social interactions of the adult male and adult females of a patas monkey group. *In* "Social Communication among Primates" (S. A. Altmann, ed.), pp. 261–280. Univ. of Chicago Press, Chicago, Illinois.

Hall, K. R. L. (1968). Behaviour and ecology of the wild patas monkey, *Erythrocebus patas*, in Uganda. *In* "Primates" (P. Jay, ed.), pp. 32–119. Holt, New York.

Hall, K. R. L., and DeVore, I. (1965). Baboon social behavior. *In* "Primate Behavior" (I. DeVore, ed.), pp. 53–110. Holt, New York.

Hall, K. R. L., and Gartlan, J. S. (1965). Ecology and behaviour of the vervet monkey, *Cercopithecus aethiops*, Lolui Island, Lake Victoria. *Proc. Zool. Soc. London* **145**, 37–56.

Hall, K. R. L., and Mayer, B. (1967). Social interactions in a group of captive patas monkeys (*Erythrocebus patas*). *Folia Primatol.* **5**, 213–236.

Hall, K. R. L., Boelkins, R. C., and Goswell, M. J. (1965). Behaviour of patas monkeys, *Erythrocebus patas,* in captivity, with notes on the natural habitat. *Folia Primatol.* **3,** 22–49.

Hampton J. K., Hampton, S. H., and Landwehr, B. T. (1966). Observations on a successful breeding colony of the marmoset, *Oedipomidas oedipus. Folia Primatol.* **4,** 265–287.

Hansen, E. (1966). The development of maternal and infant behavior in the rhesus monkey. *Behaviour* **27,** 107–149.

Harlow, H. F. (1964). Early social deprivation and later behavior in the monkey. *In* "Unfinished Tasks in the Behavioral Sciences" (A. Abrams, H. H. Garner, and J. E. P. Toman, eds.), pp. 154–173. Williams & Wilkins, Baltimore, Maryland.

Harlow, H. F. (1969). Age-mate or peer affectional system. *In* "Advances in the Study of Behavior" (D. S. Lehrman, R. A. Hinde, and E. Shaw, eds.), Vol. 2, pp. 333–383. Academic Press, New York.

Harlow, H. F., Rowland, G. L., and Griffin, G. A. (1964). The effect of total social deprivation on the development of monkey behavior. *Psychiat. Res. Rep.* **19,** 116–135.

Heron, J. (1970). The phenomenology of social encounter: The gaze. *Phil. Phenomenol. Res.* **31**(2), 243–264.

Hill, W. C. O. (1957). "Primates: Comparative Anatomy and Taxonomy. III. Pithecoidea." Univ. of Edinburgh Press, Edinburgh.

Hill, W. C. O. (1970). "Primates: Comparative Anatomy and Taxonomy. VIII. Cynopithecinae." Univ. of Edinburgh Press, Edinburgh.

Hinchliffe, M., Lancashire, M., and Roberts, F. J. (1970). Eye-contact and depression: A preliminary report. *Brit. J. Psychiat.* **117,** 571–572.

Hinde, R. A., and Rowell, T. E. (1962). Communication by posture and facial expression in the rhesus monkey. *Proc. Zool. Soc. London* **138,** 1–21.

Hinde, R. A., and Spencer-Booth, Y. (1967). The behaviour of socially living rhesus monkeys in their first two and a half years. *Anim. Behav.* **15,** 169–196.

Hinde, R. A., Rowell, T. E., and Spencer-Booth, Y. (1964). Behaviour of socially living rhesus monkeys in their first six months. *Proc. Zool. Soc. London* **143,** 609–649.

Hutt, C., and Ounsted, C. (1966). The biological significance of gaze aversion with particular reference to the syndrome of infantile autism. *Behav. Sci.* **11,** 346–356.

Jay, P. (1965a). The common langur of North India. *In* "Primate Behavoir" (I. DeVore, ed.), pp. 197–249. Holt, New York.

Jay, P. (1965b). Field studies. *In* "Behavior of Nonhuman Primates: Modern Research Trends" (A. M. Schrier, H. F. Harlow, and F. Stollnitz, eds.), Vol. 2, pp. 525–591. Academic Press, New York.

Jay, P., ed. (1968). "Primates." Holt, New York.

Jensen, G. D., and Bobbitt, R. A. (1965). On observational methodology and preliminary studies of mother-infant interaction in monkeys. *In* "Determinants of Infant Behaviour" (B. M. Foss, ed.), Vol. 3, pp. 47–65. Methuen, London.

Jensen, G. D., and Gordon, B. N. (1970). Sequences of mother-infant behavior following a facial communicative gesture of pigtail monkeys. *Biol. Psychiat.* **2,** 267–272.

Jolly, A. (1966). "Lemur Behavior." Univ. of Chicago Press, Chicago, Illinois.

Kaufman, I. C., and Rosenblum, L. A. (1966). A behavioral taxonomy for *Macaca nemestrina* and *Macaca radiata*: Based on longitudinal observation of family groups in the laboratory. *Primates* **7,** 205–258.

Kaufmann, J. H. (1966). Behavior of infant rhesus monkeys and their mothers in a free-ranging band. *Zoologica (New York)* **51,** 17–27.

Kaufmann, J. H. (1967). Social relations of adult males in a free-ranging band of rhesus monkeys. *In* "Social Communication among Primates" (S. A. Altmann, ed.), pp. 73–98. Univ. of Chicago Press, Chicago, Illinois.

Kawabe, M. (1970). A preliminary study of the wild siamang gibbon (*Hylobates syndactylus*) at Fraser's Hill, Malaysia. *Primates* **11**, 285–291.

Klopfer, P. H. (1970). Discrimination of young in Galagos. *Folia Primatol.* **13**, 137–143.

Kummer, H. (1967). Tripartite relations in hamadryas baboons. *In* "Social Communication among Primates" (S. A. Altmann, ed.), pp. 63–71. Univ. of Chicago Press, Chicago, Illinois.

Kummer, H. (1968). "Social Organization of Hamadryas Baboons. A field study," Bibliotheca Primatologica, No. 6. Karger, Basel.

Lancaster, J. B. (1967). Communication systems of Old World monkeys and apes. *Int. Social Sci. J.* **19**, 28–35.

LeCompte, W. F., and Rosenfeld, H. M. (1971). Effects of minimal eye contact in the instruction period on impressions of the experimenter. *J. Exp. Social Psychol.* **7**, 211–220.

Leventhal, H., and Sharp, E. (1965). Facial expressions as indicators of distress. *In* "Affect, Cognition, and Personality: Empirical Studies" (S. S. Tomkins and C. E. Izard, eds.), pp. 296–318. Springer, New York.

Lewis, M. (1969). Infants' responses to facial stimuli during the first year of life. *Develop. Psychol.* **1**, 75–86.

Lindburg, D. G. (1971). The rhesus monkey in North India: An ecological and behavioral study. *In* "Primate Behavior: Developments in Field and Laboratory Research" (L. A. Rosenblum, ed.), Vol. 2, pp. 1–106. Academic Press, New York.

Loizos, C. (1969). Play behaviour in higher primates: A review. *In* "Primate Ethology" (D. Morris, ed.), pp. 226–282. Anchor Doubleday, Garden City, New York.

Lorenz, K. (1966). "On Aggression." Bantam Books, New York.

Macdonald, J. (1965). "Almost Human. The Baboon: Wild and Tame—In Fact and in Legend." Chilton, Philadelphia, Pennsylvania.

MacKinnon, J. (1971). The orang-utan in Sabah today. *Oryx* **11**, 141–191.

Marcus, N. N. (1969) A psychotherapeutic corroboration of the meaning of the smiling response. *Psychoanal. Rev.* **56**, 387–401.

Marler, P. (1965). Communication in monkeys and apes. *In* "Primate Behavior" (I. DeVore, ed.), pp. 544–584. Holt, New York.

Marler, P. (1968). Aggregation and dispersal: Two functions in primate communication. *In* "Primates" (P. Jay, ed.), pp. 420–438. Holt, New York.

Mason, W. A. (1960). The effects of social restriction on the behavior of rhesus monkeys: I. Free social behavior. *J. Comp. Physiol. Psychol.* **53**, 582–589.

Mason, W. A. (1961a). The effects of social restriction on the behavior of rhesus monkeys: II. Tests of gregariousness. *J. Comp. Physiol. Psychol.* **54**, 287–290.

Mason, W. A. (1961b). The effects of social restriction on the behavior of rhesus monkeys: III. Dominance tests. *J. Comp. Physiol. Psychol.* **54**, 694–699.

Mason, W. A., and Green, P. C. (1962). The effects of social restriction on the behavior of rhesus monkeys: IV. Responses to a novel environment and to an alien species. *J. Comp. Physiol. Psychol.* **55**, 363–368.

Mason, W. A., Green, P. C., and Posepanko, C. J. (1960). Sex differences in affective-social responses of rhesus monkeys. *Behaviour* **16**, 74–83.

Maxim, P. E., and Buettner-Janusch, J. (1963). A field study of the Kenya baboon. *Amer. J. Phys. Anthropol.* **21**, 165–180.

Michael, R. P. (1969). The role of pheromones in the communication of primate behaviour. *Proc. Int. Congr. Primatol., 2nd, Atlanta, Ga.* **1**, 101–107.

Michael, R. P., and Keverne, E. B. (1968). Pheromones in the communication of sexual status in primates. *Nature (London)* **218**, 746–749.

Michael, R. P., and Keverne, E. B. (1970). Primate sex pheromones of vaginal origin. *Nature (London)* **225**, 84–85.

Michael, R. P., and Welegalla, J. (1968). Ovarian hormones and the sexual behaviour of the female rhesus monkey (*Macaca mulatta*) under laboratory conditions. *J. Endocrinol.* **41**, 401–420.

Michael, R. P., and Zumpe, D. (1971). Patterns of reproductive behavior. *In* "Comparative Reproduction of Nonhuman Primates" (E. S. E. Hafez, ed.), pp. 205–242. Thomas, Springfield, Illinois.

Michael, R. P., Herbert, J., and Welegalla, J. (1967). Ovarian hormones and the sexual behaviour of the male rhesus monkey (*Macaca mulatta*) under laboratory conditions. *J. Endocrinol.* **39**, 81–98.

Michael, R. P., Keverne, E. B., and Bonsall, R. W. (1971). Pheromones: Isolation of male sex attractants from a female primate. *Science* **172**, 964–966.

Miller, R. E. (1967). Experimental approaches to the physiological and behavioral concomitants of affective communication in rhesus monkeys. *In* "Social Communication among Primates" (S. A. Altmann, ed.), pp. 43–54. Univ. of Chicago Press, Chicago, Illinois.

Miller, R. E. (1971). Experimental studies of communication in the monkey. *In* "Primate Behavior: Developments in Field and Laboratory Research" (L. A. Rosenblum, ed.), Vol. 2, pp. 139–175. Academic Press, New York.

Miller, R. E., Murphy, J. V., and Mirsky, I. A. (1959a). Non-verbal communication of affect. *J. Clin. Psychol.* **15**, 155–158.

Miller, R. E., Murphy, J. V., and Mirsky, I. A. (1959b). Relevance of facial expression and posture as cues in communication of affect between monkeys. *Arch. Gen. Psychiat.* **1**, 480–488.

Miller, R. E., Banks, J. H., and Ogawa, N. (1963). Role of facial expression in "cooperative-avoidance conditioning" in monkeys. *J. Abnorm. Social Psychol.* **67**, 24–30.

Miller, R. E., Caul, W. F., and Mirsky, I. A. (1967). Communication of affects between feral and socially isolated monkeys. *J. Pers. Social Psychol.* **7**, 231–239.

Missakian, E. A. (1969). Effects of social deprivation on the development of patterns of social behavior. *Proc. Int. Congr. Primatol., 2nd, Atlanta, Ga.* **2**, 50–55.

Mitchell, G. D. (1968). Persistent behavior pathology in rhesus monkeys following early social isolation. *Folia Primatol.* **8**, 132–147.

Mitchell, G. (1970). The development of abnormal behavior in monkeys. Paper presented at the 1970 Behav. Modif. Workshop, Stockton, Calif.

Mitchell, G. (1972). Looking behavior in the rhesus monkey. *J. Phenomenol. Psychol.* **3**, 53–67.

Mitchell, G. D., and Clark, D. L. (1968). Long-term effects of social isolation in nonsocially adapted rhesus monkeys. *J. Genet. Psychol.* **113**, 117–128.

Mitchell, G. D., Raymond, E. J., Ruppenthal, G. C., and Harlow, H. F. (1966). Long-term effects of total social isolation upon behavior of rhesus monkeys. *Psychol. Rep.* **18**, 567–580.

Mitchell, G., Stevens, C. W., and Lindburg, D. G. (no date). General Behavioral Differences between Five Macaques. Unpublished paper, Univ. of California at Davis.

Modigliani, A. (1971). Embarrassment, facework, and eye contact: Testing a theory of embarrassment. *J. Pers. Social Psychol.* **17**, 15–24.

Møller, G. W., Harlow, H. F., and Mitchell, G. D. (1968). Factors affecting agonistic communication in rhesus monkeys (*Macaca mulatta*). *Behaviour* **31**, 339–357.

Mordkoff, A. M. (1967). A factor analytic study of the judgment of emotion from facial expression. *J. Exp. Res. Pers.* **2**, 80–85.

Morris, D., ed. (1969). "Primate Ethology." Anchor Doubleday, Garden City, New York.

Moynihan, M. (1964). Some behavior patterns of platyrrhine monkeys. I. The night monkey (*Aotus trivirgatus*). *Smithson. Misc. Collect.* **146**(5).

Moynihan, M. (1966). Communication in the titi monkey, *Callicebus*. *J. Zool.* **150**, 77–127.

Moynihan, M. (1969). Comparative aspects of communication in New World primates. *In* "Primate Ethology" (D. Morris, ed.), pp. 306–342. Anchor Doubleday, Garden City, New York.

Moynihan, M. (1970a). Some behavior patterns of platyrrhine monkeys. II. *Saguinus geoffroyi* and some other tamarins. *Smithson. Contrib. Zool.* No. 28.

Moynihan, M. (1970b). Control, suppression, decay, disappearance and replacement of displays. *J. Theor. Biol.* **29**, 85–112.

Myers, R. E. (1969). Neurology of social communication in primates. *Proc. Int. Cong. Primatol., 2nd, Atlanta, Ga.* **3**, 1–9.

Napier, J. R., and Napier, P. H. (1967). "A Handbook of Living Primates." Academic Press, New York.

Napier, J. R., and Napier, P. H., eds. (1970). "Old World Monkeys. Evolution, Systematics, and Behavior." Academic Press, New York.

Nishida, T. (1970). Social behavior and relationship among wild chimpanzees of the Mahali Mountains. *Primates* **11**, 47–87.

Osman Hill, W. C., and Bernstein, I. S. (1969). On the morphology, behaviour and systematic status of the assam macaque (*Macaca assamensis*). *Primates* **10**, 1–17.

Petter, J. J. (1965). The lemurs of Madagascar. *In* "Primate Behavior" (I. DeVore, ed.), pp. 292–319. Holt, New York.

Ploog, D., and Melnechuk, T. (1969). Primate communication: A report based on an NRP work session. *Neurosci. Res. Program, Bull.* **7**(5), 419–510.

Poirier, F. E. (1970). Dominance structure of the Nilgiri langur (*Presbytis johnii*) of South India. *Folia Primatol.* **12**, 161–186.

Polak, P. R., Emde, R. N., and Spitz, R. (1964). The smiling response to the human face. I.: Methodology, quantification and natural history. *J. Nerv. Ment. Dis.* **139**, 103–109.

Rahaman, H., and Parthasarathy, M. D. (1968). The expressive movements of the bonnet macaque. *Primates* **9**, 259–272.

Ralls, K. (1971). Mammalian scent marking. *Science* **171**, 443–449.

Ransom, T. (1971). A field study of forest-dwelling baboons (*Papio anubis*) in the Gombe Stream Reserve. Lecture presented at the Univ. of California at Davis.

Redican, W. K., and Mitchell, G. (1974). Play between adult male and infant rhesus monkeys. *Amer. Zool.* **14**, 295–302.

Redican, W. K., Kellicutt, M. H., and Mitchell, G. (1971). Preferences for facial expressions in juvenile rhesus monkeys (*Macaca mulatta*). *Develop. Psychol.* **5**, 539.

Reynolds, P. C. (1970). Social communication in the chimpanzee: A review. *In* "The Chimpanzee. III. Immunology, Infections, Hormones, Anatomy, and Behavior of Chimpanzees" (G. H. Bourne, ed.), pp. 369–394. Univ. Park Press, Baltimore, Maryland.

Richard, A. (1970). A comparative study of the activity patterns and behavior of *Alouatta villosa* and *Ateles geoffroyi*. *Folia Primatol.* **12**, 241–263.

Riemer, M. D. (1949). The averted gaze. *Psychiat. Quart.* **23**, 108–115.

Riemer, M. D. (1955). Abnormalities of the gaze—A classification. *Psychiat. Quart.* **29**, 659–672.

Robson, K. S. (1967). The role of eye-to-eye contact in maternal-infant attachment. *J. Child Psychol. Psychiat.* **8**, 13–25.

Rowell, T. E. (1963). The social development of some rhesus monkeys. *In* "Determinants of Infant Behaviour" (B. M. Foss, ed.), Vol. 2, pp. 35–49. Methuen, London.

Rowell, T. E. (1966). Hierarchy in the organization of a captive baboon group. *Anim. Behav.* **14**, 430–443.

Rowell, T. E. (1967). A quantitative comparison of the behaviour of a wild and a caged baboon group. *Anim. Behav.* **15**, 499–509.

Rowell, T. E., and Hinde, R. A. (1963). Responses of rhesus monkeys to mildly stressful situations. *Anim. Behav.* **11**, 235–243.

Rowell, T. E., Hinde, R. A., and Spencer-Booth, Y. (1964). "Aunt"-infant interaction in captive rhesus monkeys. *Anim. Behav.* **12**, 219–226.

Rowell, T. E., Din, N. A., and Omar, A. (1968). The social development of baboons in their first three months. *J. Zool.* **155**, 461–483.

Rowland, G. L. (1964). The Effects of Total Social Isolation upon Learning and Social Behavior in Rhesus Monkeys. Doctoral dissertation, Univ. of Wisconsin, Madison.

Saayman, G. S. (1970). The menstrual cycle and sexual behaviour in a troop of free ranging chacma baboons (*Papio ursinus*). *Folia Primatol.* **12**, 81–110.

Saayman, G. S. (1971). Behaviour of the adult males in a troop of free-ranging chacma baboons (*Papio ursinus*). *Folia Primatol.* **15**, 36–57.

Sackett, G. P., (1965). Response of rhesus monkeys to social stimulation presented by means of colored slides. *Percept. Mot. Skills* **20**, 1027–1028.

Sackett, G. P. (1966). Monkeys reared in isolation with pictures as visual input: Evidence for an innate releasing mechanism. *Science* **154**, 1471–1473.

Sade, D. S. (1967). Determinants of dominance in a group of free-ranging rhesus monkeys. *In* "Social Communication among Primates" (S. A. Altmann, ed.), pp. 99–114. Univ. of Chicago Press, Chicago, Illinois.

Salzen, E. A. (1963). Visual stimuli eliciting the smiling response in the human infant. *J. Genet. Psychol.* **102**, 51–54.

Salzen, E. A. (1967). Imprinting in birds and primates. *Behaviour* **28**, 232–254.

Schaller, G. B. (1963). "The Mountain Gorilla: Ecology and Behavior." Univ. of Chicago Press, Chicago, Illinois.

Schaller, G. B. (1964). "The Year of the Gorilla." Ballantine Books, New York.

Schaller, G. B. (1965). The behavior of the mountain gorilla. *In* "Primate Behavior" (I. DeVore, ed.), pp. 324–367. Holt, New York.

Schlosberg, H. (1941). A scale for the judgment of facial expressions. *J. Exp. Psychol.* **29**, 497–510.

Schlosberg, H. (1952). The description of facial expression in terms of two dimensions. *J. Exp. Psychol.* **44**, 229–237.

Schlosberg, H. (1954). Three dimensions of emotion. *Psychol. Rev.* **61**, 81–88.

Schmidt, W. H., and Hore, T. (1970). Some nonverbal aspects of communication between mother and preschool child. *Child Develop.* **41**, 889–896.

Simonds, P. E. (1965). The bonnet macaque in South India. *In* "Primate Behavior" (I. DeVore, ed.), pp. 175–196.

Sjöberg, L. (1968). Unidimensional scaling of multidimensional facial expressions. *J. Exp. Psychol.* **78**, 429–435.

Southwick, C. H., Beg, M. A., and Siddiqui, M. R. (1965). Rhesus monkeys in North India. *In* "Primate Behavior" (I. DeVore, ed.), pp. 111–159. Holt, New York.

Spitz, R., and Wolf, K. M. (1946). The smiling response: A contribution to the ontogenesis of social relations. *Genet. Psychol. Monogr.* **34**, 57–125.

Stephenson, G. M., and Rutter, D. R. (1970). Eye-contact, distance and affiliation: A re-evaluation. *Brit. J. Psychol.* **61**, 385–393.

Struhsaker, T. T. (1967a). Auditory communication among vervet monkeys (*Cercopithecus aethiops. In* "Social Communication among Primates" (S. A. Altmann, ed.), pp. 281–324. Univ. of Chicago Press, Chicago, Illinois.

Struhsaker, T. T. (1967b). Behavior of vervet monkeys and other cercopithecines. *Science* **156**, 1197–1203.

Struhsaker, T. T. (1967c). Behavior of vervet monkeys (*Cercopithecus aethiops*). *Univ. Calif., Berkeley, Publ. Zool.* **82**, 1–74.

Struhsaker, T. T., and Gartlan, J. S. (1970). Observations on the behaviour and ecology of the patas monkey (*Erythrocebus patas*) in the Waza Reserve, Cameroon, *J. Zool.* **161**, 49–63.

Thompson, D. F., and Meltzer, L. (1964). Communication of emotional intent by facial expression. *J. Abnorm. Social Psychol.* **68**, 129–135.

Tokuda, K., Simons, R. C., and Jensen, G. D. (1968). Sexual behavior in a captive group of pigtailed monkeys (*Macaca nemestrina*). *Primates* **9**, 283–294.

Tomkins, S. S., and Izard, C. E., eds. (1965). "Affect, Cognition, and Personality: Empirical Studies." Springer, New York.

van Hooff, J. A. R. A. M. (1962). Facial expressions in higher primates. *Symp. Zool. Soc. London* **8**, 97–125.

van Hooff, J. A. R. A. M. (1969). The facial displays of the catarrhine monkeys and apes. *In* "Primate Ethology" (D. Morris, ed.), pp. 9–88. Anchor Doubleday, Garden City, New York.

van Lawick, H., Marler, P., and van Lawick-Goodall, J. (1971). "Vocalizations of Wild Chimpanzees," Film. Rockefeller Univ. Film Serv., New York.

van Lawick-Goodall, J. (1967). "My Friends the Wild Chimpanzees." Nat. Geogr. Soc., Washington, D.C.

van Lawick-Goodall, J. (1968a). A preliminary report on expressive movements and communication in the Gombe Stream chimpanzees. *In* "Primates" (P. Jay, ed.), pp. 313–374. Holt, New York.

van Lawick-Goodall, J. (1968b). The behaviour of free-living chimpanzees in the Gombe Stream Reserve. *Anim. Behav. Monogr.* **1**, 161–311.

van Lawick-Goodall, J. (1971). "In the Shadow of Man." Houghton, Boston, Massachusetts.

Vine, I. (1970). Communication by facial-visual signals. *In* "Social Behaviour in Birds and Mammals" (J. H. Crook, ed.), pp. 279–354. Academic Press, London.

von Cranach, M. (1971). The role of orienting behavior in human interaction. *In* "Behavior and Environment: The Use of Space by Animals and Men" (A. H. Esser, pp. 217–237. Plenum, New York.

Wada, J. A. (1961). Modification of cortically induced responses in brain stem by shift of attention in monkeys. *Science* **133**, 40–42.

Washburn, S. L., and Hamburg, D. A. (1968). Aggressive behavior in Old World monkeys and apes. *In* "Primates" (P. Jay, ed.), pp. 458–478. Holt, New York.

Woodworth, R. S., and Schlosberg, H. (1954). "Experimental Psychology," 2nd Ed. Holt, New York.

Zaidel, S. F., and Mehrabian, A. (1969). The ability to communicate and infer positive and negative attitudes facially and vocally. *J. Exp. Res. Pers.* **3**, 233–241.

Zuckerman, S. (1932). "The Social Life of Monkeys and Apes." Harcourt, New York.

Zumpe, D., and Michael, R. P. (1968). The clutching reaction and orgasm in the female rhesus monkey (*Macaca mulatta*). *J. Endocrinol.* **40**, 117–123.

Zumpe, D., and Michael, R. P. (1970). Redirected aggression and gonadal hormones in captive rhesus monkeys (*Macaca mulatta*). *Anim. Behav.* **18**, 11–19.

The Behavior of Marmoset Monkeys (Callithricidae)*

GISELA EPPLE

Monell Chemical Senses Center
University of Pennsylvania
Philadelphia, Pennsylvania

I. INTRODUCTION

Until recently marmoset monkeys, like many other South American primates, were considered delicate animals that would not survive in captivity for a very long time. Quite early, however, single individuals were kept more-or-less successfully in captivity. The French naturalist Buffon, for instance, examined a live lion marmoset which was in the possession of Madame Pompadour (cf. Hill, 1957). Among later reports on

* Part of my own studies were supported by the National Science Foundation, Grant GB 12 660 and were undertaken while I was a Biomedical Fellow of the Population Council of the Rockefeller University. The technical assistance of Mary C. Alveario is gratefully acknowledged.

maintenance and behavior of captive marmosets are those of Desmarest (1818), Heck (1892), Hornung (1896, 1899), Meeter von Zorn (1903), Neill (1829), and Paris (1908). Most of these early publications refer to a few individuals kept as private pets. Lucas *et al.* (1927, 1937) were among the first to report long-term success in maintaining an active breeding group of common marmosets in the laboratory. Their studies and those of Ditmars (1933), Fitzgerald (1935), Marik (1931), and Schreitmüller (1930) considerably increased our knowledge of the basic dietary and environmental requirements of these primates. Thus marmoset monkeys proved to be quite hardy when kept properly and will live many years in captivity.

In recent years some species of marmosets have been imported to Europe and the United States in large numbers and are now used increasingly in biomedical research. Their extended use as laboratory primates has stimulated a large number of studies, many of them specifically concerned with their requirements in a laboratory, their reproductive behavior, parental care, and development of offspring. The present review will not deal with these aspects of marmoset behavior in detail. The reader is referred to the reports cited above and to the publications of Altmann-Schönberner (1965), Benirschke and Richart (1963), Christen (1968, 1974), Coimbra-Filho (1965), Epple (1967, 1970a), Franz (1963), Grüner and Krause (1963), Hampton *et al.* (1966), Heinemann (1970), Kingston (1969), Langford (1963), Levy and Artecona (1964), Lorenz (1969), Lorenz and Heinemann (1967), Mallinson (1964, 1969), Muckenhirn (1967), Rabb and Rowell (1960), Roth (1960), Shadle *et al.* (1965), Snyder (1972), Stellar (1960), Ulmer (1961), and Wendt (1964).

Our knowledge of the behavior of wild members of the Callithricidae is extremely limited (see Section III, A), and almost all data are derived from animals studied in captivity. Only a few of the numerous species of marmosets are regularly available from animal dealers and are used for laboratory studies in large numbers. Among these are the common marmoset (*Callithrix jacchus jacchus*), the cottontop or pinché (*Saguinus oedipus oedipus*), and members of the white-faced tamarins (*Saguinus nigricollis* group; see Section II). Therefore, most of our present knowledge of the Callithricidae is derived from these very few representatives of the family, whereas most species have never been studied in the wild or in captivity. This fact, as well as many observations showing that the strongly impoverished and crowded laboratory environment might seriously affect the behavior of primates (Gartlan, 1968), has to be kept in mind when discussing and generalizing the results presented in this review.

II. TAXONOMY AND DISTRIBUTION

The family Callithricidae contains the tamarins and marmosets of Central and South America. The family is divided into two main groups according to the morphology of the lower canines: all tamarins (genera *Leontopithecus* and *Saguinus*; Table I) have lower canines which are longer than the adjoining incisors, whereas all true marmosets (genera *Callithrix* and *Cebuella*; Table I) have lower canines that are equal in length to the incisors.

The taxonomy of the Callithricidae has been a subject of discussion for many years. Hill (1957) reviewed the earlier work. In his monograph on the anatomy and taxonomy of the Callithricidae, he divides the long-tusked tamarins into five genera and the short-tusked marmosets into three genera (Table I). Hershkovitz (1958, 1966, 1968, 1969, 1972) has recently published a series of papers on this subject and has announced the forthcoming publication of a monograph on marmoset monkeys. He employs *Saguinus* Hoffmannsegg 1807 as the oldest valid name for all tamarins which Hill (1957) lists as the genera *Tamarin, Tamarinus, Marikina,* and *Oedipomidas.* The maned tamarins or lion marmosets are recognized as a separate genus (*Leontopithecus* Lesson 1840) (see also Coimbra-Filho and Mittermeier, 1972). The true marmosets, with the exception of the pygmy marmoset (*Cebuella* Gray 1866), according to Hershkovitz (1968, 1972), are members of the genus *Callithrix* Erxleben 1777.

Goeldi's monkey or Goeldi's marmoset (*Callimico goeldii* Thomas 1904) is the single genus and species of a distinct group of the Platyrrhini. Because of certain anatomical characteristics that place it close to the marmoset monkeys (e.g., small size, manus and pes tamarin-like), it has been treated as the single representative of a subfamily (Callimiconinae) of the Callithricidae (cf. Hill, 1959). The dentition, which resembles that of the Cebidae, has prompted other workers to place the Callimiconinae into the family Cebidae (cf. Fiedler, 1956). Hershkovitz (1970) and Hill (1957), on the other hand, place it in a family (Callimiconidae) of its own.

Table I gives a brief overlook over the groups of tamarins and marmosets, comparing their classification according to Hershkovitz (1958, 1966, 1968, 1969, 1970, 1972) and Hill (1957). In the present paper, I shall follow Hershkovitz's classification.

Figure 1, adapted from Perkins (1969d), schematically indicates the distribution of the eight genera of Callithricidae recognized by Hill

TABLE I

Classification of Marmosets, Tamarins, and Goeldi's Monkey According to Hill (1957) and Hershkovitz (1966, 1968, 1969, 1972)

Vernacular names	Genera according to Hill (Hapalidae Wagner 1840)	Genera according to Hershkovitz (Callithricidae Gray 1821)	Species according to Hershkovitz
Black-faced tamarins	*Tamarin* Gray 1870	*Saguinus* Hoffmannsegg 1807	*S. midas*
White-faced tamarins	*Tamarinus* Trouessart 1899	*Saguinus* Hoffmannsegg 1807	*S. nigricollis*
			S. fuscicollis
			S. mystax
			S. labiatus
			S. imperator
Barefaced tamarins	*Marikina* Lesson 1840	*Saguinus* Hoffmannsegg 1807	*S. bicolor*
Crested barefaced tamarins or Pinchés	*Oedipomidas* Reichenbach 1862	*Saguinus* Hoffmannsegg 1807	*S. oedipus*
			S. leucopus
			S. inustus
Maned tamarins or lion marmosets	*Leontocebus* Wagner 1840	*Leontopithecus* Lesson 1840	*L. rosalia*
Naked-eared marmosets	*Mico* Lesson 1840	*Callithrix* Erxleben 1777	*C. argentata*
Tassel-eared marmosets	*Hapale* Illiger 1811	*Callithrix* Erxleben 1777	*C. humeralifer*
Tufted-eared marmosets	*Hapale* Illiger 1811	*Callithrix* Erxleben 1777	*C. jacchus*
Pygmy marmosets	*Cebuella* Gray 1866	*Cebuella* Gray 1866	*C. pygmaea*
Goeldi's monkey	(*Callimiconidae* Dollman 1933) *Callimico* Ribeiro 1912	(*Callimiconidae* Dollman 1933) *Callimico* Ribeiro 1912	*C. goeldii*

(1957). Detailed information on the distribution of the species and subspecies of *Saguinus* and *Callithrix* is provided by Avila-Pires (1969), Coimbra-Filho (1971), and Hershkovitz (1966, 1968, 1969, 1972).

It must be reported here that the distribution of *Leontopithecus* is consistently becoming more restricted. Destruction of the forest habitat of these primates and excessive harvesting by hunters have seriously endangered the survival of the three subspecies of *Leontopithecus*. Detailed discussions of the present situation of the genus and of its chances to survive in the wild are given by Coimbra-Filho (1969) and Coimbra-Filho and Mittermeier (1972).

All members of the Callithricidae are arboreal and diurnal. According to our very limited information on the ecology and the behavior of wild marmosets, these small animals seem mainly to forage in the gallery forest or at forest edges, only occasionally ascending to more than 20 ft. high (Bates, 1863, DuMond, 1971; Graetz, 1968; Hladik and Hladik, 1969; Mazur and Baldwin, 1968; Muckenhirn, 1967; Thorington, 1968). *Leontopithecus* apparently prefers the upper tree canopy of small elevated tree groups surrounded by swampy meadows. The selection of high elevations, however, might be forced upon the animals by a continuous clearing and destruction of their natural habitat (Coimbra-Filho, 1969).

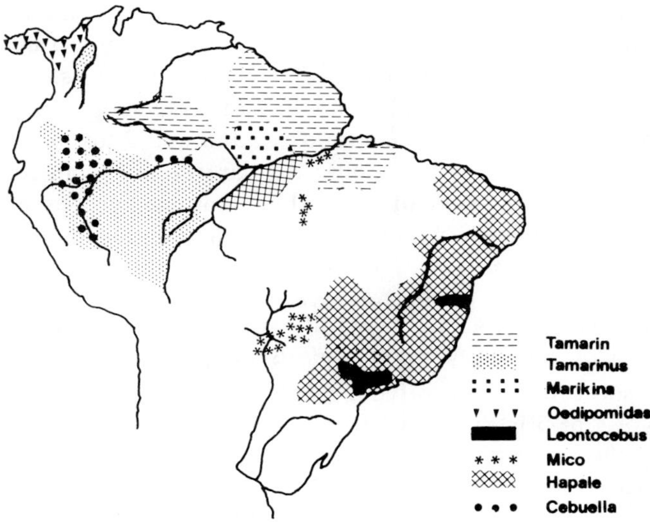

FIG. 1. Schematic diagram of the distribution of Callithricidae. (Adapted from Perkins, 1969d.)

III. SOCIAL STRUCTURE AND SOCIAL ORGANIZATION

A. Social Structure of Wild Groups

To date we know very little of the social structure and social behavior of any species of the Callithricidae in the wild. Almost all earlier observers, such as Bates (1863), Chapman (1929), Cruz Lima (1945), Enders (1930), Humboldt (1805), Krieg (1930), Miller (1930), Sanderson (1945) and others, report that the species they saw traveled in groups of between 2 and 12 individuals. Thorington (1968) briefly observed *Saguinus midas* in the field in Brazil. He saw groups numbering between 2 and 6 individuals. He mentions, however, that, in Surinam, Geikesks observed groups of *Saguinus midas* that numbered up to 20 animals. Hladik and Hladik (1969) and Moynihan (1970) studied *Saguinus oedipus geoffroyi* in Panama, in the wild and under seminatural conditions. They found wild groups composed of 2 to 9 animals and also observed single individuals. Most of these groups seemed to be quite stable over a period of several months (Moynihan, 1970). Moynihan (1970), as well as Thorington (1968) (for *Saguinus midas*), however, report incidences of merging between small groups.

Muckenhirn (1967), who observed *Saguinus oedipus geoffroyi* on Barro Colorado (Panama), reports that there were at least two stable groups in that part of the island she studied intensively. One group contained a minimum of 3 adults and 2 juveniles. The second one consisted of at least 9 individuals representing a subgroup of 3 adults, a subgroup of 2 adults and 1 juvenile, and a third subgroup of 1 adult and 2 juveniles. She also saw apparent feeding aggregations, the largest one consisting of 13 animals which separated in different directions into groups of 3 and 4 adults. It is not known whether the feeding aggregations contained members from the two stable groups referred to above.

Mazur and Baldwin (1968) observed a free-ranging group of 7 *Saguinus fuscicollis* in a 4-acre, enclosed, simulated, South American rain forest in Florida (U.S.A.). DuMond (1971) reported that, after several years and successive births of young, this original group split and a new subgroup of 4 or 5 individuals left the rain forest permanently.

Leontopithecus rosalia rosalia was studied by Coimbra-Filho (1969) in the field. He saw small groups, usually containing 2 or 3 individuals, but occasionally numbering up to 8. Other observers, he states, have reported groups of 10 or more monkeys in the wild. Coimbra-Filho (1969) assumes that at the time of estrus a mated pair separates from the social group and returns to it only after the offspring have been born.

Moynihan (1970) and Thorington (1968) provide evidence suggesting that the marmoset groups they studied occupied and defended territories. In both species, these territories seemed to be rather large, comparable in size to those of *Aotus trivirgatus*, *Cebus capucinus*, and *Alouatta palliata* (cf. Moynihan, 1970) and much larger than those of *Callicebus moloch* (Mason, 1966). DuMond (1971) reported that a 4-acre, enclosed, rain forest area in Florida accommodated just one group of *Saguinus fuscicollis*, but the same area was sufficient to harbor over 200 squirrel monkeys. His marmosets made little use of the vertical dimensions of their habitat, seldom ascending to more than 20 ft. in height, but they used the two-dimensional area of the 4 acres more than the squirrel monkeys did. As already pointed out in Section II, most marmoset species seem to be typical foragers of the gallery forests and forest edges. As DuMond (1971) suggests, one would expect a larger territory in species that use the available forest space in two dimensions rather than three.

When summarizing the very limited field observations, it becomes clear that the size and composition of wild marmoset groups vary considerably, even within the same species. Captive marmoset monkeys frequently display surprisingly strong aggressiveness against adults of their own sex (see Section III, B) which often limits the group size to one adult pair and its progeny. This has prompted several observers (Epple, 1967, 1970a; Hampton *et al.*, 1966) to conclude that, under natural conditions, marmosets live in typical family groups, consisting of a mated pair and its young, from which the offspring are driven by the parents as they reach reproductive age. The observation of small groups of 3 or 4 animals in the wild suggests that family units of this type might represent the basic "breeding nucleus" in many marmoset species. However, because large stable groups, sometimes obviously containing several subgroups, as well as merging small units and even temporary aggregations were observed, it can be concluded that these small breeding nuclei form larger permanent as well as temporary aggregations of whose organization we know nothing.

One wonders whether the large groups might consist of several small family units closely related to each other through parents or grandparents, which temporarily or permanently traveled together. A number of field studies on Old World monkeys and apes report that the ties between mothers and their offspring often persist into adulthood, resulting in close and long-term association, especially among females (e.g., Imanishi, 1960; Koford, 1963a,b; van Lawick-Goodall, 1969; Sade, 1965; Yamada, 1963). Although all marmosets studied in captivity so far are extremely intolerant of conspecifics belonging to strange groups, affectional ties between parents and offspring are often quite permanent (see Section III, B). In some individuals of our laboratory colony of *Saguinus fusciollis,* strong attraction between parents and late juvenile or adult offspring persisted for many weeks after the offspring had been separated from their parents to be mated with a nonrelated adult. These animals deliberately escaped from their new home cages in a large colony room containing fifteen family cages. Without hesitation they picked out the cages containing their parents and siblings and tried to rejoin this group, well tolerated by the family within. Siblings, months after being separated from the parents and each other, often remain completely peaceful when reintroduced to each other for experimental purposes.

These observations suggest that even in such relatively primitive species as marmosets, affectional ties among kin can be quite strong and permanent. Under laboratory conditions, however, more frequently than not, we observed strong aggression between adult offspring and the parent of the same sex, once the animals had been separated from each other for extended periods of time. This, on the other hand, does not necessarily imply that in a similar situation, under natural conditions, aggression between parents and offspring would occur. Moynihan (1970), for instance, never saw serious fighting in wild *Saguinus oedipus geoffroyi.* It is very likely that the crowding stress to which captive marmosets are submitted results in a strong increase of aggressive behavior. One could imagine that in the wild, affectional ties between members of a family prompt them to rejoin permanently or temporarily even after having formed their own breeding units.

Under the conditions of captivity, as will be discussed in Section III, B, the dominant female seems to inhibit all other females in her group from reproducing. It is well possible that temporary spatial separation from the parental breeding unit is a necessary prerequisite to release the mature female offspring from this inhibition exerted upon her by the mother and to allow the establishment of a pair bond and a new breeding unit.

B. Social Structure and Social Organization of Laboratory Groups

It will not advance our knowledge of the biology of the Callithricidae very much to speculate further about the social organization and territorial behavior of wild marmosets on the basis of our present knowledge of wild representatives of this family. Only field studies can give us the much needed information. Our knowledge of the social behavior of some species of marmosets in captivity is a little more complete. Caution, however, must be employed in generalizing information on the behavior of captive marmosets. Recent field studies in Old and some New World primates have demonstrated how strongly the social structure and social behavior in different populations of a single species of primate can vary. Jay (1965), for instance, found stable groups of the langur *Presbytis entellus* in northern India consisting of several adult males and females. The studies of Sugiyama (1967) and Yoshiba (1968), on the other hand, showed that in southern India many of the *Presbytis entellus* groups were one-male units or male bachelor groups. Gartlan (1968) discusses in detail differences observed in the social behavior of captive primates and wild representatives of the same species. He also reported many examples of differences observed in the social and aggressive behaviors between populations of one single species adapted to different habitats in the wild. These studies very clearly demonstrate that in many species of higher primates social behavior is highly adaptive. It is, of course, less likely to find striking differences between the social behavior of different populations of the same species and of captive and wild groups in such relatively primitive arboreal primates as marmosets, especially since the habitat of most species, the vast expanses of unbroken tropical rain forests, seems to be quite uniform. However, the few field data, and the data obtained from captive groups, suggest that the frequency of social and sexual behavior patterns might be higher in laboratory groups and that aggression and dominance interactions are enhanced under the crowded conditions of captivity.

In the following pages, I shall mainly discuss our laboratory studies in two genera of marmosets: *Callithrix j. jacchus* and *Saguinus fuscicollis* ssp. (We maintain *Saguinus f. fuscicollis, Saguinus f. illigeri,* and *Saguinus f. lagonotus,* if possible, in all one subspecies groups.) All our marmosets were either kept in family groups consisting of one adult pair and successive sets of its offspring or in groups artificially formed in the laboratory by placing several unrelated adults together. The *Callithrix j. jacchus* groups lived in small rooms (approximately 10 ft × 8 ft × 8 ft) connected to one or two slightly larger outdoor enclosures. The *Saguinus fuscicollis* were kept in indoor cages (approximately 4 ft × 3 ft × 6 ft) or in small rooms (approximately 10 ft × 8 ft × 8 ft).

Almost all reports on the social behavior of marmosets in captivity stress the aggressiveness of these primates. Severe fighting between adults, even causing injuries and death, has been reported in *Cebuella pygmaea* and *Saguinus midas midas* (Christen, 1968), *Callithrix j. jacchus* (Epple, 1967; Fitzgerald, 1935), *Leontopithecus r. rosalia* (Franz, 1963; DuMond, 1971), *Saguinus o. oedipus* (Hampton *et al.*, 1966), and *Saguinus oedipus geoffroyi* (Graetz, 1968).

In our colony aggression among group members was more frequent in artificial groups than among members of families. All pairs of *Callithrix j. jacchus* (4), *Saguinus fuscicollis* (6), and our only pair of *Saguinus oedipus geoffroyi* lived in complete harmony with their offspring up to 2½ to 3 years. Aggressive interactions between parents and offspring were only observed in the *Saguinus oedipus geoffroyi* group between the mother and her 3-year-old daughter, which was then removed from the group. Even in groups formed from several nonrelated adults, the animals born into the group were tolerated by the adults up to a period of 2½ years. Christen (1968) also reports that *Cebuella pygmaea* parents do not interact aggressively with their adult offspring while they are incompatible with nonrelated adults.

According to Rothe (personal communication) * family groups may increase considerably in size without breaking apart. He maintained three *Callithrix j. jacchus* families, two of which had increased from a mated pair to 16 individuals and the third one to 12 individuals within 4 years. In all three groups the mothers were pregnant again at the time I received Rothe's communication. It is noteworthy that the parents in the three groups never showed a tendency to drive any of their offspring out of the group. On the other hand, Rothe observed fights among the male offspring in two groups, resulting in the removal of the losers from the family unit. Other observers, however, report incidences of strong aggression between parents and adult offspring in *Leontopithecus r. rosalia* (DuMond, 1971), *Saguinus o. oedipus* (Wendt, personal communication), and *Saguinus oedipus geoffroyi* (Graetz, 1968) which even caused fatalities.

We did not observe any obvious rank order among the members of the family groups in our laboratory. Rothe (personal communication), however, reports that a linear hierarchy existed in his large *Callithrix j. jacchus* families mentioned above. In these groups the parents maintained an unchallenged alpha position, whereas the oldest offspring had to defend their beta position against the later-born young. The rank relationships in these families were even more complex than this, however, since the animals also appeared to be holding different ranks within the different age classes. Studies of their hierarchies are still in progress in Rothe's laboratory.

In *Cebuella pygmaea* a female usually seems to be the most dominant individual, maintaining her social status after her offspring grow up. Among her progeny, the older offspring dominate the younger ones (Christen, 1974).

---

* Dr. H. Rothe's help in making his unpublished data available to me is gratefully acknowledged.

In our colony, groups that consisted of several nonrelated adults established a nonlinear hierarchy. This was especially true for *Callithrix j. jacchus*. In *Callithrix j. jacchus* groups formed from adult nonrelated animals, one adult male dominated all other adult males but usually did not interact aggressively with any of the females or with the juveniles of both sexes. One adult female dominated the other adult females of the group, tolerating the males and all juveniles. No hierarchy seemed to exist among the submissive males and females. Quantitative data, however, which might have provided more information on ranking among submissive group members were not collected. Rothe (personal communication) reports the existence of a strictly linear rank order in groups of *Callithrix j. jacchus* consisting of several nonrelated adults of both sexes.

In our *Callithrix j. jacchus* groups, dominance interactions were most frequently observed during the day of group formation and the days immediately following it. During this period, in some but not all of our groups, dominance was established by means of overt fighting both between males and between females. In those groups where no fights occurred, the monkeys' behavior indicated the existence of rank relations a few hours after group formation (Epple, 1967). The direction of aggressive threats as displayed in facial expressions, genital presenting, scent marking, and vocalizations (Epple, 1967, 1968) and avoidance of the dominant animal by the inferior as well as inferiority calls were indicators of rank in the group. Once the social order was established, aggressive interactions in some groups, but not in all, occurred infrequently. In other groups a high frequency of aggressive interactions both among males and females persisted. Even in groups where no overt aggression was observed for relatively long periods of time (e.g., 1–2 years), serious fights betwen the dominant male or the dominant female and inferior animals of the same sex suddenly occurred, and in some cases made the removal of the inferior from the group necessary.

In *Callithrix j. jacchus* groups, the dominant male and the dominant female tended to form a stable pair. Although no quantitative data were collected, our observations indicate that the dominant male and the dominant female spent more time in contact with each other than with other adult group mates. They also engaged in sexual interactions with each other most frequently, though not exclusively. We have repeatedly observed that the dominant male interfered with copulations between the dominant female and another male of the group, threatening the male who immediately withdrew. The dominant female showed a similar behavior when "her" male engaged in copulation with other females. Thus, *Callithrix j. jacchus*, at least in captivity, shows a strong tendency to establish permanent pair bonds.

In our *Saguinus fuscicollis* groups, formed of adult, nonrelated animals, dominance interactions were of a more serious nature among females than males. In five out of a total of six groups containing more than 1 adult female, vicious fighting between the females occurred within a maximal period of 13 months after group formation and made the removal of the inferior animal from the group necessary. On the other hand, eight groups, formed from 1 adult female and 2 or more adult males, have been stable over periods up to 2 years and no fighting has occurred. Very little aggressive behavior was observed among the males of these groups. The occurrence and direction of mild threats, however, indicated that the males did not hold equal status in the group. Moreover, the single female of the group seemed to associate more closely with one of her male group mates than with the others.

In order to gain some quantitative information on social interactions and on the nature of male–female association, we studied the behavior of four groups, each containing 1 adult female and 2 adult males (Epple, 1972). Three groups were formed from *Saguinus f. illigeri*. One group contained 2 *Saguinus f. fuscicollis* males and a female hybrid between *Saguinus f. lagonotus* and *Saguinus f. illigeri*. A total of 174 observational sessions (= 87 hours) were performed on the four groups. Each session consisted of three 10-minute periods. During each 10-minute period the observer concentrated on only 1 of the 3 group members, recording the behavior of the subject as well as the behavior directed at the subject by its group mates. The total time of 10 minutes was divided into 40 intervals of 15 seconds each. For each interval the subject received a score of 1 for performing or receiving a variety of selected behavioral patterns. The scores were analyzed by the Mann Whitney U-Test.

A detailed analysis of the data from these groups as well as from some larger groups of the same species is still in progress. However, data on the number of contacts between group members, on huddling (resting in close contact) and grooming activities, and on the occurrence of copulations are presented in Table II. During the tests almost no overt aggression and a very low frequency of threat displays were observed. Therefore, the category "total contacts" (Table II) mostly reflects nonspecific contacts as well as sexual and friendly social contacts and is a good measure of social affinity between group members.

In all four groups the female tended to associate more closely with 1 of the 2 males. The female of group 1 had significantly more contacts with male 72 than with male 73 ($p = 0.028$). She was more frequently engaged in huddling ($p = 0.004$) and grooming encounters with the same male and 96% of all copulations were performed by him. (No statistical analysis was performed on the very low frequencies of grooming and

TABLE II

Saguinus fuscicollis: PERCENTAGE OF TOTAL CONTACTS, HUDDLING, GROOMING, AND COPULATIONS OCCURRING BETWEEN TWO MEMBERS OF GROUPS 1 TO 4 [a]

Group 1	♀ 40 – ♂ 72	♀ 40 – ♂ 73	♂ 72 – ♂ 73	*N*
Total contacts	39.7%	17.6%	42.7%	30
Huddling	55%	5%	40%	
Grooming	67%	26.2%	6.8%	
Copulations	96%	4%	–	
Group 2	♀ 6 – ♂ 12	♀ 6 – ♂ 13	♂ 12 – ♂ 13	
Total contacts	53.7%	20.7%	25.6%	
Huddling	43%	22%	35%	45
Grooming	85.8%	12%	2.2%	
Copulations	100%			
Group 3	♀ 21 – ♂ 26	♀ 21 – ♂ 25	♂ 26 – ♂ 25	
Total contacts	39.1%	38.5%	22.4%	
Huddling	45.5%	33.2%	21.3%	60
Grooming	31.6%	54.9%	13.5%	
Copulations	88.5%	11.5%		
Group 4	♀ 18 – ♂ 15	♀ 18 – ♂ 32	♂ 15 – ♂ 32	
Total contacts	42.9%	41.7%	15.4%	
Huddling	45.5%	42%	12.5%	39
Grooming	55.2%	41.4%	3.4%	
Copulations	55%	45%		

[a] From Epple (1972).

copulating in this group.) In group 2, there was close sexual and social association between the female and 1 of the males (male 12) but not the other (male 13) (contacts $p = 0$; grooming $p = 0$, huddling $p = 0.002$). In group 3, the female was about equally frequent in contact with both males. However, she huddled significantly more frequently with male 26 who also performed 88% of all copulations. The tendency of the female to prefer 1 of the 2 males is also apparent in group 4. However, only the number of grooming encounters between her and the males showed a significant difference ($p = .038$).

It would be interesting to obtain information on whether the closer sexual and social association between the female and one of the males reflects an *active* preference for one of the males by the female. It would probably not be established if the female would refuse the male. The males, however, certainly have an active part in it. In group 1 and in group 3, we observed both males actively competing for the female at times when she was apparently in estrus.* In both groups both males followed the female very closely wherever she went. Both constantly sniffed her body, her genitals, and her scent marks (see Section IV, A), tried to establish contact with her, sit by her side, and groom her. In both groups the male with whom the female normally associated more closely was frequently succesful in replacing the second male by her side. He would simply jump between the female and the second male, pushing him away from her. In one case male 26 maneuvered the female (female 21) into one corner of the cage and sat in front of her holding onto the wire of the cage so that his arms encircled the female and prevented her from moving away. It was surprising for us that during both of these periods of active competition for the apparently receptive female, we did not observe overt aggression or even an unusually high amount of threats between the two males.

We have, however, observed overt aggression between the males of group 1 after the experiment was terminated. Male 73 began to associate more closely with the female than he did during the tests. After several weeks of this, both males engaged in a prolonged fight during which they injured each other quite severely. The fight resulted in a reversal of social dominance. During the days following it, the formerly dominant male 72 showed all behavioral patterns and vocalizations of a clearly submissive animal. He was frequently threatened by male 73 who did everything to prevent him from contacting the female. During the hours immediately following the fight, the female repeatedly threatened male 73, the victor

* High sexual activity indicated that the female was in estrus. Vaginal washings contained a large amount of sperm.

of the fight, when he squeezed himself between her and the now submissive male 72. This might or might not have been an expression of her unaltered preference for male 72. Overt sexual interest in the female was not noticed at that time in either of the males.

It was not possible for us to decide which of the 3 animals had initiated the shift in their relationship and whether the female changed her preference from one male to another. No quantitative data were taken on this group after the 2 males fought, since male 72 was removed from the group shortly afterward. In another *Saguinus fuscicollis* group with 2 males and 1 female, we have observed a change from permanent association with 1 male to permanent association with the second male. In this group no overt aggression was observed in the males while the change took place. Again we have no information on whether the female changed her preference of males, actively ceasing to associate with one male and starting to associate more frequently with the other one.

Although overt aggression among males did occur on occasion, relationships between the males in our groups were quite subtle. Even under the relatively crowded conditions of captivity the males did not direct a high frequency of aggressive behavioral patterns at each other. It seems that in groups of this structure, social dominance among males is mainly expressed as the relative frequency with which a male associates with the female.

So far, we have not discussed the behavior of females in groups containing more than 1 nonrelated adult female. We have studied *Saguinus fuscicollis* groups containing 2 males and 2 females (2), 1 male and 3 females (1), and 3 females and 4 males (1). As already pointed out, none of these groups was stable over a long period of time. The data on these groups are still being analyzed, but a preliminary interpretation suggests that even in groups containing more males than females, one of the females actively inhibits the other females from establishing a pair bond. In the groups containing 2 males and 2 females, the animals showed a tendency to establish two permanent pairs. However, in all groups, 1 female finally eliminated all other females from the group by attacking them so viciously that they had to be removed in order to save their lives. Hampton *et al.* (1966) reported similar behavior in a group of *Saguinus o. oedipus*, where 1 female finally succeeded in eliminating 3 other females, tolerating the 2 males of the group.

Our observations on *Callithrix j. jacchus* and *Saguinus fuscicollis*, those of Hampton *et al.* (1966) on *Saguinus o. oedipus*, and those of Shadle *et al.* (1965) on *Saguinus fuscicollis*, *Saguinus o. oedipus*, and *Saguinus midas* show that males and females often (but not always) attempt to mate with members of adjacently caged groups if given a chance.

This suggests that a pair bond is not based on an exclusive sexual or social attachment to one partner. We agree with Hampton *et al.* (1966) that the fierce competition among females probably is one of the important factors in the establishment and maintenance of a pair bond. Under laboratory conditions the aggression of adult females against other adult females effectively inhibits pair formation in all but 1 female in the group. This seems certainly to be true for *Saguinus fuscicollis, Saguinus o. oedipus,* and probably also for *Callithrix j. jacchus.* As our observations on these species have shown, males to a certain extent aggressively compete for females. This competition seems quite mild in *Saguinus fuscicollis* and *Saguinus o. oedipus* (Hampton *et al.*, 1966), while it might be more frequently associated with overt aggression in some other species (e.g., *Callithrix j. jacchus*). However, our studies on aggressive behavior in *Saguinus fuscicollis* (Epple, unpublished data) have shown considerable individual variability in aggressiveness. This observation should caution against making generalizations concerning species differences in behavior on the basis of our limited knowledge.

In *Saguinus fuscicollis*, males as well as females aggressively compete for the "possession" of their mates with conspecifics which do not belong to their own group. This became very obvious in a series of experiments during which we introduced strange adults of both sexes to a mated pair for a test period of 10 minutes (Epple, unpublished data). Very frequently, we observed that the strange individual was fiercely attacked by the member of the pair which had the same sex as the stranger as soon as its mate established any sort of contact with the stimulus animal. Thus males aggressively prevented their females from establishing contact with strange males by attacking the strange male furiously, and females prevented their mates from establishing contact with strange females by attacking female intruders. Similar observations were made in *Callithrix j. jacchus* (Epple, 1970b).

The findings on pair bonding in marmosets are of interest when compared with obseravtions on another South American primate, the titi monkey. In *Callicebus moloch*, Mason (1966, 1971) reported the establishment of permanent pair bonds, both in the field and in the laboratory. He states that in this species "the bond between a particular male and female seemed to be unusually intimate, enduring and intense." His laboratory studies have shown that the major bond existing between the male and the female is mutual permanent attraction rather than aggressive competition for the attention of the mate. Although in marmosets there is probably a good deal of mutual attraction between partners of a pair, aggressive competition for the mate seems to be one of the major mechanisms that stabilizes the pair bond, and, at least under the conditions of captivity, limits the group size.

The pair bond usually is stable over long periods of time, including, for instance, the period of pregnancy, parturition, and infancy of the offspring. A pair bond that is stable for extended periods of time might even be an adaptation which is essential for the survival of the offspring. It has been known for a long time that in marmosets the male actively participates in the care of the offspring, carrying them most of the time and transferring them to their mother for nursing only (Epple, 1967; Fitzgerald, 1935; Franz, 1963; Hampton *et al.*, 1966; Roth, 1960; and others). Marmosets usually give birth to twins. A female alone, burdened with the care and protection of 2 infants, would hardly have a chance to raise them successfully under natural conditions. In our groups of *Callithrix j. jacchus* and *Saguinus fuscicollis*, the dominant male, that is, the male who associated most closely with the breeding female, tended to carry the infants for more frequent and longer periods of time than other members of the group. All members of the group, however, participated in their care.

There is some evidence that reproduction in the female is affected by the presence of other adult females in the same group (Epple, 1967, 1970a). In all breeding groups (*Callithrix j. jacchus, Saguinus oedipus geoffroyi,* and *Saguinus fuscicollis* ssp.) maintained in our laboratory during the last 10 years (approximately 200 individuals) only the dominant female of each group reproduced. Although inferior females copulated with the alpha male and the other group males, none of them ever became visibly pregnant. Christen (1974), on the other hand, reports that both the dominant and the submissive female in a *Cebuella pygmaea* and in a *Saguinus m. midas* group delivered full-term young. Not one of the females, however, raised her offspring. In our colony, young females, born in the group, never reproduced when they remained with their parental group into adulthood (up to 3 years). These females, as well as inferior adult females, however, almost immediately became pregnant when they were removed from their group and paired with an adult male or when the dominant female was removed from the group and the formerly inferior animal gained the alpha position in her group.

These observations strongly suggest that at least in some species the presence of a dominant female inhibits reproduction in submissive females and that the presence of the mother inhibits it in her daughters. Obviously the observations reported above raise more questions than can be discussed here. So far, one can only speculate about the mechanisms involved in an inhibition of reproduction in all but the dominant female of each group. Our behavioral studies (see above) indicate that the dominant female or mother prevents pair formation in all other females under the conditions of captivity. This might even be true under seminatural conditions. DuMond (1971) reported that in a semi-free-ranging group of *Saguinus fuscicollis* in the Florida Monkey Jungle, only 1 female seemed to be reproductively active. Another behavioral mechanism that might contribute to the failure of adult female offspring to reproduce when remaining within the family group is an inhibition of pair formation between siblings as suggested by DuMond (1971). However, the report on reproduction by sibling pairs *after* they had been removed from the parental group (Epple, 1970a) suggests that an inhibition of pair formation between siblings may only take place when both remain with the parents and that it is not the only reason why female offspring fail to reproduce within the parental breeding unit. There remains, of course, the question in how far the dominant male affects the reproductive capacities of other males within the group including his sons. There are no pertinent observations that indicate whether or not such effects exist.

The behavioral phenomena discussed above might result in endocrinological adaptations in the submissive and young mature individuals which finally prevent reproduction. These changes probably are reversible by permanent or even temporary separation from the parental breeding unit.

A considerable number of studies report failure to reproduce and retardation of sexual maturation in correlation with an increase in population density and with social stress in rodents. The results of these studies and the possible endocrinological mechanisms that cause them have recently been discussed by Brain (1971). There is good evidence that in rodents social stress results in an increase of adrenal weight and plasma corticosterone and a decrease of sex steroid levels in the blood. Social stress seems to affect submissive animals earlier and more severely than dominant individuals.

It is, of course, impossible to discuss the vast literature on the endocrine regulation of reproductive and aggressive behavior and of population control in this review. The publications of Leshner and Candland (1972) and Rose *et al.* (1971, 1972) discuss the problems involving studies in rodents as well as primates in more detail.

IV. PATTERNS OF SOCIAL AND SEXUAL COMMUNICATION

A. Communication by Olfactory Signals

Many observational studies and fewer experimental investigations have demonstrated that communication by odors is of considerable importance in the sexual and social life of primates. Patterns of sniffing and mouthing objects in the environment as well as the bodies of sexual and social partners are common in all primates. Scent marking behavior, that is, behavior resulting in the emission and deposition of urine, feces, saliva, and the secretions of specialized skin glands has been described mainly in prosimians and neotropical monkeys. Three species of nocturnal prosimians (*Nycticebus coucang, Loris tardigradus,* and *Perodicticus potto*) use urine marking as a means of orientation in their environment (Seitz, 1969). Further experimentation would probably show that olfactory orientation is widespread among nocturnal prosimians. In other species, e.g., *Galago crassicaudatus* (Eibl-Eibesfeldt, 1953), *Loris tardigradus* (Ilse, 1955), *Aotus trivirgatus* (Moynihan, 1964), *Callicebus moloch* (Mason, 1966), it has been suggested that scent marking has territorial function. Body odors and scent marking behavior seem to play a role in sexual communication and in the correlation of sexual arousal of many species such as, for instance, *Galago senegalensis* (Doyle *et al.*, 1967), *Lemur catta* (Evans and Goy, 1968), and perhaps in *Saimiri sciureus* (Baldwin, 1970). Moreover, scent marking apparently has important functions in the regulation of social relationships within the group in *Tupaia spp.* (von Holst, 1969; Kaufmann, 1965; Sprankel, 1961), *Lemur catta* (Evans and Goy, 1968; Jolly, 1966) *Saimiri sciureus* (Baldwin, 1968; Kirchshofer, 1963), and perhaps in *Lagothrix cana* and *Ateles geoffroyi* (Epple and Lorenz, 1967). The short list given here by no means reviews all functions of body odors and scent marking in all primate species. Many more observations have been published. The studies cited above refer to them.

During recent years a strong interest in mammalian pheromones has developed. Although this mainly resulted in numerous studies on rodents (cf. Bronson, 1968; Whitten and Bronson, 1970), it is to be expected that primate pheromones will be studied more closely in the future. To date the most extensive experimental study of primate pheromones has been conducted in *Macaca mulatta* by Michael and his co-workers. These authors demonstrated the existence of sex attractants in the females, isolated the attractants from vaginal washings of estrogenized females, and described their chemical nature (cf. Curtis *et al.*, 1971; Michael and Keverne, 1970; Michael *et al.*, 1971).

In marmoset monkeys, communication by scent is important in several areas of sexual and social life. In our laboratory we have recently started to study details of the biological function of chemical communication in two genera of the Callithricidae, namely, *Callithrix j. jacchus* and *Saguinus fuscicollis.* We shall discuss these studies and relate them to our present knowledge of primate pheromones.

1. *Scent Glands and Scent Marking*

Many marmoset species possess odor-producing skin glands in the circumgenital and sternal areas (Pocock, 1920; Sonntag, 1924; Epple and Lorenz, 1967; and others). Christen (1974), Perkins (1966, 1968, 1969a,b,c), Starck (1969), and Wislocki (1930) studied the histology of these skin glands in detail. They found dense accumulations of large apocrine and sebaceous glands in the circumgenital, suprapubic, and sternal regions of *Cebuella pygmaea*, *Callithrix j. jacchus*, *Callithrix argentata*, *Saguinus fuscicollis*, *Saguinus o. oedipus*, *Saguinus oedipus geoffroyi*, and *Callimico goeldii*. The glands are present in adult males and in adult females of all species. Details of their histological composition vary from species to species.

Marmoset monkeys show a variety of behavioral patterns that are directly or indirectly concerned with the application of glandular secretions and urine to the environment, their own body, or that of group mates. Scent marking, a behavioral pattern characterized by rubbing by the animal of glandular skin areas against items of its environment, has been observed in males and females of many species (*Callithrix j. jacchus*, *Callithrix jacchus geoffroyi*, *Callithrix argentata*, *Cebuella pygmaea*, *Leontopithecus r. rosalia*, *Saguinus o. oedipus*, *Sanguinus oedipus geoffroyi*, *Saguinus imperator*, *Saguinus fuscicollis*, and *Callimico goeldii*). Its wide distribution suggests that it probably exists in all species of the family.

The scent glands develop at the time of puberty (Wislocki, 1930, and personal observations). At this time the animals also begin to scent mark regularly. In hand-raised *Saguinus fuscicollis*, we noticed the typical body odor of adults at a time coinciding with the development of the scent glands, scent marking behavior, and other patterns of adult sexual and social behavior. These observations suggest that the development of the glands and of scent marking behavior are under the control of gonadal hormones.

Three basic scent marking patterns occur in both sexes of many species. They are described below.

a. *Marking with Circumgenital Glands.* All species we observed (see above) perform the lowest-intensity marking in a sitting position. The animal sits, presses the circumgenital and circumanal areas against the substrate, and either rubs back and forth or from side to side. The monkeys do not only apply the secretions of the glands covering the labia pudendi and scrotum to the substrate but at the same time produce a few drops of urine. In females, vaginal secretions may also be mixed into the scent mark. We observed an opening of the cleft of the vulva in intensively rubbing *Saguinus fuscicollis* females.

b. *Marking with Suprapubic Glands.* In *Leontopithecus r. rosalia*, *Saguinus oedipus geoffroyi*, *Saguinus o. oedipus*, and *Saguinus fuscicollis*, a thick cushion of glandular skin extends anteriorly from the labia pudendi and scrotum across the suprapubic region. Secretions from the suprapubic glands are applied when the animal lies flat on its stomach pressing the suprapubic glands to the substrate and either pulling itself forward with its hands and/or pushing the body with the feet (Fig. 2). *Saguinus fuscicollis* also rubs the suprapubic glands against small protruberances while in a sitting position. The motor pattern seems not to be strictly fixed but is adaptable to the characteristics of the substrate.

c. *Marking with Sternal Glands.* In all the species mentioned above, rubbing of the patch of glandular skin above the sternum against items of the environment has been observed. Sternal marking is a variable behavioral pattern. A monkey may rub the chest against the substrate while lying flat on its stomach. In another typical sternal marking pattern, the animal only presses its sternal area to the item being marked while the belly and rear end are elevated. If the item to be marked is located above the surface supporting the monkey, the animal frequently reaches it standing on its hind legs as shown in Fig. 3.

The three patterns of marking are quite distinct. Each may be performed alone but frequently all three are given in succession. Marking with the suprapubic and sternal glands is also combined into one pattern during which the animal rubs the entire ventral surface over the substrate.

Most scent marking is performed on items in the monkeys' environment.

FIG. 2. *Saguinus oedipus geoffroyi* ♂ marks with the suprapubic glands. From Epple (1967).

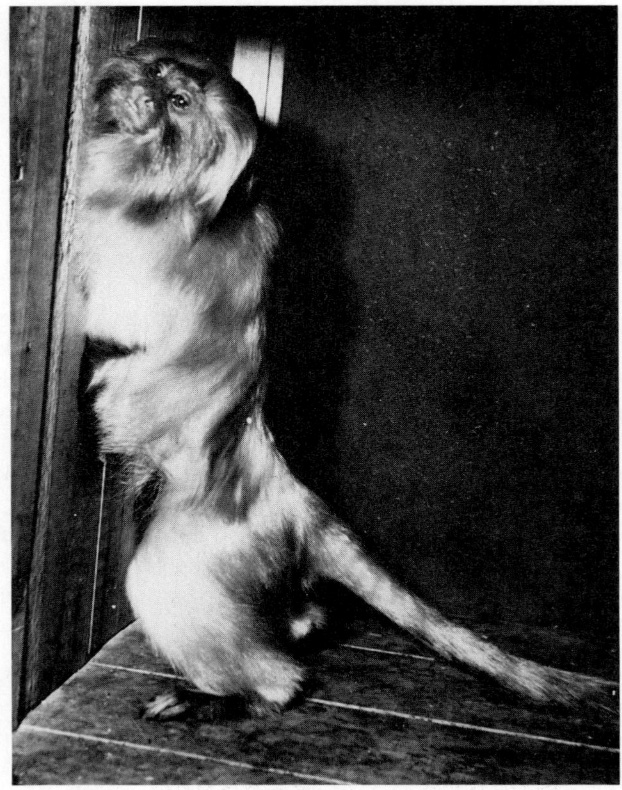

FIG. 3. Sternal marking in *Leontopithecus r. rosalia* ♀.

Callithrix j. jacchus, *Saguinus fuscicollis*, and *Saguinus oedipus* spp., however, also scent mark the body of conspecifics with the circumgenital and suprapubic glands. Most of the partner marking observed in our animals was performed when the monkeys were excited by a change in their environment or by aggressive encounters.

Some other patterns, which have not been developed specifically for the purpose of applying odorous substances to the body or the environment, nevertheless result in the spreading of these substances. During scratching and self-grooming, small amounts of odorous material are doubtlessly spread over the body surface. The monkeys groom their lower abdomen, and the area around the scent glands, thus automatically picking up and spreading secretions from the glands.

Contacts between two or more monkeys, such as grooming, huddling, or sleeping in one nest box, will certainly also result in the transfer of small amounts of odorous substances from one animal to another. It is not known if this creates a common "group odor" as is the case in the flying phalanger (Schultze-Westrum, 1965).

During aggresive encounters, *Saguinus fuscicollis* very conspicuously scratch their chests and suprapubic areas. This scratching is a little slower than the usual "casual" scratching in undisturbed situations. It is probably not a "grooming" activity or a displacement behavior but serves to stimulate the secretions of sternal and suprapubic glands and to spread these secretions into the animal's coat. Moynihan (1970) reports manual stimulation of the suprapubic glands in *Saguinus oedipus goeffroyi* in much the same way as described above.

2. *Communicatory Function of Olfactory Signals*

The well-developed, active scent glands found in both sexes of many species, and the frequent marking behavior observed in captive marmosets, suggest that chemical signals are of considerable importance in the life of these primates. In social species such as marmosets, chemical signals may be classified as involving:

1. Intragroup communication—(*a*) sexual communication; (*b*) regulation of social relationships among adults; and (*c*) infant–adult relationships.

2. Intergroup communication—(*a*)territorial defense; and (*b*) formation of new groups.

3. Orientation in the environment.

In our laboratory we have recently started to study the nature and function of the signals transmitted by scent marks. Odor might play an important role in territorial defense and group formation in wild marmosets. However, its role in this context cannot easily be studied in the laboratory. Moreover, the behavior of *Callithrix j. jacchus* and *Saguinus fuscicollis* suggested that scent is particularly important in sexual and social communication. Therefore, we have so far limited our efforts to this area, and I shall discuss the results of our studies briefly.

a. *Sexual Communication.* One function of scent marking, perhaps not the most important one, seems to be sexual communication. In *Callithrix j. jacchus*, *Saguinus fuscicollis*, and *Saguinus oedipus geoffroyi* males and females regularly mark with the circumgenital glands before and after mating. The male intensively and frequently licks and sniffs the female's genitals and her scent marks. A female sniffs and licks the genitals and scent marks of the male during strong sexual excitement. While submissive and juvenile members of a group normally do not scent mark very frequently (see below), in *Callithrix j. jacchus* such activity is increased considerably in connection with copulations (Epple, 1970b). In *Callimico goeldii* the male shows a large amount of olfactory exploration of the female (Lorenz, 1972). Lorenz (1972) suggests that by smelling and licking the scent glands as well as urine samples of the female, the male keeps a regular check on her reproductive condition and that courtship and copulation are in part initiated and controlled by chemical signals produced by the female.

A very distinct pattern of self-marking was seen in 1 adult female *Saguinus oedipus geoffroyi* in connection with sexual behavior. This animal coiled up her tail into a loop behind her rear end. The looped tail was then actively wiped back and forth across the labia pudendi several times while the female remained standing. As soon as the female finished the wiping pattern the male took her tail in both hands and sniffed it intensively. It is very likely that the wiping across the labia served to impregnate the tail with glandular secretions, urine and, maybe, vaginal secretions. The female showed this pattern only at times when she seemed sexually receptive. It was often followed by copulations. Since this behavior was only seen in the 1 sexually active female of our colony, it is yet too early to make any generalizations. Moynihan (1970) describes "tail coiling" as an expression of copulatory motivation in female *Saguinus oedipus geoffroyi*. The coiling pattern seems to be identical to the one seen in our female. His females, however, almost always sat down after showing tail coiling.* He did not see the wiping of the tail across the circumgenital area, although he briefly mentions that the pattern "may also provide olfactory information and stimulation." Maybe the motor pattern of tail coiling serves as a visual stimulus or sexual solicitation by which the female attracts the male. From Moynihan's (1970) observations, one would suspect that the motor pattern also has an arousal function. The scent, which is very likely applied to the tail by wiping it across the glands, may serve to communicate receptivity of the female in a closeup situation and also arouse the male.

We have not seen tail coiling associated with tail wiping in *Callithrix j. jacchus*. R. Lorenz (personal communication), however, observed a very similar pattern in *Saguinus imperator* and *Callimico goeldii*. According to Lorenz (1972) impregnation of the tail in female *Callimico goeldii* increases in frequency during estrus. In this species, however, the behavior is seen in both sexes and is not limited to times of high sexual activity, suggesting that it might have other functions beside sexual communication.

---

* A similar behavior is sometimes seen in our *Saguinus fuscicollis* females during encounters with strange conspecifics of both sexes. The pattern is quite rare and we cannot interpret it yet. It seems, however, in part sexually motivated.

So far, we have no information on the type of messages that are transmitted between partners during sexual marking. The tail wiping behavior of *Saguinus oedipus geoffroyi* and *Callimico goeldii*, the increased scent marking frequency of female *Callithrix j. jacchus* and *Saguinus fuscicollis* at times of behavioral sexual receptivity, and the very high frequency with which males sniff the genitals and the scent marks of receptive females, suggest that females produce an odor that informs the male about their estrus state. Such a sex attractant probably not only communicates receptivity but also arouses the male and correlates copulatory behavior. Michael and his co-workers have recently identified a mixture of volatile aliphatic acids in the vaginal secretions of female rhesus monkeys under estrogenic stimulation. These substances, named "copulins" by Michael *et al.*, serve as olfactory releaser pheromones, inducing mounting activity and ejaculations in males when applied to castrated females (Curtis *et al.*, 1971; Michael and Keverne, 1968; Michael *et al.*, 1971).

Apart from the possibility that the scent of female marmosets serves as a sex attractant for the male, there is the possibility that odor of females as well as males produced during sexual marking serves to maintain and strengthen the pair bond and in that way fulfills a function similar to some of the very elaborate visual displays in some birds (Wynne-Edwards, 1962).

Mated pairs of *Saguinas fuscicollis* showed an increase in the frequency of marking their mates' bodies (partner marking) during aggressive interactions with strange adults which were introduced to the pairs for a series of 10-minute tests (Epple, unpublished data). Our data are yet too few to interpret this observation. However, it might indicate that, in *Saguinus fuscicollis,* partner marking perhaps serves to demonstrate the existence of a pair bond to the stranger or to create a common group odor that strengthens the bond between mates in the prescence of the stranger. Partner marking, however, does not only occur between members of mated pairs but also between parents and offspring and nonrelated adult group members (Epple, 1967; Moynihan, 1970; Muckenhirn, 1967). This indicates that its function is not limited to the control of pair bond relationships. Even Moynihan's (1970) suggestion that it might be an artifact of the condition of captivity cannot be excluded.

In *Saguinus fuscicollis* we have observed a high frequency of scent marking in pregnant females, especially when they are close to term. Moreover, the labia pudendi, the suprapubic scent glands, and the circumgenital areas of these females are often very moist and the hair immediately surrounding them may appear matted and moist. At these times the male group mates of a pregnant female frequently licked and sniffed her marks and her free-flowing urine. They generally seemed to be quite attracted by the female. Increased sexual activity was not observed at this time. It is possible that the high marking activity of pregnant females, and the quality of their odor, function to attract the males and to strengthen the pair bond. Muckenhirn (1967) made similar observations on *Saguinus o. oedipus* and interprets them in the same way as we do.

Keeping the male close to the female when she is approaching parturition would be of advantage for the survival of the infants since the dominant male is the one who takes over much of their care. Muckenhirn (1967) observed that male *Saguinus o. oedipus* show increased interest in the marks of pregnant females. She also reports increased mounting of pregnant females by males. She interprets these observations as an indication that female scent marking is important in maintaining the pair bond in *Saguinus o. oedipus.*

b. *Regulation of Social Relationships among Adults.* Earlier, purely qualitative observations of *Callithrix j. jacchus* groups indicated that dominant males and females generally scent mark more frequently than adults of inferior social status or than juveniles (Epple, 1967). To confirm this observation and to obtain some information on the way in which scent marking might be involved in dominance interactions, we recorded the frequency of marking with the circumgenital glands and some patterns of aggressive behavior in three *Callithrix j. jacchus* groups of different social structure (Epple, 1970b). In each group the marking frequency per hour was recorded for every individual under three conditions: (A) for 40 hours of trial-free observation; (B) in 21 tests, a strange male was introduced to the group for a period of 10 minutes, and immediately after removal of the stimulus animal, the marking frequency of each group member was recorded for 1 hour; and in 14 tests, a strange female was introduced to the group for a period of 10 minutes, and after removal of the stimulus animal, the marking frequencies of the group members were recorded for 1 hour (for details of the procedures, see Epple, 1970b). The results are summarized in Table III.

During trial-free situations, the dominant male showed the highest scent marking frequency of all group members ($p = 0.01$) in group 2 that contained more than 1 adult male. In groups 1 and 3, containing only 1 adult male but 2 adult females, the dominant female showed the highest marking frequency of all group members ($p < 0.001$).

When a strange male was introduced to the group, the dominant males of all three groups were the most active of all group members in threatening, attacking, and fighting the stranger. The strange male was the loser of all encounters. He never attacked any group member, although he defended himself when being attacked. He showed no aggressive

TABLE III

Callithrix j. jacchus: MEAN SCENT MARKING FREQUENCY PER HOUR IN TRIAL-FREE SITUATIONS (A), AFTER AGGRESSIVE ENCOUNTERS WITH STRANGE MALES (B), AND AFTER AGGRESSIVE ENCOUNTERS WITH STRANGE FEMALES (C)[a]

Group	Individual	A Trial-free	B After encounters with strange ♂	C After encounters with strange ♀
1	Dominant ♂	4.61	52.00	2.50
	Dominant ♀	13.92	12.00	20.50
	Juvenile ♂	5.07	10.50	1.25
	Submissive adult ♀	1.84	7.16	1.50
	Juvenile ♀	1.69	0.66	1.00
2	Dominant ♂	16.58	54.87	11.50
	Dominant ♀	9.70	18.37	36.83
	Submissive adult ♂	3.40	1.00	2.20
	Juvenile ♀	5.17	11.34	2.16
3	Dominant ♂	9.09	43.28	21.70
	Dominant ♀	13.27	13.85	36.50
	Submissive adult ♀	5.45	8.42	12.50

[a] Adapted from Epple (1970b).

threat patterns and did not scent mark (Epple, 1970b). During the hour immediately following aggressive encounters with strange males, the dominant males in all three groups scent marked more frequently than in trial-free situations ($p < 0.01$; $p < 0.05$; $p = 0.05$) and (in groups 1 and 2) more frequently than any other group member ($p < 0.02$; $p = 0.05$). Among the other members of the group, only the juvenile male and the submissive female of group 1, and the dominant female of group 2 showed increased marking.

When a strange female was placed with the groups, the dominant group females threatened, attacked, and fought the stimulus female very actively. The strange stimulus female was submissive in all encounters, showing only a minimum number of attacks, patterns of aggressive threat, and scent marks. During the hour immediately following the removal of the stimulus female, the dominant females of all three groups scent marked more frequently than in trial-free situations ($p < 0.01$; $p > 0.01$; $p < 0.01$) or after encounters with strange males. The dominant females of groups 1 and 2 showed the highest marking activity of all group members ($p < 0.001$; $p > 0.02$).

As our data show, adult *Callithrix j. jacchus* display strong aggression against adult strange conspecifics of their own sex, whereas they more-or-less tolerate adult strangers of the opposite sex. In undisturbed groups (A) the presence of an adult group mate of the same sex stimulates scent marking activity in dominant males and females but not in group members of low social status or in juveniles. That scent marking in these situations is motivated by aggression against the stimulus animal is strongly suggested by the finding that aggressive interactions (B, C) result in an even higher frequency of marking. Thus the frequency with which the scent is applied to the monkeys' environment is an indication of aggressive motivation. It might be regarded as a kind of "triumph ceremony" expressing personal success, as has been suggested by Moynihan (1964) for the scent marking of *Aotus trivirgatus*. The frequency of marking, however, and perhaps even the quality of the scent might also maintain social dominance in intergroup and intragroup encounters.

During recent years, several studies have demonstrated similar tendencies in a variety of social mammals, e.g., the flying phalanger (Schultze-Westrum, 1965), the rabbit (Mykytowycz, 1965, 1970), Maxwell's duiker (Ralls, 1971), the mountain goat (Geist, 1964), the tree shrew (Sprankel, 1961), and *Lemur catta* (Jolly, 1966). Ralls (1971) recently reviewed the studies on mammalian scent marking and pointed out that all species studied experimentally "tend to mark frequently in any situation where they are *both intolerant* of and *dominant* to other members of the same species." The demonstration of aggressiveness and social dominance by scent marking might well be a rule among social mammals, from marsupials to primates.

Much more information, of course, is needed before we can determine whether and how odors are able to affect the social status of an individual. A basic step in this direction is a detailed study of the amount and character of messages that are transmitted by scent marking. We have recently started to study these questions in *Saguinus fuscicollis*.

In these studies we test the monkeys' ability to discriminate between various types of odors derived from the scent marks of conspecifics (e.g., between the odor of a male and that of a female). Series of tests are given to adult males and females during which their preference for wooden perches scent marked by conspecifics is checked. During all these tests the animals are given the choice between two pieces of white pine (2 ft × 1½ inches × ¾ inches) impregnated with different odors. To obtain these odors, we allow donor monkeys to place their scent marks on the wood. Fresh "perches" are placed into the cages of two donor monkeys for 30 minutes, during which they usually scent mark them repeatedly. After impregnation with the scent of the donors, the two perches are presented to the subject for a 15-minute test. The total test period is divided into intervals of 5 seconds. During the test the total time the subject spends in contact with either of the perches is recorded with stopwatches and the subject receives a score of 1 per interval for the performance of sniffing and scent marking.

The following results were obtained:

1. When the animals were given the choice between a wooden perch scent marked by a male with whom they had no previous contact and a perch scent marked by a strange female, they significantly preferred the stimulus carrying the odor of a male. Their preference was expressed in a higher frequency of scent marking on the perch carrying the odor of a male ($p = 0.012$). Some individuals also spent more time in contact with the male stimulus than with that carrying female odor (Epple, 1971). The preference for the odor of males was not a preference for the perch carrying the larger or smaller amount of scent. Male odor was preferred when the male stimulus had been marked by 1 male only, whereas the female stimulus was marked by 2 females. A preference for male odor was also shown when the male stimulus was marked by 2 males, but the female stimulus only by 1 female or when both stimuli carried the odor of one individual (Epple, 1971).

The preference for the odor of males over that of females shows that the scent marks of *Saguinus fuscicollis* carry information on the sex of the animal who produced them and that this information is independent of the total amount of scent applied to the environment.

2. In another experiment, we tested the ability of the monkeys to discriminate between the odors of two individuals of the same sex. A strange male or female (the stimulus animal) was introduced to the subject for 10 minutes. This resulted in a high frequency of aggressive interactions between the subjects and the stranger. When the subjects later were given the choice between a wooden perch scent marked by the stimulus animal and one scent marked by an animal of the same sex as the stimulus animal with whom they did not have a recent aggressive encounter, they highly preferred the perch carrying the odor of their recent "enemy" (Epple, 1973) (Table IV). This preference shows that the subjects were able to discriminate between the odors of individuals. The preference for the odor of their recent enemy was shown when the scent sample was presented on the same day as the subjects had encountered the stimulus animal. It persisted even when the subjects were presented with the choice of marked perches 3 or more days after their most recent encounter with the stimulus animal. This shows that the marmosets can remember the odor of other individuals for several days. It also controls for the possibility that the subjects reacted to the scent of a donor monkey who had been recently stressed by an aggressive encounter versus that

TABLE IV

Saguinus fuscicollis: MEAN SCORE FOR SNIFFING, MARKING, AND CONTACTING PERCHES CARRYING THE ODORS OF 2 INDIVIDUAL MALES AND 2 INDIVIDUAL FEMALES [a]

Time (min) spent in contact with:				Scent marking score on:				Sniffing score on:			
Familiar ♂ perch	Strange ♂ perch	*d*	*p*	Familiar ♂ perch	Strange ♂ perch	*d*	*p*	Familiar ♂ perch	Strange ♂ perch	*d*	*p*
4.634	3.426	1.208	0.027	8.030	5.090	2.940	0.027	14.492	8.997	5.494	0.008
Time (min) spent in contact with:				**Scent marking score on:**				**Sniffing score on:**			
Familiar ♀ perch	Strange ♀ perch	*d*	*p*	Familiar ♀ perch	Strange ♀ perch	*d*	*p*	Familiar ♀ perch	Strange ♀ perch	*d*	*p*
2.942	2.364	0.578	0.012	5.198	3.323	1.875	0.004	9.317	6.249	3.068	0.004

[a] Adapted from Epple (1973).

of an undisturbed animal rather than to a truly individual odor (Epple, 1973).

3. The preference of the monkeys for perches scent marked by dominant males versus those scent marked by subordinate males was also tested. Five ranked pairs of males served as donors. They lived in groups consisting of 1 female and 2 males each. Their rank had been determined by the quantitative observational tests discussed in Section III. The dominant male was defined as being the one who associated more closely with the single female of their group.

As Table V shows, the animals spent about equal amounts of time investigating both perches during the whole 15 minutes of the test. However, during the first 5 minutes they spent significantly more time with the perch carrying the odor of dominant males. They also sniffed and scent marked the stimulus carrying the odor of dominant males more frequently than that carrying the odor of submissive males. The results indicate that the scent marks of animals of different social status contain some information that enables the monkeys to discriminate between them. They do not, of course, demonstrate the existence of an odor specifically signaling social dominance or submissiveness. As pointed out above, dominant *Callithrix j. jacchus* mark more frequently than submissive individuals (Epple, 1970b). This observation is presently being confirmed for *Saguinus fuscicollis*. It is possible that during the present study the subjects preferred the perches marked by dominant males because they carried larger amounts of odor than those marked by submissive males. More experiments are necessary to control this possibility.

The studies reported above show that the scent marks of *Saguinus fuscicollis* are very complex chemical signals. They transmit detailed information on the individual, its sex, and social status. Studies now in progress in our laboratory will doubtlessly reveal even more complexity. Thus, what Evans and Goy (1968) suggested for *Lemur catta* seems to apply also to at least one species of marmoset monkeys: these primates "may prove to possess an olfactory repertoire whose complexity rivals the most sophisticated visual and acoustic systems of larger brained primates."

B. Communication by Auditory Signals

All species of the *Callithricidae* observed so far are fairly vocal, producing quite a variety of different calls. Many of the earlier papers on marmosets cited in Section I contain verbal transcriptions of the calls of various species. However, only the recent development of techniques for

TABLE V

Saguinus fuscicollis: MEAN SCORE FOR SNIFFING, MARKING, AND CONTACTING PERCHES CARRYING THE ODOR OF DOMINANT MALES AND SUBMISSIVE MALES[a]

Time (min) spent in contact with:				Contact score during first 5 min of test:				Scent marking score on:				Sniffing score on:			
Dominant ♂ perch	Submissive ♂ perch	d	p	Dominant ♂ perch	Submissive ♂ perch	d	p	Dominant ♂ perch	Submissive ♂ perch	d	p	Dominant ♂ perch	Submissive ♂ perch	d	p
2.710	2.213	0.497	>0.05	16.512	14.217	2.295	<0.025	5.280	3.582	1.698	0.005	9.974	7.382	2.592	0.005

[a] Adapted from Epple (1973).

the recording and analysis of acoustic signals has made an objective study of marmoset vocalizations possible.

Recently, Andrew (1963), Christen (1974), Epple (1968), Le Roux (1967), Moynihan (1970), and Muckenhirn (1967) have studied the vocal signals of several species of marmosets. A very limited account of the vocal repertoire and its physical characteristics is given for *Callithrix a. argentata, Callithrix j. jacchus geoffroyi, Leontopithecus r. rosalia,* and *Callimico goeldii* by Andrew (1963) and Epple (1968). A more detailed analysis of the calls of *Callithrix j. jacchus, Cebuella pygmaea, Saguinus o. oedipus,* and *Saguinus oedipus geoffroyi* is presented by Christen (1974), Epple (1968), Le Roux (1967), Moynihan (1970), and Muckenhirn (1967). The vocalizations of all species studied so far have some features in common. The majority of all calls are high pitched, almost birdlike in character, and tonal in structure. Few vocalizations are low pitched and harsh or noisy in structure. Many of the high-pitched calls contain ultrasonic frequencies. Epple (1968) showed that the high-pitched harmonics of some calls of *Callithrix j. jacchus, Callithrix a. argentata, Leontopithecus r. rosalia,* and *Saguinus oedipus geoffroyi* reach 60 kc.

It is not possible to discuss here in detail the extensive reports of Epple (1968), Le Roux (1967), Moynihan (1970), and Muckenhirn (1967) on marmoset vocalization. In this review I shall limit myself to two groups of vocal signals present in *Callithrix j. jacchus* and *Saguinus oedipus.* One group of calls is involved in the maintenance of social contact. The calls of the second group seem mainly to function as alarm signals in the presence of predators. I follow my former practice (Epple, 1968) of assigning some of the vocalizations of marmosets to functional categories such as contact calls or mobbing calls. This is done because tests on the situations that elicit the calls in the laboratory and the reaction of the marmosets when recordings of calls were played back to them indicated that their major function is in the area they have been assigned to. This does not mean that they should be regarded as simple releaser signals that are strictly bound to one function only. Furthermore, it should be pointed out that names such as "long-distance contact call" or "mobbing call," when given to the vocal signals of two different species, only imply that the calls have a similar function in both species. They do not imply that, for instance, the long-distance contact calls of two species are necessarily similar in tonal character or physical structure or that they must be homologous.

In adult *Callithrix j. jacchus, Callithrix jacchus geoffroyi, Callithrix a. argentata,* and *Leontopithecus r. rosalia,* Epple (1968) found quite distinct vocal patterns that were relatively closely linked to specific social situations. The calls of infant *Callithrix j. jacchus* are much more variable than those of adult animals. Most of the infantile calls form a graded system. In small infants they serve as contact calls, communicating different levels of distress when the baby looses bodily contact with an adult who is carrying it or when it is hungry or cold (for further details, see Epple, 1968).

In adult *Callithrix j. jacchus* and other species of *Callithrix*, Epple (1968) described a continuum of high-pitched whistles, all very similar in tonal character, which seems to maintain contact between the members of a group. The animals utter short, soft whistles at infrequent intervals when they are undisturbed and in close visual contact with their group mates. It is very difficult for a human observer to locate these low-intensity whistles and to attribute them to a specific animal. One might assume that a predator would have the same difficulties in locating the animals who utter the calls. Thus the soft whistles have all the characteristics postulated for a contact call by Marler (1955). They announce the presence of the caller to its nearby mates and at the same time avoid announcing it to predators. When members of a group temporarily lose visual contact with each other, when they are hungry, or otherwise in mild distress, they give whistles which are longer, louder, and more frequently produced than the soft whistles given in close contact. Now rhythmical twitters are also given. The twitters and the longer whistles can be heard over longer distances than the soft whistles given in close contact. They are also much more easily localized.

The most intense contact calls, very loud whistles of up to 1.1 seconds duration, are given when animals become totally isolated from their group mates. One of our animals, who escaped from the laboratory into an adjacent park, kept in contact with the rest of the colony for 3 days by long-distance whistles.

The graded complex of whistles and the twitters seem mainly to be motivated by the feeling of uneasiness and mild distress. The soft whistles express just a slight uneasiness which an individual might always feel when moving around. As adverse stimuli such as social isolation or hunger increase, intensified whistles and the twitters may communicate increasing feelings of distress.

The long loud whistles of individuals who become completely isolated from their group mates are quite obviously motivated by strong distress. However, they do not seem to function as long-distance contact calls alone. In our laboratory, several groups of *Callithrix j. jacchus*, housed in different parts of the building, regularly engaged in long vocal dialogues during which these long-distance calls were uttered, sometimes mixed with twitters. These dialogues were regularly accompanied by visual threat displays (Epple, 1968).

The behavior accompanying the vocalizations strongly suggested that the long whistles and twitters under these conditions were involved in the defense or acoustic marking of the territory. In these situations distress caused by social isolation can hardly have been their motivation. In wild *Saguinus oedipus geoffroyi,* Moynihan (1970) reported that "long whistles" (which are quite different from those of *Callithrix j. jacchus*) serve not only as long-distance contact calls among members of the same group but also may fulfill the function of a territorial call. That means that they attract members of the same group or isolated individuals seeking a mate but repel members of strange groups. Evidence for the existence of territorial calls is also given for wild *Saguinus midas* by Thorington (1968).

Moynihan (1970) reported that aggressive chasing among wild *Saguinus oedipus geoffroyi* was associated with long whistles. We have repeatedly observed *Callithrix j. jacchus* giving several long-distance contact calls during aggressive encounters with members of strange groups. This vocalization immediately induced their own group members to join them.* It seems that under these circumstances the long-distance contact call also serves as a call for assistance.

In *Callithrix jacchus,* and all other species studied by Epple (1968), any potential predator that comes within sight of the animals without representing an imminent danger elicits a typical mobbing response. Stimuli that elicit mobbing were studied in detail by Epple (1968) and Muckenhirn (1967). Mobbing can be stimulated by a wide variety of objects, including harmless objects such as a small rolling ball.

The mobbing response is a very conspicuous behavioral pattern. Like the mobbing behavior of many birds (Marler, 1955), it serves to alarm all conspecifics to the potential danger and helps to locate this danger. The vocalizations uttered during mobbing are very "contagious." When the animals hear these calls, even when they are unable to see the caller, they usually stop all other ongoing activities and join into the mobbing calls.

The mobbing response includes visual displays such as swaying and piloerection and high-intensity vocalizations. In *Callithrix j. jacchus,* long series of loud, very sharp "tsik, tsik, tsik, . . ." calls are given. At low-intensity mobbing, these are intermingled with softer, lower-pitched crackles and coughs. As the animal becomes more excited, the crackles and coughs are omitted from the series of tsik calls. Details of the physical properties of the calls and of the species–specific mobbing vocalizations of other marmosets are given by Epple (1968).

* This behavior was shown whenever the door between two animal rooms was open and neighboring groups could mingle.

In *Callithrix j. jacchus* and all other species studied by Epple (1968) and Muckenhirn (1967), an alarm response, including freezing in place followed by withdrawal or immediate flight, is stimulated by predators that unexpectedly appear in an animal's field of vision. In *Callithrix j. jacchus* the alarm response includes a few very sharp, high-pitched calls of relatively low intensity. The situations under which these calls were given and also their low intensity made it difficult to obtain a sufficient number of recordings for analysis (Epple, 1968).

When a predator appears suddenly on the ground, *Callithrix j. jacchus* utters three to four notes, which sound very much like a sharper and abbreviated form of the tsik calls used in mobbing. They might actually represent a variation of these tsik calls. For human ears, the alarm calls given to birds of prey and other overhead movements (for details, see Epple, 1968; Muckenhirn, 1967) are slightly different from those given to ground predators. They might represent a further and more extreme variation of the tsik calls used in mobbing.

The vocal patterns of *Saguinus o. oedipus* and *Saguinus oedipus geoffroyi* were studied by Epple (1968), Moynihan (1970), and Muckenhirn (1967). The vocalizations of infants were discussed in detail by Moynihan (1970). Some of the vocalizations of adults are much more variable than the calls of *Callithrix j. jacchus* and *Leontopithecus r. rosalia.* In captivity both species of *Saguinus oedipus* vocalize much more frequently than the other marmosets mentioned above. Muckenhirn (1967) also reports that wild *Saguinus oedipus geoffroyi* are highly vocal, whereas Moynihan (1970) points out that they are actually quite silent.

In *Saguinus oedipus geoffroyi,* a continuum of highly variable calls is stimulated by any slight disturbance of the animals. These vocalizations, which Epple (1968) called "monosyllabic calls in loose visual contact and when disturbed," are identical with the twitters and "short whines" of Moynihan (1970) and the "bar whines" of Muckenhirn (1967). They seem to be elicited by a wide variety of mildly distressing stimuli such as a new environment, temporary loss of close visual contact with group mates, or the appearance of a potential predator of which the animals are not very much afraid. Moynihan (1970) gives a detailed discussion of the motivation of the variants of these vocalizations. From the situations in which they are given and from the reaction of the marmosets to tape recordings of the calls, one would assume that they serve a double function: they seem to serve as contact calls and at the same time alert all members of a group to the disturbing stimulus that elicited them. Since usually a great number of these highly variable calls are given, the animal has the possibility of expressing details of its motivation both by the number of vocal signals and by the quality of a single signal.

Individuals who become totally isolated from their groups give one, two, or three monosyllabic calls, each call being slightly longer and louder than the one before. Moynihan (1967) calls the long-distance contact notes "long whistles," Muckenhirn (1967) "long call whines." As Moynihan (1970) reports, the long whistles of wild *Saguinus oedipus geoffroyi* are not only used as contact calls by isolated individuals but also function as territorial calls (see above).

A mobbing response is elicited in *Saguinus oedipus geoffroyi* by much the same stimuli as in *Callithrix j. jacchus.* It is a conspicuous display that alerts all conspecifics to potential danger. The vocalizations given during mobbing include very loud sharp trills and series of sharp tsik, tsik, tsik, . . . calls, the "loud sharp notes" of Moynihan (1970). Whereas Epple (1968) suggested that the trills combine the function of contact calls with that of a mobbing call, Moynihan (1970) thinks that they are mainly alarm notes. The tsik calls or loud sharp notes resemble the mobbing tsik calls of *Callithrix j. jacchus, Callithrix a. argentata,* and *Leontopithecus r. rosalia* in character. Mobbing tsik calls are probably homologous within the Callithricidae but show species–specific structural differences. According to Moynihan (1970), the loud sharp notes of *Saguinus oedipus geoffroyi* are also homologous with the alarm patterns of some other Platyrrhini, e.g., *Cebus* sp.

Similarly to *Callithrix j. jacchus,* the *Saguinus oedipus geoffroyi* react with an alarm response to predators that appear very suddenly on the ground and to overhead movements. Only Muckenhirn (1967) provides a sonagram of an alarm call. It shows a distinctive trill given by her animals in reaction to overhead movements. In our own laboratory, we were not able to record and analyze any of the alarm calls of *Saguinus oedipus geoffroyi.*

Lancaster (1968) and Marler (1965) have recently discussed the complexity of communication systems in monkeys and apes. As these authors pointed out, communication in many higher primates relies heavily on multimodal signals. In many cases a single display, be it olfactory, vocal, visual or tactile, does not represent a releaser in itself but functions as a part of a complex of signals. Communication systems of this type are present in some species of Platyrrhine primates such as *Saimiri* (Winter *et al.,* 1966), *Ateles* (Eisenberg and Kuehn, 1966; Moynihan, 1964), and *Callicebus moloch* (Moynihan, 1966). The vocal patterns of these species are quite variable. Calls integrate with each other, many of them forming graded systems. As already pointed out by Moynihan (1964), graded systems of vocal signals can communicate much information about the sender in a short period of time. They have the disadvantage, however, that they may be easily misunderstood by the receiver. Therefore there is a tendency to develop less variable, more stereotyped vocal signals wherever there is the necessity for a call to be interpreted by the receiver very clearly or very fast. This is the case, for instance, of long-distance signals that cannot be accompanied by other patterns of communication behavior, and of alarm calls.

Stereotyped vocal patterns are also found in species that do not possess a great variability of visual displays and, therefore, have to rely more on means of vocal communication, or in species living in an environment that makes visual communication difficult. *Aotus trivirgatus*, which is nocturnal and arboreal, for instance, has a vocal repertoire of stereotyped signals. The calls are precise releaser signals that can be described quite legitimately in terms such as threat or appeasement (Moynihan, 1964). The vocal repertoire of the small arboreal Callithricidae contains both stereotyped releaser signals and graded systems of calls. It seems to be halfway between the complex systems of some Platyrrhini and the stereotyped one of the night monkey. In *Callithrix j. jacchus* and *Leontopithecus r. rosalia,* most of the calls show a tendency to be stereotyped and quite closely associated with specific social situations. In *Saguinus oedipus geoffroyi,* on the other hand, some of the calls form quite complex graded systems. The calls of Goeldi's monkey displayed the greatest variability of all the species studied by Epple (1968). Although most of the vocal signals of this species are high-pitched and birdlike, as in the Callithricidae, they are distinguished from the typical vocalizations of the Callithricidae by a very high degree of variability.

C. Communication by Visual Signals

Compared with most other species of simian primates, especially with Old World monkeys and apes, marmosets have few facial expressions and visual displays. This is not surprising since the Callithricidae might well be the most primitive simians in existence (Hershkovitz, 1969). The lack of a high number of displays and especially of facial expressions might also, to some degree, be due to the small body size of all Callithricidae. In an arboreal environment, where the animals spend at least a part of their time out of close visual contact with their group mates, visual signals cannot always be used effectively by small-sized primates such as marmosets. This does not mean, however, that all members of the Cebidae show a higher number of visual signals than the small and probably primitive Callithricidae. Moynihan (1970), for instance, reports that *Saguinus oedipus geoffroyi* has more types of visual displays than the larger-sized *Aotus trivirgatus* and *Callicebus moloch,* which live in much the same environment. Moynihan (1970) lists thirty-two major displays ("give or take a half dozen") for adult *Saguinus oedipus geoffroyi*. Various combinations of these displays result in a huge number of recognizably different signals.

One wonders how many of these displays of *Saguinus oedipus geoffroyi* and the body postures of other species of marmosets have been evolved specifically for the purpose of communication and are releaser signals in a strict sense. Especially some of the tail movements of *Saguinus oedipus geoffroyi* described by Moynihan (1970) seem to be nothing more than by-products of emotional states such as uneasiness or alarm. However, it is reasonable to assume that a primate can learn to associate body or tail postures of conspecifics with the situations in which they usually occur. Thus many body postures that were not evolved as releaser signals may become valuable means of communication simply through the personal experience of the animal that perceives them.

Marmoset monkeys show some facial expressions and body movements that undoubtedly represent ritualized displays. These displays involve patterns of piloerection and smoothing, flattening of the external ear and the hairs surrounding it, swaying movements of the whole body, extreme arching of the back, presentation of the genitalia and the circumgenital area. A few facial expressions have also been reported. Since the displays and facial expressions of Callithricidae have been described in detail and are illustrated by drawings and photographs in the papers of Christen (1974), Epple (1967, 1970b), Le Roux (1967), Moynihan (1970), and Muckenhirn (1967), they will not be further discussed in this review.

References

Altmann-Schönberner, D. (1965). Beobachtungen über Aufzucht und Entwicklung des Verhaltens beim grossen Löwenäffchen, *Leontocebus rosalia*. *Zool. Garten* **31**, 227–239.

Andrew, R. J. (1963). The origin and evolution of the calls and facial expressions of the primates. *Behaviour* **20**, 1–109.

Avila-Pires, F. (1969). Taxonomia e zoogeografia do gênero "Callithrix" Erxleben, 1777 (Primates, Callithricidae). *Rev. brasil. Biol.* **29**, 49–64.

Baldwin, J. D. (1968). The social behavior of adult male squirrel monkeys (*Saimiri sciureus*) in a seminatural environment. *Folia Primatol.* **9**, 281–314.

Baldwin, J. D. (1970). Reproductive synchronization in squirrel monkeys (*Saimiri*). *Primates* **11**, 317–326.

Bates, H. W. (1863). "The Naturalist on the River Amazon." Murray, London.

Benirschke, K., and Richart, R. (1963). Establishment of a marmoset breeding colony. *Lab. Anim. Care* **13**, 70–83.

Brain, P. F. (1971). The physiology of population limitation in rodents—a review. *Commun. Behav. Biol.* **6**, 115–123.

Bronson, F. H. (1968). Pheromonal influences on mammalian reproduction. *In* "Perspectives in Reproduction and Sexual Behavior" (M. Diamond, ed.), pp. 341–361. Univ. of Indiana Press, Lafayette.

Chapman, F. M. (1929). "My Tropical Air Castle. Nature Studies in Panama." Appleton, New York.

Christen, A. (1968). Haltung und Brutbiologie von *Cebuella*. *Folia Primatol.* **8**, 41–49.

Christen, A. (1974). Fortpflanzungsbiologie und Verhalten bei *Cebuella pygmaea* und *Tamarin tamarin*. Fortschritte der Verhaltensforschung. *Beih. Z. Tierpsychol.* No. 14.

Coimbra-Filho, A. F. (1965). Breeding lion marmosets, *Leontideus rosalia* at Rio de Janeiro Zoo. *Int. Zoo Yearb.* **5**, 109–110.

Coimbra-Filho, A. F. (1969). Mico-leão, *Leontideus rosalia* (Linnaeus, 1766), situação atual de espécie no Brasil (Callithricidae – Primates). *An. Acad. Brasil. Ciênc.* **41**, 29–52.

Coimbra-Filho, A. F. (1971). Os sagüis do gênero *Callithrix* da região oriental brasileira e um caso de duplo-hibridismo entre três de suas formas (Callithricidae, Primates). *Rev. brasil. Biol.* **31**, 377–388.

Coimbra-Filho, A. F., and Mittermeier, R. A. (1972). Taxonomy of the genus *Leontopithecus* Lesson, 1840. *In* "Saving the Lion Marmoset: Proceedings of the WAPT Golden Lion Marmoset Conference" (D. D. Bridgwater, ed.), pp. 7–22. Wild Animal Propagation Trust, Oglebay Park, Wheeling, West Virginia.

Cruz Lima, E. (1945). "Mammals of Amazonia," Vol. 1. Livraria Agir Editora, Belém do Para.

Curtis, R. F., Ballantine, J. A., Keverne, E. B., Bonsall, R. W., and Michael, R. P. (1971). Identification of primate sexual pheromones and the properties of synthetic attractants. *Nature (London)* **232**, 396–398.

Desmarest, A. G. (1818). Ouistiti, Jacchus et Midas. *Nouv. Dict. Hist. Natur., Paris* **24**, 237–244.

Ditmars, R. L. (1933). Development of the silky Marmoset. *N.Y. Zool. Soc. Bull.* **36**, 175–176.

Doyle, G. A., Pelletier, A., and Bekker, T. (1967). Courtship, mating and parturition in the lesser bushbaby (*Galago senegalensis moholi*) under semi-natural conditions. *Folia Primatol.* **7**, 169–197.

DuMond, F. (1971). Comments on Minimum Requirements in the Husbandry of the Golden Marmoset (*Leontopithecus rosalia*). *Lab. Primate Newsl.* **10**, 30–37.

Eibl-Eibesfeldt, I. (1953). Eine besondere Form des Duftmarkierens beim Riesengalago, *Galago crassicaudatus* (E. Geoffroy 1812). *Säugetierkundl. Mitt.* **1**, 171–173.

Eisenberg, J. F., and Kuehn, R. E. (1966). The behavior of *Ateles geoffroyi* and related species. *Smithson. Misc. Collect.* **151**, 1–60.

Enders, R. K. (1930). Notes on some mammals from Barro Colorado Island, Canal Zone. *J. Mammal.* **11**, 280–292.

Epple, G. (1967). Vergleichende Untersuchungen über Sexual- und Sozialverhalten der Krallenaffen (*Hapalidae*). *Folia Primatol.* **7**, 37–65.

Epple, G. (1968). Comparative studies on vocalization in marmoset monkeys (*Hapalidae*). *Folia Primatol.* **8**, 1–40.

Epple, G. (1970a). Maintenance, breeding, and development of marmoset monkeys (Callithricidae) in captivity. *Folia Primatol.* **12**, 56–76.

Epple, G. (1970b). Quantitative studies on scent marking in the marmoset (*Callithrix jacchus*). *Folia Primatol.* **13**, 48–62.

Epple, G. (1971). Discrimination of the odor of males and females by the marmoset *Saguinus fuscicollis* ssp. *Proc. Int. Congr. Primatol., Zürich, 1970* **3**, 166–171.

Epple, G. (1972). Social behavior of laboratory groups of *Saguinus fuscicollis*. *In* "Saving the Lion Marmoset: Proceedings of the WAPT Golden Lion Marmoset Conference" (D. D. Bridgwater, ed.), pp. 50–58. Wild Animal Propagation Trust, the American Association of Zoological Parks and Aquariums. Oglebay Park, Wheeling, West Virginia.

Epple, G. (1973). The role of pheromones in the social communication of marmoset monkeys (Callithricidae). *J. Reprod. Fert. Suppl.* **19**, 447–454.

Epple, G., and Lorenz, R. (1967). Vorkommen, Morphologie und Funktion der Sternaldrüse bei den Platyrrhini. *Folia Primatol.* **7**, 98–126.

Evans, C. S., and Goy, R. W. (1968). Social behaviour and reproductive cycles in captive Ring-tailed lemurs (*Lemur catta*). *J. Zool.* **156**, 181–197.

Fiedler, W. (1956). Übersicht über das System der Primaten. *In* "Primatologia" (H. Hofer, A. H. Schultz, and D. Starck, eds.), Vol. 1, pp. 1–266. Karger, Basel.

Fitzgerald, A. (1935). Rearing marmosets in captivity. *J. Mammal.* **16**, 181–188.

Franz, J. (1963). Beobachtungen bei einer Löwenäffchen-Aufzucht. *Zool. Garten* **28**, 115–120.

Gartlan, J. S. (1968). Structure and function in primate society. *Folia Primatol.* **8**, 89–120.

Geist, V. (1964). On the rutting behavior of the mountain goat. *J. Mammal.* **45**, 551–568.

Graetz, E. (1968). Studien über das mittelamerikanische Krallenäffchen *Oedipomidas spixi*. *Sitzungsber. Ges. Naturforsch. Freunde, Berlin* **8**, 29–40.

Grüner, M., and Krause, P. (1963). Biologische Beobachtungen an Weisspinseläffchen, *Hapale jacchus* (L., 1758) im Berliner Tierpark. *Zool. Garten* **28**, 108–114.

Hampton, J. K., Jr., Hampton, S. H., and Landwehr, B. T. (1966). Observations on a successful breeding colony of the marmoset. *Oedipomidas oedipus*. *Folia Primatol.* **4**, 265–287.

Heck, L. (1892). Affernzwerge. *Natur Haus* **1**, 105.

Heinemann, H. (1970). The breeding and maintenance of captive Goeldi's monkeys *Callimico goeldii*. *Int. Zoo Yearb.* **10**, 72–78.

Hershkovitz, P. (1958). A geographic classification of neotropical mammals. *Fieldiana: Zool.* **36**, 581–620.

Hershkovitz, P. (1966). Taxonomic notes on tamarins, genus *Saguinus* (Callithricidae, Primates), with descriptions of four new forms. *Folia Primatol.* **4**, 381–395.

Hershkovitz, P. (1968). Metachromism or the principle of evolutionary change in mammalian tegumentary colors. *Evolution* **22**, 556–575.

Hershkovitz, P. (1969). The evolution of mammals on southern continents. VI. The recent mammals of the neotropical region: a zoogeographic and ecological review. *Quart. Rev. Biol.* **44**, 1–70.

Hershkovitz, P. (1970). Notes on tertiary platyrrhine monkeys and description of a new genus from the late Miocene of Colombia. *Folia. Primatol.* **12**, 1–37.

Hershkovitz, P. (1972). Notes on new world monkeys. *Int. Zoo Yearb.* **12**, 3–12.

Hill, W. C. O. (1957). "Primates, Vol. III. *Pithecoidea*," Edinburgh Univ. Press, Edinburgh.

Hill, W. C. O. (1959). The anatomy of *Callimico goeldii* (Thomas), a primitive American primate. *Trans. Amer. Phil. Soc.* **49**, Part 5.

Hladik, A., and Hladik, C. M. (1969). Rapports trophiques entre vegetation et primates dans la foret de Barro Colorado (Panama). *Terre Vie* **1**, 25–117.

Hornung, V. (1896). Der Pinselaffe (*Hapale penicillata*). *Zool. Garten* **37**, 273–277.

Hornung, V. (1899). Weitere Mitteilungen über den Pinselaffen (*Hapale penicillata*). *Zool. Garten* **40**, 208–209.

Humboldt, A. V. (1805). "Vom Orinoko zum Amazonas." Brockhaus, Wiesbaden. (Reprinted, 1959).

Ilse, D. R. (1955). Olfactory marking of territory in two young male Loris, *Loris tardigradus lydekkerianus*, kept in captivity in Poona. *Brit. J. Anim. Behav.* **3**, 118–120.

Imanishi, K. (1960). Social organization of subhuman primates in their natural habitat. *Curr. Anthropol.* **1**, 393–407.

Jay, P. (1965). The common langur of North India. *In* "Primate Behavior: Field Studies of Monkeys and Apes" (I. DeVore, ed.), pp. 197–249. Holt, New York.

Jolly, A. (1966). "Lemur Behavior." Univ. of Chicago Press, Chicago, Illinois.

Kaufmann, J. H. (1965). Studies on the behavior of captive tree shrews (*Tupaia glis*). *Folia Primatol.* **3**, 50–74.

Kingston, W. R. (1969). Marmosets and tamarins. *Lab. Anim. Handb.* **4**, 243–250.

Kirchshofer, R. (1963). Einige bemerkenswerte Verhaltensweisen bei Saimiris im Vergleich zu anderen Arten. *Z. Morphol. Anthropol.* **53**, 77–91.

Koford, C. B. (1963a). Rank of mothers and sons in bands of rhesus monkeys. *Science* **141**, 356–357.

Koford, C. B. (1963b). Group relations in an island colony of rhesus monkeys. *In* "Primate Social Behavior" (C. H. Southwick, ed.), pp. 136–152. Van Nostrand-Reinhold, Princeton, New Jersey.

Krieg, H. (1930). Die Affen des Gran Chaco und seiner Grenzgebiete. *Z Morphol. Ökol. Tiere* **18**, 760–785.

Lancaster, J. B. (1968). Primate communication systems and the emergence of human language. *In* "Primates: Studies in Adaptation and Variability" (P. C. Jay, ed.), pp. 439–457. Holt, New York.

Langford, J. B. (1963). Breeding behavior of *Hapale jacchus* (common marmoset). *S. Afr. J. Sci.* **59**, 299–300.

Le Roux, G. (1967). Contribution à l'Étude des Moyens d'Intercommunication chez le Ouistiti à Pinceaux (*Hapale jacchus*). Thesis, Fac. Sci., Univ. Rennes.

Leshner, A. J., and Candland, D. K. (1972). Endocrine effects of grouping and dominance rank in squirrel monkeys. *Physiol. Behav* **8**, 441–445.

Levy, B. M., and Artecona, J. (1964). The marmoset as an experimental animal in biological research: care and maintenance. *Lab. Anim. Care* **14**, 20–27.

Lorenz, R. (1969). Notes on the care, diet and feeding habits of Goeldi's monkey (*Callimico goeldii*). *Int. Zoo Yearb.* **9**, 150–155.

Lorenz, R. (1972). Management and reproduction of the Goeldi's monkey *Callimico goeldii* (Thomas, 1904) Callimiconidae, Primates. *In* "Saving the Lion Marmoset: Proceedings of the WAPT Golden Lion Marmoset Conference" (D. D. Bridgwater, ed.), pp. 92–109. Wild Animal Propagation Trust, Oglebay Park, Wheeling, West Virginia.

Lorenz, R., and Heinemann, H. (1967). Beitrag zur Morphologie und körperlichen Jugendentwicklung des Springtamarin *Callimico goeldii* (Thomas, 1904). *Folia Primatol.* **6**, 1–27.

Lucas, N. S., Hume, E. M., and Smith, H. H. (1927). On the breeding of the common marmoset (*Hapale jacchus* Linn.) in captivity when irradiated with ultra-violet rays. *Proc. Zool. Soc. London* **30**, 447–451.

Lucas, N. S., Hume, E. M., and Smith, H. H. (1937). On the breeding of the common marmoset (*Hapale jacchus* Linn.) in captivity when irradiated with ultra-violet rays. II. A ten years' family history. *Proc. Zool. Soc. London, Ser. A* **107**, 205–211.

Mallinson, J. (1964). Notes on the nutrition, social behaviour and reproduction of *Hapalidae* in captivity. *Int. Zoo Yearb.* **5**, 137–140.

Mallinson, J. (1969). The breeding and maintenance of marmosets at Jersey Zoo. *Annu. Rep. Jersey Wildl. Preserv. Trust, 6th.*

Marik, M. (1931). Beobachtungen zur Fortpflanzungsbiologie der Uistiti (*Callithrix jacchus* L.). *Zool. Garten* **4**, 347–349.

Marler, P. (1955). The characteristics of some animal calls. *Nature (London)* **176**, 6–8.

Marler, P. (1965). Communication in monkeys and apes. *In* "Primate Behavior: Field Studies of Monkeys and Apes" (I. DeVore, ed.), pp. 544–584. Holt, New York.

Mason, W. A. (1966). Social organization of the South American monkey, *Callicebus moloch*: A preliminary report. *Tulane Stud. Zool.* **13**, 23–28.

Mason, W. A. (1971). Field and laboratory studies of social organization in *Saimiri* and *Callicebus*. *In* "Primate Behavior: Developments in Field and Laboratory Research" (L. A. Rosenblum, ed.), Vol. 2, pp. 107–138. Academic Press, New York.

Mazur, A., and Baldwin, J. D. (1968). Social behavior of semi-free ranging white-lipped tamarins. *Psychol. Rep.* **22**, 441–442.

Meeter von Zorn. (1903). Affenzwerge: Nach langjährigen Beobachtungen. *Natur Haus* **12**, 35–38.

Michael, R. P., and Keverne, E. B. (1968). Pheromones in the communication of sexual status in primates. *Nature (London)* **218**, 746–749.

Michael, R. P., and Keverne, E. B. (1970). Primate sex pheromones of vaginal origin. *Nature (London)* **225**, 84–85.

Michael, R. P., Keverne, E. B., and Bonsall, R. W. (1971). Pheromones: isolation of male sex attractants from a female primate. *Science* **172**, 964–966.

Miller, F. W. (1930). Notes on some mammals of southern Mato Grosso. *Brazil J. Mammal.* **11**, 10–22.

Moynihan, M. (1964). Some behavior patterns of platyrrhine monkeys. I. The night monkey (*Aotus trivirgatus*). *Smithson. Misc. Collect.* **146**, (5), 1–84.

Moynihan, M. (1966). Communication in the Titi monkey, *Callicebus*. *J. Zool.* **150**, 77–128.

Moynihan, M. (1970). Some behavior patterns of platyrrhine monkeys. II. *Saguinus geoffroyi* and some other tamarins. *Smithson. Contrib. Zool.* **28**, 1–77.

Muckenhirn, N. A. (1967). The behavior and vocal repertoire of *Saguinus oedipus* (Hershkovitz, 1966) (Callithricidae, Primates). Thesis, Univ. of Maryland, College Park.

Mykytowycz, R. (1965). Further observations on the territorial function and histology of the submandibular cutaneous (chin) glands in the wild rabbit, *Oryctolagus cuniculus* (L.). *Anim. Behav.* **13**, 400–412.

Mykytowycz, R. (1970). The role of skin glands in mammalian communication. *In* "Communication by Chemical Signals" (J. W. Johnston, D. G. Moulton, and A. Turk, eds.), Vol. 1, pp. 327–360. Appleton, New York.

Neill, P. (1829). Account of the habits of a specimen of the *Simia jacchus*, Lin., or *Jacchus vulgaris* Geoff., now in the possession of Gavin Milroy, Esq. *Edinburgh. Mag. Natur. Hist.* **1**, 18–20.

Paris, P. (1908). Un cas de reproduction du Ouistiti (*Hapale jacchus.*) *Bull. Soc. Zool., Paris* **31/33**, 147.

Perkins, E. M. (1966). The skin of the black-collared tamarin (*Tamarinus nigricollis*). *Amer. J. Phys. Anthropol.* **25**, 41–69.

Perkins, E. M. (1968). The skin of the pygmy marmoset – *Callithrix* (=*Cebuella*) *pygmaea. Amer. J. Phys. Anthropol.* **29**, 349–364.

Perkins, E. M. (1969a). The skin of the cottontop pinché *Saguinus* (=*Oedipomidas*) *oedipus. Amer. J. Anthropol.* **30**, 13–28.

Perkins, E. M. (1969b). The skin of Goeldi's marmoset (*Callimico goeldii*). *Amer. J. Phys. Anthropol.* **30**, 231–250.

Perkins, E. M. (1969c). The skin of the silver marmoset – *Callithrix* (=*Mico*) *argentata. Amer. J. Phys. Anthropol.* **30**, 361–388.

Perkins, E. M. (1969d). Monkeys of the New World. *Primate News* **7**, 4–15.

Pocock, R. J. (1920). On the external characters of the South American monkeys. *Proc. Zool. Soc. London* **1**, 91–113.

Rabb, G. B., and Rowell, T. E. (1960). Notes on reproduction in captive marmosets. *J. Mammal.* **41**, 401.

Ralls, K. (1971). Mammalian scent marking. *Science* **171**, 443–449.

Reynolds V., and Reynolds, F. (1965). Chimpanzees of the Budongo Forest. *In* "Primate Behavior" (I. DeVore, ed.), pp. 368–424. Holt, New York.

Rose, R. M., Holaday, J. W., and Bernstein, I. S. (1971). Plasma testosterone, dominance rank and aggressive behaviour in male rhesus monkeys. *Nature* (*London*) **231**, 366–368.

Rose, R. M., Gordon, T. P., and Bernstein, I. S. (1972). Plasma testosterone levels in the male rhesus: Influence of sexual and social stimuli. *Science* **178**, 643–645.

Roth, H. H. (1960). Beobachtungen an *Tamarin spec. Zool. Garten* **25**, 166–182.

Rowell, T. E., and Hinde, R. A. (1962). Vocal communication by the Rhesus monkey (*Macaca mulatta*). *Proc. Zool. Soc. London* **138**, 279–294.

Sade, D. S. (1965). Some aspects of parent-offspring and sibling relations in a group of Rhesus monkeys, with a discussion of grooming. *Amer. J. Phys. Anthropol.* **23**, 1–18.

Sanderson, I. T. (1945). "Caribbean Treasure." Viking Press, New York.

Schaller, G. B. (1963). "The Mountain Gorilla: Ecology and Behavior." Univ. of Chicago Press, Chicago, Illinois.

Schreitmüller, A. (1930). Einiges über das Marmosetäffchen. *Aquarium* (*Berlin*) pp. 42–44.

Schultze-Westrum, T. (1965). Innerartliche Verständigung durch Düfte beim Gleitbeutler *Petaurus breviceps papuanus* THOMAS (Marsupialia, Phalangeridae). *Z. Vergl. Physiol.* **50**, 151–220.

Seitz, E. (1969). Die Bedeutung geruchlicher Orientierung beim Plumplori, *Nycticebus coucang* Boddaert 1785 (*Prosimii, Lorisidae*). *Z. Tierpsychol.* **26**, 73–103.

Shadle, A. R., Mirand, E. A., and Grace, J. T. (1965). Breeding responses in tamarins. *Lab. Anim. Care* **15**, 1–10.

Snyder, P. A. (1972). Behavior of *Leontopithecus rosalia* (the golden lion marmoset) and related species: A review. *In* "Saving the Lion Marmoset: Proceedings of the WAPT Golden Lion Marmoset Conference" (D. D. Bridgwater, ed.), pp. 23–49. Wild Animal Propagation Trust, Oglebay Park, Wheeling, West Virginia.

Sonntag, C. F. (1924). "Morphology and Evolution of the Apes and Man." Bale Sons & Daniels Son, London.

Sprankel, H. (1961). Histologie und biologische Bedeutung eines jugosternalen Duftdrüsenfelds bei *Tupia glis* Diard 1820. *Verh. Deut. Zool. Ges.* (*Saarbrücken*) pp. 198–206.

Starck, D. (1969). Die circumgenitalen Drüsenorgane von *Callithrix* (*Cebuella*) *pygmaea* (Spix 1823). *Zool. Garten* **36**, 312–326.

Stellar, E. (1960). The marmoset as a laboratory animal: maintenance, general observations of behavior and simple learning. *J. Comp. Physiol. Psychol.* **53**, 1–10.

Sugiyama, Y. (1967). Social organization of Hanuman langurs. *In* "Social Communication among Primates" (S. A. Altmann, ed.), pp. 221–236. Univ. of Chicago Press, Chicago, Illinois.

Thorington, R. W., Jr. (1968). Observations of the tamarin *Saguinus midas*. *Folia Primatol.* **9**, 95–98.

Ulmer, F. A., Jr. (1961). Gestation period of the lion marmoset. *J. Mammal.* **42**, 253–254.

van Lawick-Goodall, J. (1969). Mother-offspring relationships in free-ranging chimpanzees. *In* "Primate Ethology" (D. Morris, ed.). pp. 365–436. Doubleday, Garden City, New York.

von Holst, D. (1969). Sozialer Stress bei Tapajas (*Tupaia belangeri*). *Z. Vergl. Physiol.* **63**, 1–58.

Wendt, H. (1964). Erfolgeiche Zucht der Baumwollköpfchens oder Pincheäffchens, *Leontocebus* (*Oedipomidas*) *oedipus* (Linné 1758), in Gefangenschaft. *Säugetierkundl. Mitt.* **12**, 49–52.

Whitten, W. K., and Bronson, F. H. (1970). The role of pheromones in mammalian reproduction. *In* "Advances in Chemoreception" (J. W. Johnston, D. G. Moulton, and A. Turk, eds.), Vol. 1, pp. 309–325. Appleton, New York.

Winter, P., Ploog, D., and Latta, J. (1966). Vocal repertoire of the squirrel monkey (*Saimiri scirueus*), its analysis and significance. *Exp. Brain. Res.* (*Berlin*) **1**, 359–384.

Wislocki, G. B. (1930). A study of scent glands in the marmosets, especially *Oedipomidas geoffroyi*. *J. Mammal.* **11**, 475–482.

Wynne-Edwards, V. C. (1962). "Animal Dispersion in Relation to Social Behaviour." Oliver & Boyd, Edinburgh.

Yamada, M. (1963). A study of blood-relationship in the natural society of the Japanese macaque. *Primates* **4**, 43–65.

Yoshiba, K. (1968). Local and intertroop variability in ecology and social behavior of common Indian langurs. *In* "Primates: Studies in Adaptation and Variability" (P. C. Jay, ed.), pp. 217–242, Holt, New York.

Behavior and Malnutrition in the Rhesus Monkey*

ROBERT R. ZIMMERMANN,† DAVID A. STROBEL,
PETER STEERE,‡ AND CHARLES R. GEIST§

University of Montana
Missoula, Montana

I. INTRODUCTION

Nutritional deficiencies have long been suspected as being one of the primary sources for many of the psychological and social problems that

* Preparation and original research presented in this manuscript was supported in part by Grant No. 401 from the Nutrition Foundation Inc., and Grant No. HD-04863 from the National Institute of Human Growth and Development.

† Present address: Central Michigan University, Mt. Pleasant, Michigan.
‡ Present address: University of Georgia, Athens, Georgia.
§ Present address: University of Alaska, Fairbanks, Alaska.

appear in the course of human development. However, in the recent past, the relationship between nutrition and psychology has not been conducive to fruitful research. During World War II, there arose considerable interest concerning the effects of starvation upon the behavior of human beings (Keys *et al.*, 1950). Much of this research has revealed that the psychological changes could be accounted for on the basis of temporary fatigue or motivational effects resulting from the conditions of starvation. The general disinterest apparent in most psychologists with respect to these early studies in undernutrition was the product of the failure to find any significant intellectual impairment concomitant with rather severe manipulations of nutritional conditions.

Concurrent with investigations of human nutritional deficiencies were studies of animal behavior employing deprivation conditions in order to motivate animals to operate in special situations. Motivation as a basic process within the organism was the proper domain of nutritional variables in this research. Investigations of nutritional deprivations were often concerned with the effects generated by manipulation of homeostatic mechanisms and the behavior associated with the maintenance of the homeostatic drives. Thus, considerable effort was directed toward the study of specific hungers and how such hungers effectuated distinctive behavior in lower animals. The academic controversy centered around the mechanisms involved in the production of behavior specific to the hunger developed (Young, 1955). Recently, however, there has been a growing body of information from observations of human behavior and investigations of lower animals indicating that organisms that have suffered from malnutrition early in life may sustain permanent behavioral abnormalities.

Of the myriad of forms and severities representative of deficiency states, protein–calorie malnutrition has been of prime concern to investigators. Emphasis on this variety of malnutrition has been fostered by studies that revealed that human beings may suffer permanent changes in cognitive development and in the final level of intellectual growth (Brockman and Ricciuti, 1971; Cravioto and Robles, 1965). Further, it has been estimated that 70% of the preschool children in the developing countries are, in all probability, suffering from some form of protein–calorie malnutrition (Coursin, 1965).

The designation of protein–calorie malnutrition as a clinical syndrome was first suggested by Jelliffe (1959) and later was proposed by the Joint Food and Agriculture Organization and the World Health Organization Expert Committee on Nutrition (FAO/WHO, 1962). Protein–calorie malnutrition refers to a wide range of physiological conditions that are often described as kwashiorkor, marasmic kwashiorkor, and marasmus. Although it is beyond the scope of this chapter to enter into a clinical controversy concerning the detailed definitions of these diverse syndromes and the possible biochemical distinctions, it may be asserted that such labels describe slightly different syndromes and etiologies associated with forms of protein–calorie malnutrition.

For the most part, the kwashiorkor syndrome appears to develop as a result of an unbalanced diet early in childhood, usually between the ages of 1–3 years. The diet is generally low in protein but contains adequate calories supplied by carbohydrates (Jelliffe and Welbourn, 1963). Although kwashiorkor is a variable syndrome, a number of constant features can be ascertained. Whereas irritability and resentfulness are prevalent in moderate cases (Wayburne, 1968), apathy, lethargy, and misery become manifest in extreme cases (Hansen, 1968; Latham, 1968, Wayburne, 1968). Indeed, the child exhibits a loss of interest in his environment and it is difficult to attract and maintain his attention (Burgess and Dean, 1962; see also Barnes, 1971). Fear and anxiety are normally absent (Wayburne, 1968). Retardation of growth (Gopalin, 1968; Graham *et al.*, 1963; Vis, 1968; Whitehead and Dean, 1964a, b) and motor development (Gerber and Dean, 1967; Waterlow *et al.*, 1960) is apparent in kwashiorkor. Edema frequently complicates chronic states of protein–calorie malnutrition (Srikantia, 1968). In addition, there are a number of variable features associated with kwashiorkor. Changes in the hair may include hypochromotrichia, loss of natural curl, brittleness, and sparseness (Jelliffe and Welbourn, 1963; Monckeberg, 1968; Trowell *et al.*, 1954; see also Bradfield, 1968b; Viteri *et al.*, 1964), and a reduction in the tensile strength of the hair (Latham and Velez, 1966). The skin may become light colored (Wayburne, 1968), a flaky rash may develop (Jelliffe, 1955; Trowell *et al.*, 1954; see also Bradfield, 1968b), or lesions may be present (Hansen, 1968; Trowell *et al.*, 1954; see also Bradfield, 1968b; Wayburne, 1968). Hepatocellular variations are pronounced and include fatty infiltration of the liver (Davies, 1948; Gopalin, 1968; Monckeberg, 1968). Often the child will develop a moon-faced appearance with blueberry cheeks (Jelliffe and Welbourn, 1963). Conditioning infections associated with kwashiorkor may comprise chest signs as in tuberculosis, severe anemia in hookworm infection, and dehydration in infective diarrhea (Jelliffe and Welbourn, 1963).

Nutritional marasmus appears to be due principally to diets deficient in both calorie and protein (Monckeberg, 1968), the so-called balanced starvation (Jelliffe and Welbourn, 1963). It is most prevalent during the period from 6 to 18 months of age (Gopalin, 1968) and is often the result of the failure of lactation or the feeding of diluted milk (Jelliffe and Welbourn, 1963). The consistent signs of this syndrome are retardation of growth (Gopalin, 1968), drastic loss of muscle mass (Pineda, 1968), and wasting of subcutaneous fat (Gongora and McFie, 1959). The child takes on the appearance of a "wizened little old man" (Jelliffe and Welbourn, 1963). On occasion there are alterations in hair (Latham, 1968), vitamin deficiency (McLaren, 1968), oral moniliasis (Jelliffe and Welbourn, 1963), relapsing diarrhea often with severe dehydration (Hansen, 1968), and chest signs due to tuberculosis (Jelliffe and Welbourn, 1963). There is no edema (Gopalin, 1968), but a lively interest in the environment (Wayburne, 1968), a fine appetite (Latham, 1968), and little sign of either apathy or anorexia (Jelliffe and Welbourn, 1963) may be present. Finally, since interest in food has been reported to be excellent in marasmus and poor in kwashiorkor, appetite has been suggested as an additional index for classification (Bradfield, 1968a).

Whether the developing child suffers from kwashiorkor or marasmus is apparently irrelevant to most of the behavioral deficits reported. Children that recover from either kwashiorkor or marasmus suffer some permanent alterations in mental development as measured by the Gesell Infant Scale (Cravioto and Robles, 1965; Gerber and Dean, 1957). Cravioto and Robles (1965) also reported that children with a history of undernourishment are inadequate in integrating sensory information. Although Brockman and Ricciuti (1971) are hesitant to project their findings in terms of long-range effects, it is apparent that 12 weeks of rehabilitation from protein–calorie malnutrition is not sufficient to bring stimulus categorization ability up to normal. In a breakdown of the Gesell tests, it was found that areas of motor behavior, adaptive responses, language, personal, and social development were not equally affected (Cravioto and Robles, 1965). In general, language development showed the lowest scores and motor responses were the least retarded. After recovery there was generally a gradual improvement in all measures and the differences between the chronological and mental ages were reduced in all of the subjects with the exception of those who were admitted to the hospital when they were less than 6 months of age.

One of the major problems in dealing with the results of studies with human subjects is that most of the findings cannot clearly define the role of nutrition in the formation of behavioral deficiencies. Studies with human beings will always confound cultural, social, and economic factors with the nutritional variables. For example, Stoch and Smythe (1968) found that even when families were equated for socioeconomic levels, the living conditions of children with a history of malnutrition were markedly different from adequately nourished subjects. The homes of the undernourished children were characterized by alcoholism, illegitimacy, and degradation, whereas the homes of the control subjects were more stable. Thus, cultural factors of child-rearing variables were confounded with the nutritional variable even when families were equated for income. Another factor that cannot be overlooked is the role of secondary infections in the development of behavioral deficiencies. It has been suggested that secondary infections may be synergistic with nutritional variables in producing impaired development (Behar *et al.*, 1958).

Studies with nonhuman animals provide more rigorous control over the variables that might influence behavior, but to date most of the research in behavior and nutrition has been conducted with chickens, pigs, and rats. These animals are usually adequate, but not necessarily sufficient, for studying some of the behavioral effects resulting from malnutrition. For example, they have provided valuable information on the relationship between the biochemical mechanisms in malnutrition and performance on simple tasks. However, each of these animals has a very rapid period of development and a limited developmental repertoire, exhibiting adult forms of behavior relatively early in life. Nevertheless, some valuable information concerning the role of malnutrition in development has been acquired.

Initial efforts were directed toward investigating the effects of malnutrition on learning. This probably came about as a result of equating the learning ability in animals with the mental abilities in humans as assessed by IQ tests in childhood. In the fiirst studies, few, if any, differences in learning ability were found (Barnes *et al.*, 1966) and those that did appear were attributed to motivational differences between normal and malnourished animals. In fact, most of the effects of early protein–calorie malnutrition on behavior seemed to take the form of heightened reactivity (Barnes *et al.*, 1967), loss of inhibition (Frankova and Barnes, 1968b), and an oversensitivity to the environment (Barnes *et al.*, 1969; Frankova and Barnes, 1968a). Zimmermann and Wells (1971) found that rats deprived of protein early in life were inferior to normal animals on performance in a Hebb–Williams maze. These data are consistent with those of Cowley and Griesel (1962) who found that rats with a history of protein–calorie malnutrition were inferior to controls when tested in a Hebb–Williams maze that had been modified so that escape from water served as the incentive. However, Wells *et al.* (1972) found that when rats were given additional maze experience or enriched environments in which to develop, the effects of the diet on maze performance were no longer significant.

The investigations with pigs and rats have been important for completing an accurate picture of the role of malnutrition on behavior development across species. However, the developing primate, which has a relatively slow rate of growth and a gradual development of a wide variety of behaviors, can be observed and tested in many different situations and on many different tasks over a period of years. Thus, the effects of a diversity of developmental variables, including nutrition, can be more accurately assessed. More important is the fact that subhuman primates appear to go through stages of development that are similar to those found in the human being. For example, the developing rhesus monkey (*Macaca mulatta*) does not reach a maximum level of intellectual development until sexual maturity; social behavior continues to develop and change from infancy to adolescence to adulthood. These attributes make the developing rhesus monkey an excellent subject for studying the behavioral development during malnutrition.

II. PRIMATE FEEDING BEHAVIOR AND NUTRITION

It is important to consider two completely different aspects of feeding behavior in relation to nutrition in primates: the first is feeding and nutrition within the laboratory and the second is feeding and nutrition in the natural habitat. The eating habits and nutrition of the laboratory primate have been summarized in great detail in a recent text by Harris (1971). The book presents the latest available data on the feeding and maintenance of laboratory primates and presents the known dietary requirements of the primates mentioned. It is quite clear from this book that there is much to be learned in the area of nutrition in the primate order. Most of what is known is limited to extensive research with rhesus macaques that have been captured in India and subjected to dietary controls after conditioning in the laboratory. Similar data are available on the developing rhesus monkey born in the confines of the laboratory setting. Most of the information consists of comparisons between groups of monkeys with limited intake of certain chemical variables and those animals raised on commercial diets. As Portman (1970) has pointed out, we have accumulated a considerable body of knowledge concerning the growth and development of the rhesus monkey, as well as the effects of variations of different nutrients on weight gain and the short-term health of this species. Little is known about the long-term effects on simple growth patterns, and as this book makes clear, less is known about the behavioral effects of dietary manipulations. Feeding behavior in the natural habitat presents the investigator with a different set of problems. However, it is readily ascertained from watching primates in their natural habitat that they spend a great deal of time in behavior that is associated with feeding.

Three basic methods are available for determining the dietary substances consumed by subhuman primates in their natural habitats. The first method is that of stomach analysis. Fooden (1964) has used this method for determining the dietary staples of Guianian monkeys, but only in a general sense. When quantifying his data, categories such as fruits, seeds, and insects were employed. This method has also been used to some extent by Carpenter (1940) in his study of gibbons in Siam. However, stomach analysis has some distinct disadvantages. The animal must be exterminated in order to undertake an analysis of stomach contents. This is unfeasible for some primate species due to the threat of extinction. There is also the possibility that a good cross section of the diet may not be obtained by killing the animal at any particular time. Further, the method fails to take into account habitat variation, age variation, and seasonal variation as factors that influence the diet. In order to gather sufficient data to account for variations of this nature, a large number of animals would have to be sacrificed on a year-round basis. Finally, it is impossible to study the feeding behavior of animals that are being constantly pursued and exterminated. It is also very unlikely that successful observations could be undertaken in an area where monkeys had been collected in the recent past.

The second method is that of direct observation of feeding behavior. The observer watches the animals as they eat followed by collection of specimens of the type of food consumed. Such samples can then be sent to botanists, entomologists, or mammalogists for analysis of the nutritional composition and identification. This method, at least through the collecting phase and preliminary identification, has been the most common one employed by field workers such as Aldrich-Blake *et al.* (1971); Carpenter (1934, 1935, 1940); Chalmers (1968); Crook and Aldrich-Blake (1968); DeVore and Hall (1965); DeVore and Washburn (1963); Goodall (1965); Jay (1965); Jones and Pi (1968); Kern (1964); Melville (1968); Reynolds and Reynolds (1965); Schaller (1961, 1963); and Simonds (1965). There is a serious omission in the above-mentioned works if the dietary composition and requirements are to be accurately investigated. Most of the authors have identified animal remains, insects, and plants by species, but only Schaller (1963) has attempted a nutritional breakdown of a particular food source. This was done for an herb stem which is a part of the diet of the mountain gorilla (*Gorilla gorilla*). Schaller (1963) has compiled the most extensive lists of dietary items for any primate study concerning the mountain gorilla. His primary concern was with what foods were eaten in particular habitats and how many of these food sources were found in different ecological settings.

A variety of problems exist with this method also. Insects that are consumed are usually completely devoured and the observer must identify the species as the animal is in the process of being eaten and then collect a similar species from the area for analysis. Plant foods are generally abundant enough in the area to be easily identified and samples collected. The small mammals that baboons (*Papio* sp.) and chimpanzees (*Pan* sp.) occasionally consume do not present such a problem for they can be identified more easily than can insect and plant foods. It would be of great value to become familiar with the flora and fauna of the area in which the primate study is to be undertaken.

The third method is that of fecal analysis. The remains of animals, insects, and plants which have been consumed can often be identified from the feces. Again, Schaller (1963) utilized this method for the identification of some dietary materials present in the nutritional repertoire of the mountain gorilla, although he stressed that this method cannot be utilized as the sole determinant of nutritional components. Some food materials may be decomposed to such an extent that identification is impossible.

Accurate study of the diet of nonhuman primates in their natural habitats requires long-term study to take into account the daily, seasonal, and annual variations in the diet. Species that have subgroups inhabiting different and often isolated geographic areas must all be thoroughly studied in order to obtain a complete picture of the diet for the entire species. Detailed observations of individual groups are needed to account for the age variation in diet composition and feeding behavior. Goodall (1965) has demonstrated the significance of age factors in foods consumed and behaviors associated with feeding in the chimpanzee (*Pan satyrus*) of the Gombe Stream Reserve in Tanzania.

It can be stated that with few exceptions, little attention has been given to the problem of the nutritional adequacy of the diet of nonhuman primates. Few field studies have gone beyond a simple description of the dietary components.

Table I lists most of the major genera of the primate order, the dietary preferences for each genera, and the source of the information.

TABLE I

DIETS OF NONHUMAN PRIMATES IN THEIR NATURAL HABITATS

Genera	Common name	Diet	Source
Prosimians			
Tupaia	Tree shrews	Insects, fruits	Napier and Napier (1967); Davis (1962)
Dendrogale	Smooth-tailed tree shrews	Insects	Harrison (1954)
Urogale	Philippine tree shrews	Insects, grubs, bird's eggs, young birds, fruit	Napier and Napier (1967); Wharton (1950)
Anathana	Madras tree shrews	Insects, fruit	Napier and Napier (1967)
Ptilocercus	Feather-tailed tree shrews	Insects, lizards	Napier and Napier (1967 ; Lim (1967)
Lemur	True lemurs	Fruits, flowers, leaves	Petter (1965)
Hapalemur	Gentle lemurs	Fruits, leaves, reeds, flowers	Petter (1962)
Lepilemur	Sportive lemurs	Leaves, fruits, bark	Napier and Napier (1967)
Cheirogaleus	Dwarf lemurs	Insects, fruits, rice	Petter (1965); Napier and Napier (1967)
Microcebus	Mouse lemurs	Insects; fruit, small mammals	Petter (1962); Shaw (1879)
Phaner	Fork-marked dwarf lemurs	Insects, fruit	Petter (1962, 1965)
Indri	Indri	Leaves, fruit	Napier and Napier (1967); Petter (1965)
Loris	Slender loris	Insects, birds, small lizards	Pocock (1939); Sanderson (1957)
Nycticebus	Slow loris	Insects, fruit, leaves, seeds, birds, lizards, bird's eggs	Davis (1962); Harrison (1962); Pocock (1939)
Daubentonia	Aye-aye	Insects, larvae	Petter (1965); Napier and Napier (1967)
Galago	Galago	Insects, tree gum	Sauer and Sauer (1963); Oates (1966); Booth (1956); Candale (1944)
Arctocebus	Angwantibos	Insects	Napier and Napier (1967)
Perodicticus	Potto	Insects, birds, small mammals, nuts, fruit, leaves	Napier and Napier (1967)

(*Continued*)

TABLE I (Continued)

Genera	Common name	Diet	Source
Avahi	Avahi	Leaves, bark, fruit, flowers	Napier and Napier (1967)
Propithecus	Sifakas	Fruit, rice, eucalyptus leaves, in captivity	Napier and Napier (1967)
New World monkeys			
Callithrix	Marmosets	Insects, fruit, vegetables	Hill (1957)
Cebuella	Pygmy marmosets	Insects, fruit, bird's eggs, birds	Ochs (1964)
Saguinus	Tamarins	Fruit, seeds	Fooden (1964); Enders (1930)
Leontideus	Golden lion marmosets	Fruit, lizards, insects	Napier and Napier (1967)
Callimico	Goeldi's marmosets	Unknown	
Aotus	Douroucoulis	Fruit, insects, small mammals	Hill (1960)
Callicebus	Titis	Insects, fruit, small birds, bird's eggs	Napier and Napier (1967)
Pithecia	Sakis	Fruit, birds, small mammals	Fooden (1964); Sanderson (1957)
Chiropotes	Bearded sakis	Fruit	Fooden (1964)
Cacajo	Uakaris	Fruit, buds, seeds, leaves	Napier and Napier (1967)
Cebus	Capuchins	Fruit, insects	Fooden (1964); Collias and Southwick (1952)
Saimiri	Squirrel	Fruit, insects	Fooden (1964)
Alouatta	Howlers	Leaves, fruit, buds, shoots,	Carpenter (1934)
Ateles	Spider	Fruits, nuts, flowers	Carpenter (1935)
Lagothrix	Woolly	Leaves, fruit	Napier and Napier (1967)
Brachyteles	Woolly spider	Fruit, leaves, seeds	Hill (1962)

Old World monkeys—Asia			
Presbytis	Langurs	Leaves, fruit, buds, shoots, licking of earth	Gee (1961); Jay (1965); Napier and Napier (1967)
Pygathrix	Douc langur	Not known	
Simias	Pagai Island langur	Not known	
Rhinopithecus	Snub-nosed langur	Fruit, buds, leaves, bamboo shoots	Napier and Napier (1967)
Nasalis	Proboscis	Young leaves, tips of mangrove and pedada trees, fruits, flowers	Kern (1964); Napier and Napier (1967)
Cynopithecus	Celebes black ape	Fruit	Napier and Napier (1967)
Macaca	Macaques	Fruit, roots, leaves, insects, grubs, rice, maize, potatoes, sugar cane, mollusks, crustaceans	Napier and Napier (1967)
Old World monkeys—Africa			
Macaca sylvannus	Barbary ape	Fruit, leaves, melons, vegetables	MacRoberts (1970)
Papio	Baboon	Omnivorous; fruits, grasses, roots, buds, lizards, insects occasional meat eating	Washburn and DeVore (1961); Dart (1963); Rowell (1966); DeVore and Hall (1965)
Colobus	Guerezas	Leaves	Booth (1956, 1957)
Cercopithecus	Guenons	Cattails, grasshoppers, mantis, ants, termites, grasses, herbs, young birds, vines, bird's eggs	Struhsaker (1967)
Erythrocebus	Patas	Grasses, fruit, beans, seeds, insects, lizards, bird's eggs	Napier and Napier (1967); Hall (1965)
Theropithecus	Geladas	Grasses, insects, bulb	Napier and Napier (1967); Crook (1968)
Cercocebus	Mangabeys	Fruit, leaves, buds, insects, shoots, dead bark	Chalmers (1968); Napier and Napier (1967)
Mandrillus	Drills, mandrills	Omnivorous	Napier and Napier (1967)

(Continued)

TABLE I (Continued)

Genera	Common name	Diet	Source
Lesser apes—Asia			
Hylobates	Gibbons	Fruit, leaves, buds, insects, young birds, bird's eggs	Carpenter (1940)
Symphalangus	Siamangs	Fruit, leaves	McClure (1964)
Great Apes—Asia			
Pongo	Orang-Utan	Fruit, leaves, bark, bird's eggs, shoots, pulp, seeds	Schaller (1961)
Great Apes—Africa			
Pan	Chimpanzee	Fruits, leaves, palm-nuts, bark, seeds, stems, termites, ants, meat, fish	Reynolds and Reynolds (1965); Goodall (1963, 1965); Jones and Cave (1960); Kortlandt (1962); Stanley (1919); Reynolds (1967)
Gorilla	Gorilla	Pith, stalk, leaves, bamboo shoots and roots, vines, fruit, sugar cane	Schaller (1961, 1963); Bingham (1932); Donisthorpe (1958); Merfield and Miller (1956); Kawai and Mizuhara (1959)

III. AN EXPERIMENTAL PROGRAM IN PROTEIN MALNUTRITION

The traditional method for studying the effects of malnutrition on behavioral development has been to deprive an organism during development (or discover an organism that has been so deprived), rehabilitate the organism, and then compare its behavior with that of a control group that has not suffered from such deprivation. This is a tried and true experimental technique that has been employed in the area of sensory deprivation, social deprivation, and, for that matter, all manner and forms of sensory and environmental restrictions. This is, in effect, a simple comparison of groups after treatment. Rarely, however, are changes in behavior evaluated or assessed during the deprivation period. It appears as if the experimenters are concerned that measurement during this interval might in some way affect the outcome. Behavioral differences that appear at the end of the deprivation period are attributed to something that did not develop during that manipulation. Thus, animals deprived of early social experience exhibit abnormal or "neurotic" forms of behavior (Mason, 1968). The source and ontogeny of these abnormal behaviors is rarely identified and, in the end, the experimenter is left to speculate as to the nature of the change that has occurred.

The principle directing the present research program is to evaluate behavior changes during the period of protein deprivation and to reexamine these behaviors after rehabilitation. Inherent in the philosophy of this program is the notion that as behavior is altered during the dedevelopmental period of the animal, adult or maturing behavior will also be altered. Thus, it is important to evaluate how a baby monkey is changed by protein malnutrition during the developmental period in order to more fully understand alterations in adult behavior following rehabilitation.

A. Experimental Groups and Treatments

Five groups of infant rhesus monkeys were separated from their mothers at 90 days of age, housed in individual cages, and maintained according to procedures described by Blomquist and Harlow (1961). Briefly, these procedures consisted of providing the animal with a milk formula on a 2–4 hour schedule, depending on the age, weight, and strength of the infant. Bottle holders on a ramp, shown in Fig. 1, were placed in the cages in which the infants were housed. This allowed access to the nursing bottles during the feeding periods. Procedures identical to those described by Zimmermann (1969) for the introduction and weaning to solid food were employed. By 120 days of age, all of the subjects had been weaned to the experimental diets. The diets were isocaloric but contained either 25, 2, or 3.5% protein by weight. Casein provided the sole source of dietary protein in each of the nutritional preparations. The composition of the experimental diets is presented in Table II. Groups 1 ($N = 4$), 2 ($N = 6$), and 4 ($N = 5$) were introduced to the 3.5% (low) protein diet at approximately 380, 210, and 120 days of age, respectively. However, in order to maintain stable body weights, subjects in groups 1, 2, and 4, were shifted to the 2% (low) protein diet at 1,036, 728, and 500 days of age, respectively. Two age control groups, group 3 ($N = 4$) and group 5 ($N = 4$) for subjects in groups 2 and 4, were maintained on the 25% (high) protein diet throughout the experiment. Subjects in group 1 served as pilot experimental animals that had been maintained on Purina monkey chow and the vitamin sandwiches described by Sidowski and Lockard (1966) prior to the introduction of the restricted diet. Although there were no dietary controls for group 1, 4 male age

TABLE II: COMPOSITION OF EXPERIMENTAL DIETS

Components	Low protein (2% protein)	Low protein (3.5% protein)	High protein (25% protein)
Primex [a]	9.0	9.0	9.0
Fat-soluble vitamin [b]	1.0	1.0	1.0
Crude casein	2.0	3.5	25.0
Cerelose [c]	39.7	39.0	29.0
Dextrin [d]	39.9	39.2	27.7
Salts (HMW) [e]	4.0	4.0	4.0
B vitamin premix [f]	2.0	2.0	2.0
Choline dihydrogen citrate	0.3	0.3	0.3
Ascorbic acid	0.03	0.03	0.03
Alphacel [g]	2.0	2.0	2.0
Total	99.93	100.03	100.03

[a] Primex, Proctor and Gamble Co., Cincinnati, Ohio.

[b] Fat-soluble vitamins in corn oil: vitamin A acetate (0.31 mg); vitamin D (calciferol) (0.0045 mg); α — tocopherol (5.00 mg).

[c] Cerelose, Coin Products Co., Argo, Illinois.

[d] Dextrin (white tech.), Nutritional Biochemicals Corp., Cleveland, Ohio.

[e] Hubbell, R. B., Mendel, L. B., and Wakeman, A. J., Salt Mixture, Nutritional Biochemicals Corp., Cleveland, Ohio.

[f] B vitamin premix in 2 gm cerelose:

Thiamine HCP	0.40 mg	Inositol	20.00 mg
Riboflavin	0.80 mg	Biotin	0.02 mg
Pyridoxine HCP	0.40 mg	Folic acid	0.20 mg
Ca pantothenate	4.00 mg	Vitamin B^{12}	0.003 mg
Niacin	4.00 mg	Menadione	1.00 mg

[g] Alphacel (nonnutritive bulk), Nutritional Biochemicals Corp., Cleveland, Ohio.

TABLE III: GROUPS AND DIETARY TREATMENTS

Group No.	*N*	Animal No.	Sex	Days of age at onset of diet	% Protein	Days of age when shifted to 2% protein diet
Portland	4	3508	M	Age control	18	—
		3509	M	Age control	18	—
		4303	M	Age control	18	—
		4310	M	Age control	18	—
1	4	1	M	380	3.5	1036
		2	M	380	3.5	1036
		3	M	380	3.5	1036
		4	M	380	3.5	1036
2	6	5	F	210	3.5	728
		6	M	210	3.5	728
		5766	F	210	3.5	728
		5872	F	210	3.5	728
		5976	F	210	3.5	728
		5979	F	210	3.5	728
3	4	5754	M	210	25	
		5755	M	210	25	
		5756	F	210	25	
		5758	M	210	25	
4	5	8	M	120	3.5	500
		9	F	120	3.5	500
		10	F	120	3.5	500
		12	M	120	3.5	500
		13	M	120	3.5	500
5	4	15	F	120	25	—
		16	F	120	25	—
		17	F	120	25	—
		18	M	120	25	—

controls were purchased from the Oregon Primate Center and maintained on Purina monkey chow throughout the experimental period.

The groups, number of animals, and dietary treatments are summarized in Table III. For convenience, each group is referred to by the mean age of the group at the onset of the dietary treatments. Thus, there are the 380-day low-protein group, the 210-day high- and low-protein groups, and the 120-day high- and low-protein groups. The age controls for the 380-day low-protein group are designated as the Portland group. Table IV shows the age of each of the dietary groups at the onset of the various experimental treatments.

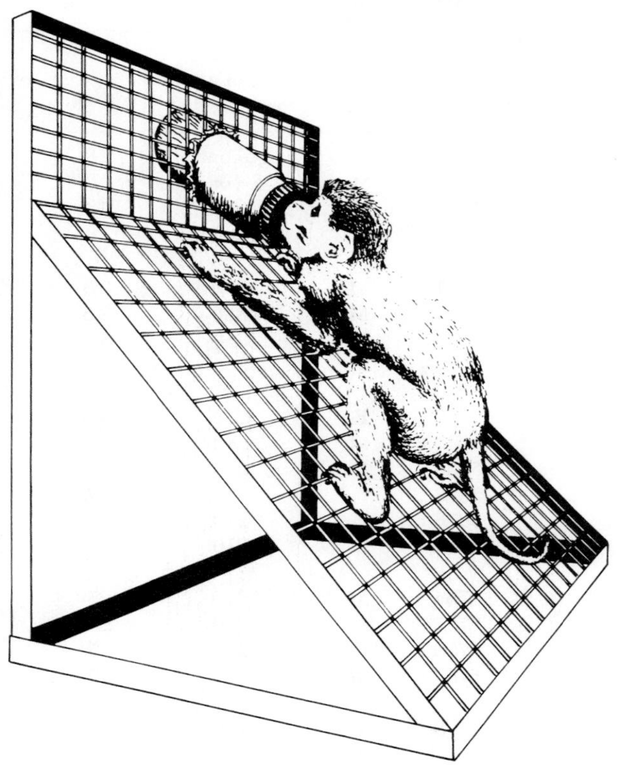

FIG. 1. Artist's impression of a bottle feeder with ramp.

B. Food Preparation

The diets were prepared according to the procedures outlined by Zimmermann and Geist (1972). Each of the ingredients were added to an 80-quart Hobart mixer and blended until a dry homogeneous powder was obtained. To differentiate each of the diets with respect to protein content, a sufficient amount of either blue or red food coloring was suspended in the fat-soluble vitamin mixture prior to blending. This procedure imparted various hues to each of the diets. Blue denoted 25% protein, yellow (the natural color of the diet) indicated 3.5% protein, and red represented 2% protein. The diets were then refrigerated until needed. From the dry dietary preparations, biscuits weighing from 150 to 200 gm were formed by adding sufficient water to the powder to make a dough. The dough was then rolled out on a cutting board and individual biscuits cut with a circular cookie cutter approximately 5 cm in diameter. Each of the animals were fed *ad libitum* quantities of their respective diets except on the occasions when high-protein subjects had to be deprived for testing purposes.

C. Housing and General Maintenance

The monkeys were housed in individual cages measuring 76.2 × 76.2 × 76.2 cm. All of the subjects were trained to enter transport cages so that handling and the trauma associated with capture was minimized. The only time the animals were handled was when they refused to enter the transport cage voluntarily and when they were restrained for taking blood samples or administering medication. The animals were tested on a wide variety of tasks so that it was difficult to assess the effects of one task from the other. However, our experience has been that the behavioral patterns of monkeys are rather situation specific. Thus, even after receiving aversive stimulation in one test situation the subjects did not fail to enter the transport cage to go to other tests.

D. Weight Gain and Blood Analysis

Every morning all of the animals were weighed before daily testing commenced. As was found in earlier studies (Ordy *et al.*, 1966), the protein-deficient diets have marked effects on the rate of weight gain in the developing rhesus monkey. Mean weights for the 380-day low-protein group and the 210-day high- and low-protein groups are illustrated in Fig. 2. The 380-day low-protein animals gained only 15% of their weight over an 800-day period. Under normal conditions these animals would have tripled in weight. The 210-day low-protein groups gained only 25% of their weight over this period of time, whereas the 210-day high-protein control group made nearly a threefold increase in body weight over the 800-day period. The mean daily weights in 10-day blocks for the 120-day high- and low-protein groups over a period of 200 days is shown in Fig. 3. During this period of time the 120-day high-protein subjects increased 4 times the weight gained by the 120-day low-protein animals. At 1070 days of age the high-protein monkeys were placed on food deprivation by limiting their nutritional intake. There was no weight loss as a result of the deprivation schedule.

Electrophoretic analysis of blood serum protein, albumin, globulin, and albumin-to-globulin ratio were determined at 12-month intervals over

TABLE IV

MEAN DAYS OF AGE AT START OF EXPERIMENTAL TREATMENTS

Experimental treatment	Portland	380–Low protein	210–Low protein	210–High protein	120–Low protein	120–High protein
Weight gain	640	310	130	180	70	92
Blood analysis	—	1090	790	840	—	—
Food consumption	—	1300	770	750	490	—
Food preference	—	—	1330	1180	—	—
Activity						
Home cage	—	380	130	180	130	—
Wheel	—	1200	980	1040	750	350
Visual curiosity	—	340	120	180	120	—
Investigative behavior–home cage manipulation						
Chains	—	500	190	240	130	120
Objects	—	530	220	260	160	130
Chains	—	560	260	320	220	160
Puzzle box						
2-Solution	—	530	220	270	100	130
3-Solution	—	720	410	460	270	450

100% Reinforcement	—	860	550	600	—	—
Extinction	—	870	560	610	—	—
Partial reinforcement	—	980	670	730	—	—
Extinction	—	1120	810	860	—	—
Partial reinforcement (cues)	—	1390	1080	1130	—	—
Learning						
Learning set retention	—	—	210	260	—	—
Delayed response	—	—	230	280	—	—
Transposition	—	—	380	580	—	—
Size–brightness discrimination	—	—	670	720	—	—
Cue–response separation	—	—	1600	1650	—	—
Response to novelty						
Neophobia	—	—	210	260	—	—
Vertical shuttle						
Adaptation	—	—	610	670	—	—
Food shuttle	—	—	660	710	—	—
Objects	—	—	820	870	—	—
Social behavior						
3 Categories	—	—	210	270	—	—
5 Categories (Old)	—	—	310	360	—	—
5 Categories (New)	—	—	590	640	—	—
Dominance						
Food competition	980	1140	830	880	600	200
Social dominance	990	1150	840	890	610	210

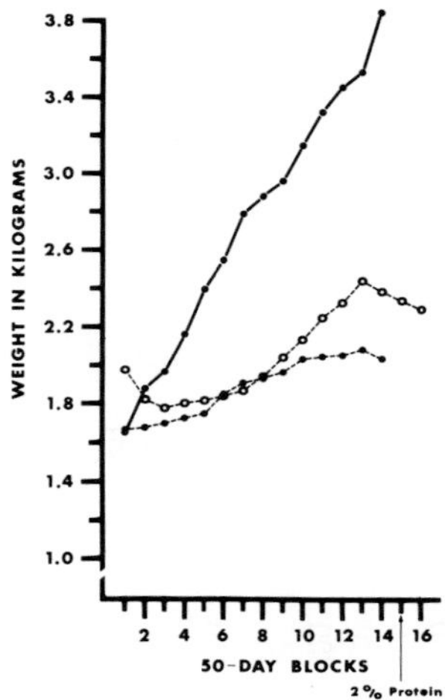

FIG. 2. Weight gain from start of diets (older monkey groups). o---o, 380-Day low-protein; ●---●, 210-day low-protein; ●—●, 210-day high-protein.

a period of 2 years. Thirty days prior to the second analysis the 380-, 210-, and 120-day low-protein groups were shifted from 3.5 to 2% protein diets. The mean electrophoretic values and standard errors for each group obtained in both years is presented in Table V. An analysis of variance

TABLE V

MEAN ELECTROPHORETIC VALUES OF BLOOD SERUM COMPONENTS

Diet (% protein)	Days of age at onset of diet	No. of subjects	Total protein (gm/100 ml + SE)	Albumin (gm/100 ml + SE)	Globulin (gm/100 ml + SE)
			Analysis 1		
3.5	380	4	6.5 + 0.5	4.0 + 0.4	2.4 + 0.5
3.5	210	6	6.6 + 0.3	3.9 + 0.3	2.6 + 0.3
25	210	4	7.4 + 0.1	4.8 + 0.2	2.6 + 0.2
			Analysis 2		
2	120	4	6.2 + 0.4	3.4 + 0.3	2.8 + 0.5
2	380	4	6.2 + 0.3	3.7 + 0.4	2.5 + 0.4
2	210	6	6.2 + 0.3	3.9 + 0.4	2.3 + 0.5
25	210	4	7.1 + 0.4	5.0 + 0.2	2.0 + 0.4

showed that the animals treated with the low-protein diets had blood serum levels of albumin and total protein that were significantly lower than the values obtained from the subjects treated with the high-protein

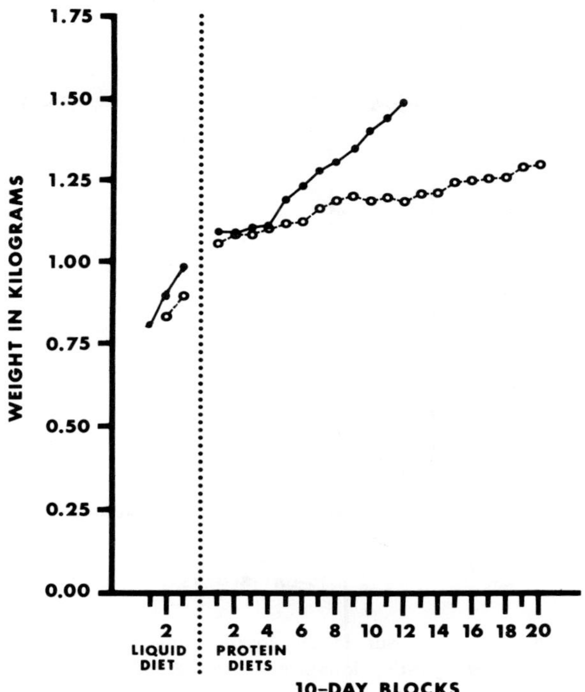

FIG. 3. Weight gain before and after start of diets (younger monkey groups). ○---○, 120-Day low-protein; ●—● 120-day high-protein.

diets. These data are consistent with those reported by Ordy *et al.* (1966) and are presented in greater detail in Geist, *et al.* (1972).

E. Food Intake

Animals equated with respect to age and size showed little difference in food consumption across dietary regimes. As a whole, monkeys on diets deficient in protein consumed less food than either those fed standard or elevated protein concentrations. In terms of absolute quantities of food consumed within the protein-deficient groups, there appeared to be an order by age effect in which the older animals consumed greater amounts of food. Although the high-protein animals ate more food than the low-protein animals, when the ratio of food consumed to grams body weight was calculated, no group differences appeared. Figure 4 shows the amount of food consumed per grams body weight. There is a considerable overlap among groups indicating that monkeys, independent of protein concentration contained in the diet, consumed a quantity of food that was proportional to their body weights.

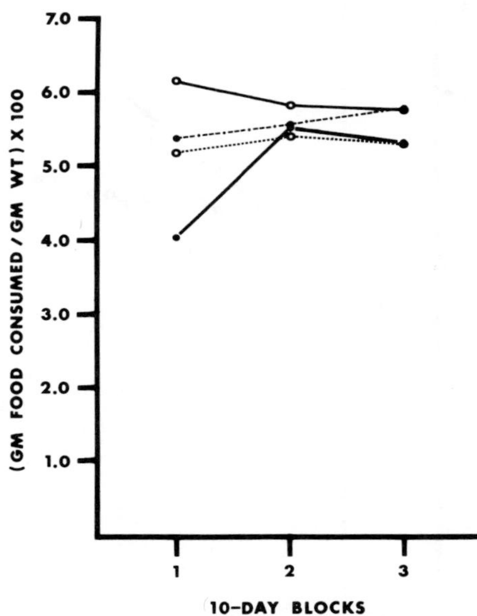

FIG. 4. Amount of food consumed per grams body weight. O····O, 380-Day low-protein; ●——·——·●, 210-day low-protein; ●—●, 210-day high-protein; O—O, low-protein.

F. Food Preferences

Although the malnourished monkeys controlled their intake of the low-protein diet to a point that was equal to the intake of animals of equal weight, their behavior in test situations involving food reward indicated that they were highly motivated for food as a source of reinforcement. In these tasks, a sugar-coated cereal was used as the reinforcement. There was the possibility that these animals might have been sensitized to prefer food that contained increasing amounts of protein. A review of the literature, however, would indicate that there is little evidence for specific protein hunger in mammals. In fact, a recent study by Hillman and Riopelle (1971) indicated that there was no evidence for a protein preference in malnourished primates. However, none of the studies report the degee of depletion of the animals. Thus, it is difficult to assess whether or not the problem of specific protein preferences do result from long-term protein deprivation.

To investigate the possibility of specific preference for high-protein foods, the 210-day high- and low-protein monkeys were given a preference test at 1180 and 1330 days of age, respectively, in which they could select bits of food that contained different quantities of protein. Standard protein diets that were color coded according to laboratory procedures were employed.

The animals were placed in the apparatus shown in Fig. 5. The preference wall consisted of a series of 7.62 × 7.62 cm compartments with clear plastic fronts that could be opened and the item secured. The animal was separated from the compartments by an opaque door. When the door was raised, the animals were free to open all of the compartments. A trial lasted 1 minute and the score was the number of items removed. The eight compartments contained 25, 3.5, and 2% protein food in 1.5 gm pieces or junk toys from our object quality series. Nine trials were run each day, which allowed the animals the opportunity to secure up to eighteen reinforcements of each of the foods and eighteen different

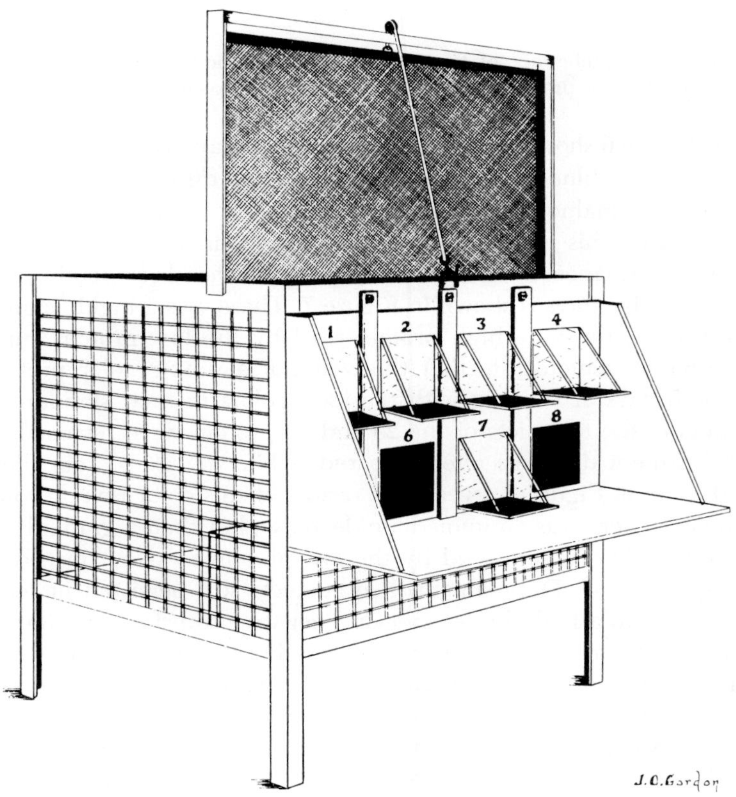

Fig. 5. Artist's impression of preference wall apparatus.

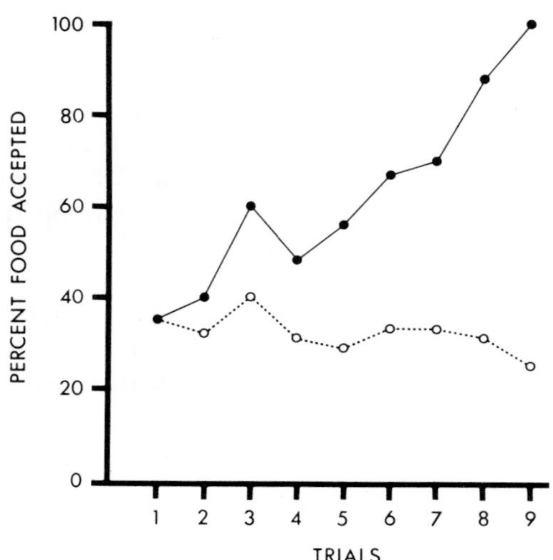

FIG. 6. Food preference of the 210-day high- and low-protein monkeys on the first day. ●—●, Low-protein group; O– – – –O, high-protein group.

objects. Figure 6 shows the responses of the 210-day high- and low-protein subjects to the blue-colored high-protein food during the first day of training. The malnourished monkeys exhibited a rapid acquisition of preferences for this food, whereas the high-protein subjects responded at chance performance levels. The responses to the three foods by both groups over five days is shown in Figure 7. There was a highly significant preference for the blue high protein food by the low protein animals.

In order to demonstrate that this was not a simple color preference, after the 5 days of training with the food colors in their standard form, the colors of the test diets of the 25 and 2% protein were reversed. Thus, the 25% protein diet was altered to red, while the 2% protein diet was altered to blue. Figure 8 reveals the results of this color reversal on food preference. There was an immediate decrease, but then a gradual shift to the 25% red high-protein food by the low-protein monkeys in this phase of the experiment. The malnourished subjects still persisted in selecting the blue diet to a high degree showing negative transfer of training. But it should be noted that the animals eventually shifted in preference to the red diet in the color reversal situation even though they were receiving red low-protein food in the home cage. Thus, at least in our animals that have had extensive periods of protein deprivation, there is a clearly expressed and rapidly learned preference for high-protein food when they have an opportunity to express this preference.

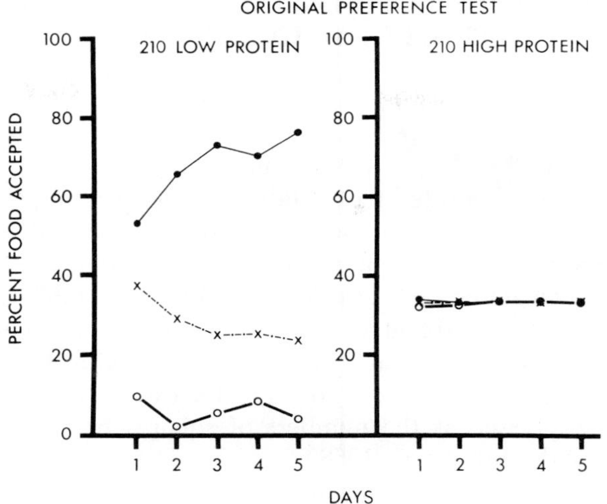

FIG. 7. Protein preference over original 5 days. ●—●, 25% Protein; ×—·—·—×, 3.5% protein; o—o, 2% protein.

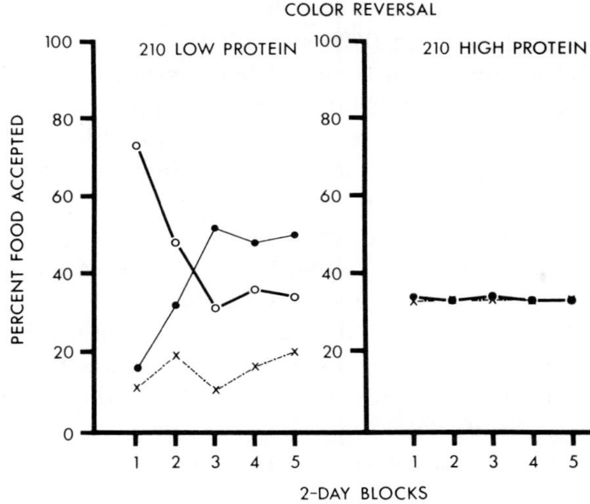

FIG. 8. Protein preference following reversal of color. ●—●, 25% Protein; ×—·—·—×, 3.5% protein; o—o, 2% protein.

IV. CHANGES IN BEHAVIOR RESULTING FROM PROTEIN MALNUTRITION

A. Activity

Nutritional variables may have an effect on a variety of motivational systems. It has been reported that children suffering from protein–calorie malnutrition were lethargic and inactive (Cravioto *et al.*, 1966). To the contrary, rats suffering from protein deficiency showed an increase in activity as measured in an activity wheel (Collier and Squibb, 1967; Collier *et al.*, 1965). Further, early protein–calorie malnutrition was found to affect vertical components of exploratory activity, expressed as the number and duration of standing movements, to a greater extent than the horizontal components, expressed as the numbers of squares traversed or entered in an open field (Frankova, 1968).

In order to investigate whether or not spontaneous activity of the developing monkey was affected by the dietary variable, the animals were subjected to two different tests of activity. In the first measure, activity was determined by measuring the general movement within a cage. Movement was detected by two photocells each mounted 38.1 cm from the floor of a cage that was 76.2 cm on a side. The breaking of the beam activated electromechanical counters. The subjects were placed singly in the cage for 1 hour per day approximately every third day, which allowed time to test other animals in the colony. In order to balance out daily cyclic factors, the time of day the animals were placed in the apparatus was varied between 0600 and 1700 hours. Figure 9 shows the results of this measure taken from 125-hour sessions over a period of 2 years for the 380-, 210-, and 120-day low-protein groups and the 210-day high-protein group. The groups were highly variable in their activity and there was no consistent trend that would indicate that there was an effect on this measure of activity.

A second measure of activity was taken by placing an animal in a running wheel which was very similar to the typical rodent activity wheel but was 91.44 cm in diameter. One revolution of the wheel in either direction activated a counter and the number of revolutions in each 1-hour session was recorded. The animals were tested on a schedule similar to that described in the cage activity measure. Again there were no differences as a function of diet.

We have tested animals under a variety of stimulus conditions with these two pieces of equipment, such as empty rooms with minimal lighting, colony rooms that were brightly illuminated, and hallways in which there was considerable human traffic. No consistent or significant bifurca-

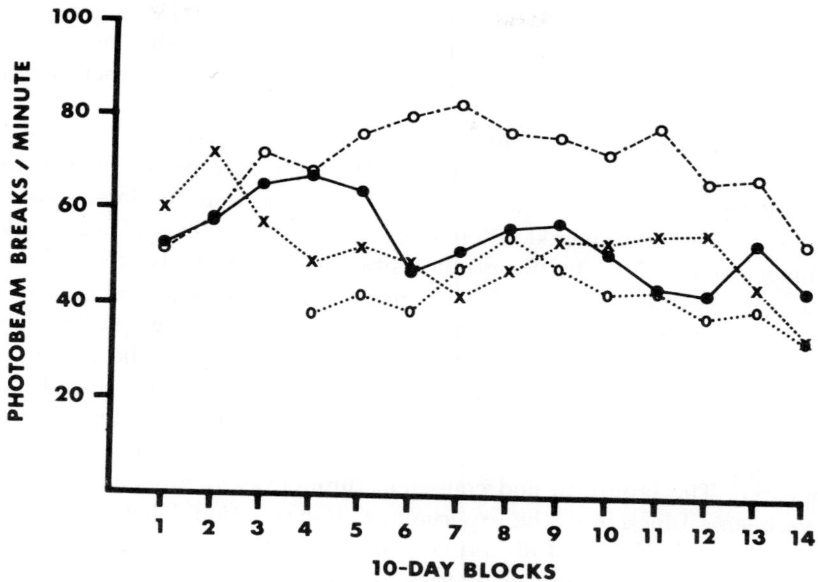

FIG. 9. Photometric measurement of activity. O — — · — — O, 380-Day low protein, × · · · · ×, 210-day low protein; ● — ●, 210-day high protein; O · · · · O, 120-day low protein.

tion of groups appeared in any of the tests. The protein-deficient monkeys continued to be as active in these situations as their high-protein controls.

B. INVESTIGATORY BEHAVIOR

Decreased investigative behavior has been identified as one of the consistent features of protein–calorie malnutrition. Children suffering from kwashiorkor or marasmus are often described as appearing apathetic, inert, lethargic, and withdrawn (Jelliffe and Welbourn, 1963). Alteration of exploratory and investigative behavior in subhuman animals have also been reported. Frankova and Barnes (1968a) found low exploratory behavior and long periods of inactivity in protein-malnourished rats. When exploratory behavior was measured by the frequency of vertical standing movements, prematurely weaned low-protein rats exhibited significantly lower values than those of normal animals (Frankova, 1968). The decrement in this type of activity could not be attributed to a general decrease in activity since the groups of animals did not differ in frequency of horizontal movements in an open field. Kerr and Waisman (1968) reported that protein-malnourished infant monkeys have apathetic behavior patterns and retarded social development. In order to test the effects of protein–calorie malnutrition on investigative behavior, groups of monkeys in our colony were subjected to a variety of experimental tasks.

One of the natural forms of investigative behavior sensitive to nutritional manipulation is visual curiosity. This is a very powerful form of motivation in the young monkey and has been used to detect visual preferences in infant monkeys separated from their mothers (Harlow and Zimmermann, 1959). In the present situation the animals were placed in a visual curiosity chamber for 1-hour sessions. Over a period of 1 year the experiments indicated that the animals raised on low protein had initially lower rates of visual exploration, especially when the responses required were on long fixed-ratio schedules such as FR6 and FR16 (Zimmermann and Strobel, 1969). However, with repeated testing and when appropriate controls for age were introduced, the differences disappeared. The failure to find consistent differences in this measure may have been a function of the experimental test situation. That is, the animals were only deprived of normal visual stimulation during the 1-hour test periods. Long periods of deprivation and a more systematic analysis of the variables that affect this measure may have yielded different results.

In order to assess investigative behavior in a free operant situation, home cage manipulatory behavior was measured. In the first experiment (Zimmermann and Strobel, 1969), manipulation was measured by hanging three long chains attached to mechanical counters in the living cages of the animals in the 380-day low-protein group. The apparatus was placed on the cage for 5 consecutive days immediately after the animals were introduced to the diet and then again 90 days later. There was a marked reduction in chain pulling behavior over the 3-month period. In the second study (Strobel and Zimmermann, 1971), the same apparatus was used in a series of experiments designed to test the before and after effects of the diet on manipulation and the effects of the introduction of novel stimuli on the chain pulling. The 210-day high- and low-protein groups were exposed to the chain manipulation apparatus for 10 days, 2 weeks before and 1 week immediately after the onset of the diets. The 120-day high- and low-protein groups first received the chains shortly after being introduced to the diets. Following this initial or base-line experience with the chains, new or novel objects were presented to all of the groups. Three aluminum cookie cutters in the shapes of animals were attached to the bottoms of the chains and suspended 7.62 cm from the tops of the cages. Following 10 days of exposure to the objects the chains were reinstated for an additional 20 consecutive days.

Figure 10 shows the results of the before and after study with the 210-

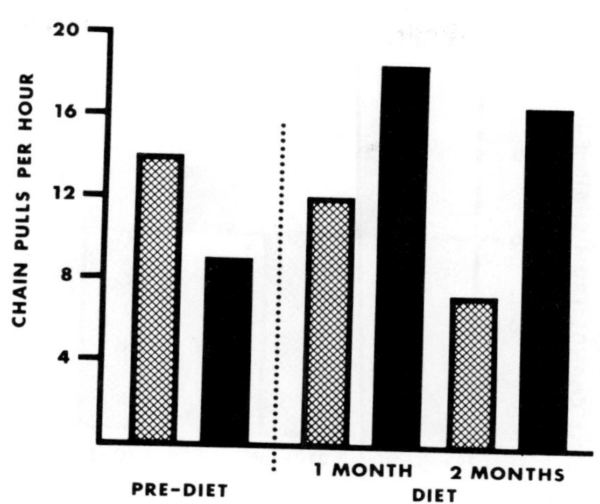

FIG. 10. Chain pulling before and after the introduction of experimental diets. The low-protein diet is represented by the hatched bar; the high-protein diet is represented by the solid bar.

day low-protein groups. The low-protein group had a higher rate of chain pulling (not statistically significant) than did the high-protein group prior to the onset of the diet. With the introduction of the diets, the 210-day low-protein animals showed a decrease in response, whereas the control subjects showed a slight increase.

The responses of the four groups, the 210- and 120-day high- and low-protein groups, to the introduction of the objects is shown in Fig. 11. Both control groups increased their rates of chain pulling over the chain only situation, whereas the two low-protein groups reduced their rates of responses. With the reinstatement of the chain only condition, the 120-day low-protein group recovered in rate of responses close to their baseline level, but the 210-day low-protein group showed only partial recovery with the chains. In contrast, the high-protein groups maintained rates that were above their original levels.

The decrement in chain manipulation (pulling) by the protein-deprived animals occurred early in the deprivation period. This period of behavioral change, the first 10 days after being introduced to the diet, correlated positively with the most rapid fall in levels of total serum protein and albumin in these animals (Geist *et al.*, 1972).

The reduction in responsiveness to the objects that were placed on the

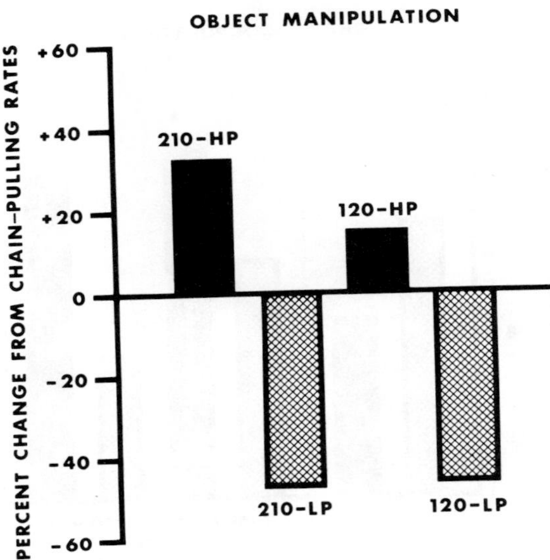

FIG. 11. Chain-pulling response to the introduction of objects. HP, high protein; LP, low protein.

chains cannot be attributed merely to a decrease in manipulatory activity, since there was a temporary increase in the response rate to the chains after the objects had been removed. Rather, it would appear that the malnourished animals had a decreased responsiveness, whereas the controls had an increased responsiveness to the novel objects. The problem of fear of new stimuli is discussed at length in Section IV, D. Although no quantitative data were available, it was observed that the low-protein monkeys showed distinct avoidance and fear reactions toward the objects during the first day of this condition. Two monkeys in the 210-day low-protein group refused to enter their cages after the introduction of these stimuli. It is apparent from this experiment that with repeated testing and with continued protein deprivation, the experimental animals showed a decrease in their investigatory and manipulatory behavior in a fashion similar to that observed in the visual curiosity task.

Harlow and co-workers (1950) first demonstrated that normal rhesus monkeys would readily manipulate and solve mechanical puzzles in the absence of any apparent intrinsic reward. Subsequent studies (Harlow, 1950; Harlow *et al.*, 1956) indicated that this was a very persistent form of behavior and that it appeared early in the life of the developing rhesus monkey. It seemed logical, therefore, to study the effects of protein-calorie malnutrition on this form of manipulatory response in the developing monkey.

Several experiments were conducted and some of them have been detailed elsewhere (Strobel and Zimmermann, 1971). The monkeys were placed in a cage containing twelve individual puzzle units that were mounted on a Masonite panel which replaced a wall of a 69.96 × 45.72 cm cage. The apparatus is shown in Fig. 12. Each puzzle unit consisted of a screen door hook and hasp in the two-part puzzle, and in the three-part puzzle a pin was added. The unit pieces were in combination so that the pin had to be removed before the hook could be released, which, in turn, permitted the hasp to be moved. Preliminary studies with the two-part puzzle indicated that 210-day low-protein monkeys showed less puzzle-solving behavior in the absence of reward than the high-protein controls. A replication of the preliminary study with the three-part puzzle produced the same trend but the differences were not significant (Strobel and Zimmermann, 1971).

The sequence of testing the monkeys on the two-part puzzle for 125 days and then the three-part puzzle was replicated with the 120-day groups. In this study there were no differences in the two-part puzzle as both groups showed lower rates of manipulation compared to the older animals. But on the three-part puzzle, which was initiated when the 120-day groups were 400 days of age, the high-protein controls showed significantly more manipulatory responsiveness than the 120-day or 210-day low-protein groups.

In the next phase of the puzzle-solving experiments, food, in the form of a sugar-coated cereal, was placed behind each of the hasps in order to reinforce the animals for solving the puzzles. Under these conditions all groups of animals immediately jumped to perfect performance. After 10 days of testing on this condition the food was removed from behind the hasps and extinction was introduced. The groups did not differ in their rates of extinction and all returned to approximately prereinforcement levels.

The animals were then put on a partial reinforcement schedule in which only one of the puzzle units was reinforced with food. A different puzzle unit was reinforced each day in a random order. Under these conditions, the low-protein subjects again jumped to perfect performance in solutions, whereas the high-protein animals came near perfection but failed to solve all of the puzzles after obtaining the reinforcement. After 70 days of partial reinforcement the food reward was no longer presented. The performance curves under both the partial reinforcement and extinction schedules is shown in Fig. 13 for the 380-day low-protein group and the 210-day high- and low-protein groups. Both low-protein groups showed more rapid extinction than the high-protein group. It was assumed that the low-protein monkey would be more highly motivated for food reward and, therefore, show greater resistance to extinction.

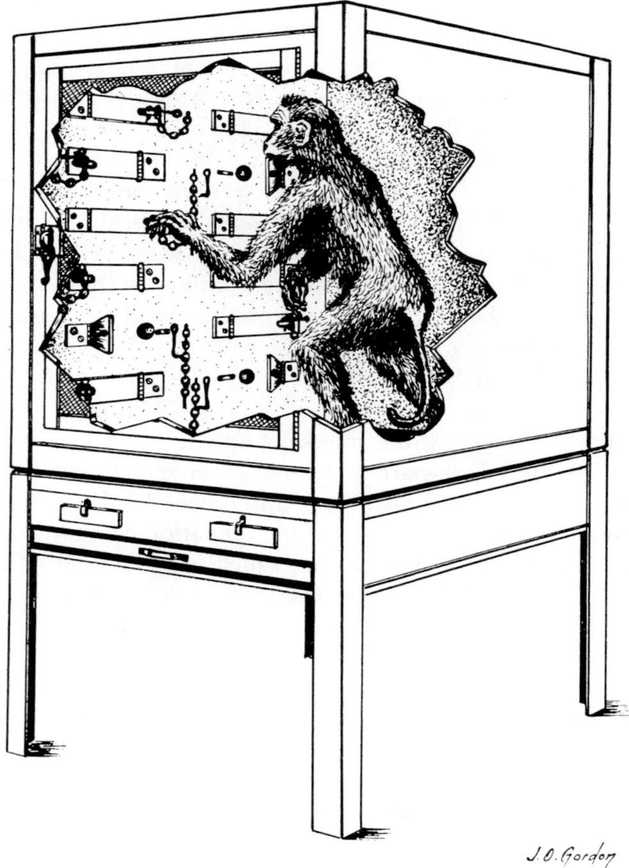

Fig. 12. Artist's impression of the puzzle box apparatus.

However, these results appear to indicate that the low-protein monkey is more sensitive to the presence and absence of reward. The high-protein monkeys, on the other hand, returned to the modestly high levels of manipulation that characterized their pre-food-reward performances. This would suggest that these animals were responding not only to the food reward contingency, but also for intrinsic manipulatory reinforcement.

To investigate the hypothesis that protein-malnourished monkeys are more sensitive to the food reward contingency, reinforcement was again introduced, but now the location of the reinforcement was cued by the presence of an orange washer on the puzzle containing the reinforcement.

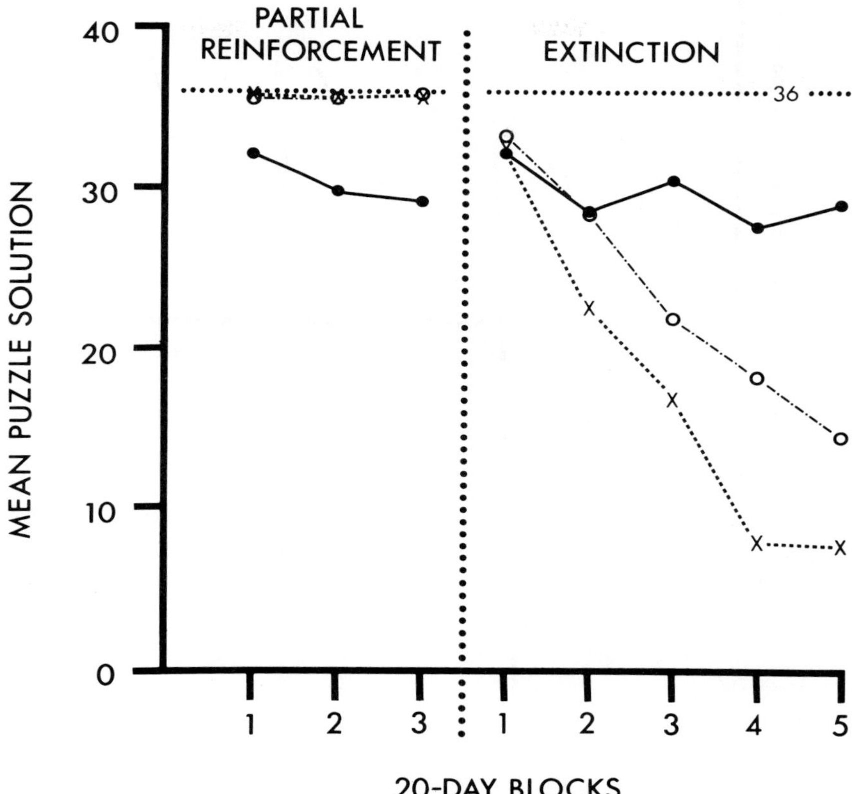

FIG. 13. Puzzle solutions during partial reinforcement and extinction. o—.—.—o, 210-Day low protein; ●—●, 210-day high protein; × · · · · ×, 14-month low protein.

It was anticipated that the low-protein animals would be more discriminating and would soon concentarte most of their attention and behavior to the cued puzzle. In the preliminary trials of this condition the low-protein subjects continued to open all of the puzzles but usually left the cued puzzle until last. Opening of the reinforced puzzle late in the 1-hour test period was contrary to our prediction. However, if malnourished monkeys have considerable fear of new or novel stimuli in their environment (refer to the section on neophobia) the subsequent data appear to make some sense. With repeated testing, the overall manipulation of the low-protein subjects decreased more rapidly than the high-protein monkeys, and the

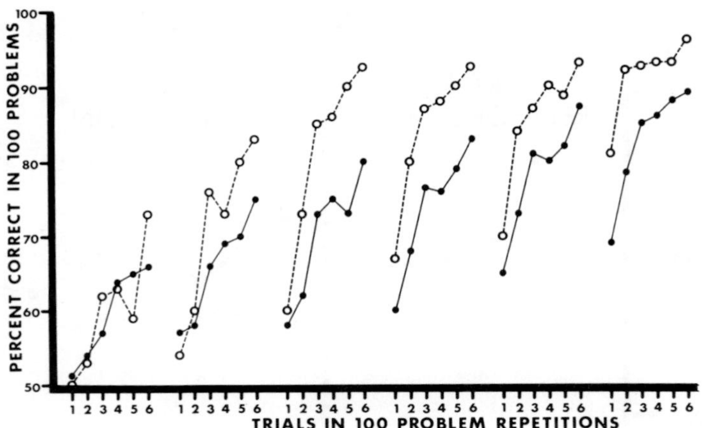

FIG. 14. Learning set retention with 100 problem repetitions. ●—●, 6-month-old normal; ○ · · · · ○, 6-month-old deprived.

malnourished animals soon began to solve the cued puzzle earlier in the test period. The high-protein subjects continued to manipulate all puzzles ostensibly in a manner similar to their performance on partial reinforcement.

C. Learning

As mentioned in the beginning of this chapter, the major interest in the area of nutrition and behavior has been the relationship between an early history of malnutrition and the development of learning. It was assumed that since there were reported deficiencies in IQ tests and other tasks in humans as a result of early malnutrition, the first place to look for differences in animal behavior was in learning studies. To date this search has resulted in contradictory evidence. When learning differences have been found the authors also reported differences in emotion or motivation of the animal. For the most part, these additional learning variables account for the basic differences in learning capacity when normal and previously malnourished animals are compared. The monkeys in our laboratory have been tested on a wide variety of learning tasks involving food reward and on a small number of tasks involving aversive stimulation. In general, it can be said that in most of the learning situations involving food rewards, the malnourished monkeys are superior to the high-protein controls, whereas, in aversive conditions the low-protein animals are often inferior.

Zimmermann (1969) found that malnourished monkeys with a history of undernutrition were superior on a learning set task that involves the memory of 100 pairs of object discrimination stimuli (Fig. 14). Three months after rehabilitation the malnourished group was no longer superior to the control animals that had been maintained on standard laboratory monkey chow. The experiment was replicated in our laboratory using a task requiring the memorization of fifty pairs of stimuli presented in a learning set paradigm of six trials per problem. The results of both experiments were almost identical, except that the high-protein animals in the second study showed no signs of significant learning or memory until a deprivation schedule was introduced. In the latter case the control animals were maintained on our high-protein diet. In subsequent studies we have found that it is absolutely necessary to deprive the high-protein animals if they are to perform in any of the learning tasks.

Under certain circumstances the low-protein subjects do not operate well in any learning situation. These conditions are best described as emotional or fear-producing situations. In particular, a number of these animals have developed strong stereotyped behaviors such as autoeroticism or crouching behavior that is typical of the social isolate. It is very difficult to train these monkeys in any of the learning situations, and it is only with much patience or therapeutic conditioning of the animal that any effective learning behavior can take place.

In aversive conditioning situations, there is no general rule that one can apply to the differences between high- and low-protein animals. The low-protein subjects do not appear to learn at a slower rate. However, in a preliminary study on threshold measurements, it was found that the low-protein animals were slightly more sensitive to the shock level. It took fewer milliamperes to produce a measurable response in the low-protein animals than in the controls. On several occasions, it was impossible to shape escape or avoidance responses in the low-protein subjects because of their stereotyped clutching or freezing behaviors in response to shock. There were also qualitative differences in learning between the high- and low-protein animals. The activity of low-protein subjects may be described as explosive and directed equally over the cage when the shock amperage reached a certain threshold level. By contrast, with a gradual increase in shock intensity, the behavior of the high-protein animals was usually an attempt to escape from the source or an attack response such as biting the bars. The diffuse manner in which the low-protein animals responded made the shaping of discrete responses, such as lever pressing, quite difficult.

That learning per se is not affected is evidenced by the fact that undesired responses that are persistent can be eliminated with a minimum amount of reinforcement (Stoffer *et al.*, 1973). Two subjects (No. 8, a low-protein S, and No. 4725, an age control) that persisted in autoeroticism were placed in a chamber and presented with response-contingent shock until they terminated the undesired response. In one animal only one shock was necessary to eliminate the habit completely throughout twenty test sessions, and in the second animal only two reinforcements were necessary. The results for both animals are shown in Fig. 15. It was

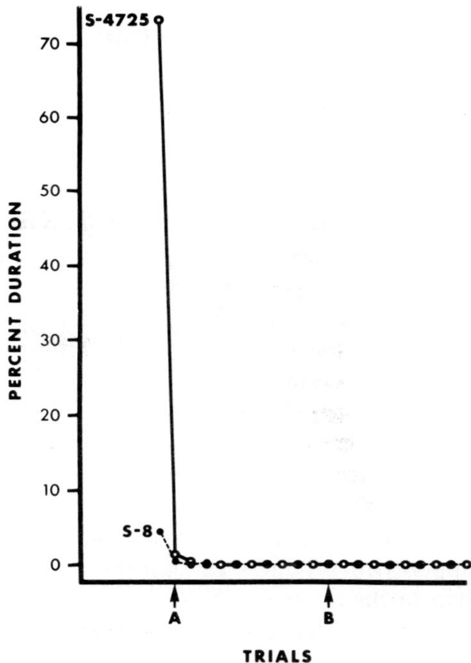

FIG. 15. Duration of autoeroticism following response-contingent shock in the shock apparatus.

found, however, that this avoidance response did not generalize to the home cage or to other test situations. Classical conditioning trials were subsequently presented by pairing the shock with a buzzer, independently of the unwanted response, while the animal was in the shock apparatus. After this procedure the buzzer was sounded when self-stimulation occurred in the home cage; Fig. 16 shows the results of this test. As can be seen there is a marked reduction of the response in the home cage after the conditioning procedure.

The 210-day high- and low-protein groups were tested on a delayed response test. The animals were allowed to view the placement of a reinforcement under one of the identical pair of stimulus objects. The subjects were prevented from responding to the objects for 0, 10, 20, 40, or 60 seconds. The low-protein animals were significantly superior to the controls on all delay intervals over a series of 400 trials. The tests were repeated 28 months later. The same conditions were employed as in the first test except that the high-protein monkeys were placed on food deprivation. Further, the animals were also tested with the stimulus objects exposed or hidden from view during the delay intervals. Under these conditions, the performance of the two diet groups were indistinguishable. The differences between the results of these two experiments may be attributed to the altered motivational levels of the high-protein animals.

In a series of studies on transposition and the effects of stimulus discriminability on rate of acquisition and transfer of the learned relational response, the performance of low-protein subjects was not inferior to the high-protein controls. This series of experiments was also run in a learning set paradigm with varying brightness or size discrimination problems presented for a fixed number of trials.

In a study of reversal learning, normal year-old animals were presented with ten discrimination learning and discrimination reversal problems, each problem being learned to a criterion. No differences were found between original and reversal learning. These results stand in contrast to an earlier study (Zimmermann, 1972) in which 180-day-old monkeys that had been on a low-protein diet for 90 days were significantly inferior to normal controls in learning a series of discrimination reversal problems. As in the test series described by Zimmermann (1972), the animals were presented with ten discrimination and reversal problems, all of which were learned to a criterion. The major differences between the two experiments were the different ages of the animals and the fact that the second experiment involved before and after measures of performance. However, another critical difference was discovered. In the second study the discrimination objects used were small toys and nonsense objects from the object-quality discrimination, learning set series that were constructed at the University of Montana. The objects employed by Zimmermann (1972) were constructed at Cornell University and, although similar to the Montana forms, were mounted on a 7.62 $\times$ 7.62 cm piece of Masonite which had been painted gray. Separating cue and response loci is known to affect the performance of monkeys on discrimination learning tasks

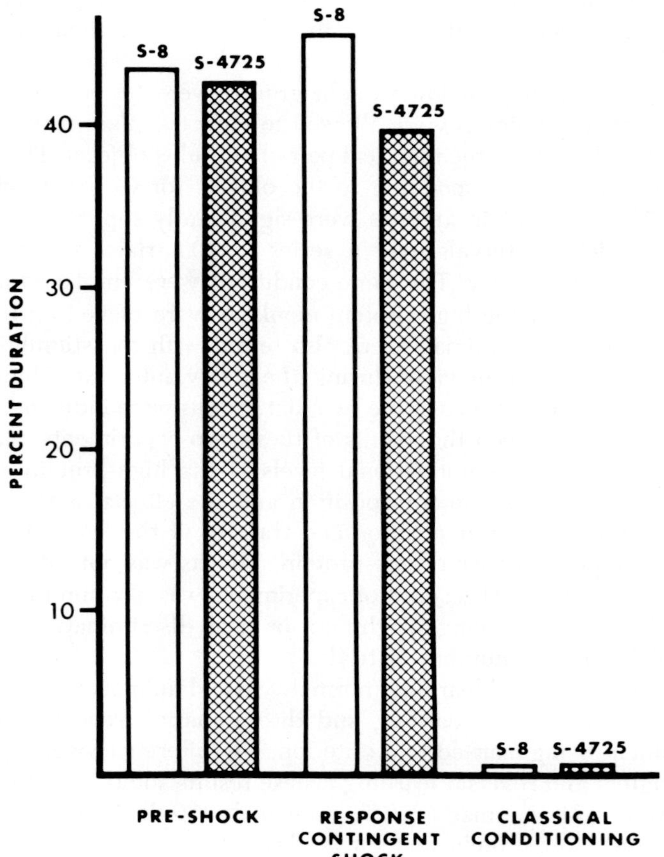

FIG. 16. Duration of autoeroticism in home cage following instrumental and classical conditioning.

(Stollnitz, 1965). If the Stollnitz analysis of discrimination learning is projected into this situation, the difference between the two sets of problems is that in the latter case the animals had to develop an observing response, attention response, or response of discrimination in order to solve the problems efficiently.

Klein *et al.* (1969) have found that children with a history of malnutrition do not differ on discrimination learning tasks but are inferior to normal controls on tasks requiring attention. These results would suggest that differences between discrimination reversal learning in the above studies conducted at Montana and Cornell were attributable to attentional deficits in malnourished monkeys. In order to test this hypothesis, stimuli were constructed to minimize or maximize the demands imposed on attentional processes in discrimination reversal problems. A sample of the stimuli are shown in Fig. 17. The stimuli were constructed to vary on two dimensions of difficulty. For one set of stimuli, a black or white discriminable cue occupied the perimeter of a 7.62 × 7.62 × 0.32 cm Masonite plaque. A second set of stimuli were similarly constructed, but the discriminable cue occupied the center portion of the plaque. In addition, the cues covered either 5, 12, 25, 50, 75, or 100% of the total area of the object. The remainder of the plaque was painted a neutral gray. Under these conditions, the low-protein subjects did not differ from controls in original learning of the different stimuli. However, the low-protein subjects were significantly inferior to the high-protein controls in reversal problems (Fig. 18).

An attempt was made to determine whether this same effect would occur in a reversal learning set paradigm by testing the animals for six trials on every combination of area and cue location. The black or white cue from each stimulus pair was randomly selected to be positively reinforced on each set of six trials. After 460 problems, there was no sign of learning in any of the groups and the experiment was discontinued. It is apparent that within the limits of the problems covered, there is an inability of the animals to overcome the problem difficulty. These findings are consistent with the results of similar studies reviewed by Stollnitz (1965) in which spatial variables were varied in a learning set paradigm. Stollnitz proposed that learning set training, in contrast to single problem learning, may be more difficult because of the repeated

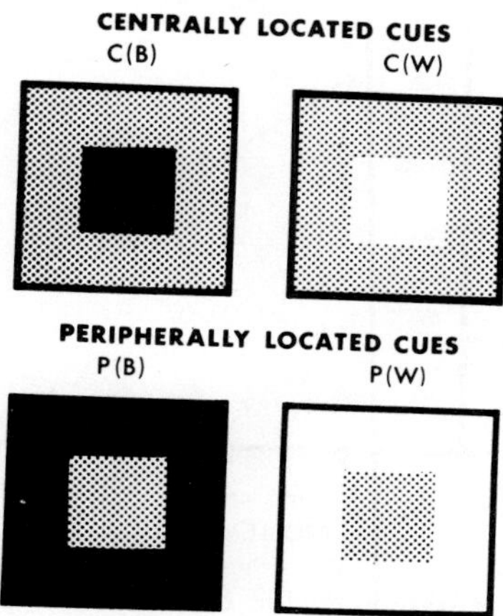

FIG. 17. Examples of centrally and peripherally located stimulus cue pairs.

extinction of within-problem observing responses to the correct cue.

In summary, it is evident that if attention mechanisms are involved in the type of problem discussed here, this mechanism is disturbed by the nutritional variable. Other experiments involving different aspects of the dimension of attention in animals will have to be measured before we can assess the role of attention per se in the performanc of these animals. However, it is important to point out that this is the first learning task in which we have found the malnourished monkey to be inferior to his normal peers. In all other tasks, the learning deficits, if they appear, are more clearly attributable to emotional or motivational changes in the malnourished animal.

D. Response to Novelty: Neophobia

Anyone who has spent any time with the developing rhesus monkey is impressed by its insatiable curiosity. Monkeys raised with adequate social and testing experience manifest a strong tendency to respond positively to new objects that are placed within their reach in familiar situations (Butler, 1965). As might be expected, there is usually an initial period of hesitancy, as shown in avoidance responses in a locomotor manner, but

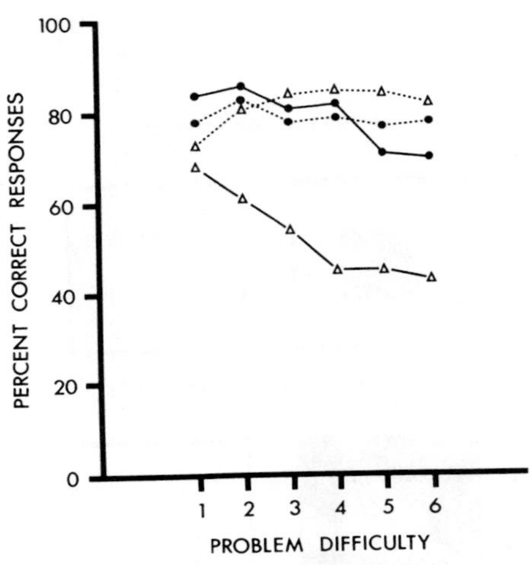

Fig. 18. Percent corect responses for centrally and peripherally located cues following brightness reversal. △—△, Central; △ - - - - △, peripheral—low protein. ● — ●, Central; ● - - - - ●, peripheral—high protein.

the animal gradually approaches, reaches for, and actively manipulates the object if it is accessible. One might say that the typical, socially reared, young rhesus monkey is neophilic, since it demonstrates strong positive approach response to novelty. The monkey raised in isolation stands in sharp contrast to the animal with social experience in response to new objects (Mason and Green, 1962; Menzel *et al.*, 1961, 1963): partial or total isolation results in the development of avoidance responses to new stimuli that might best be described as neophobic reactions.

The general responsiveness of feral or socially raised monkeys to new or novel objects is also found in traditional tests of discrimination learning. Harlow (1959) has described the response shift phenomenon and attributes it to the curiosity of the monkey. Furthermore, it has been demonstrated that rhesus monkeys will persistently select new objects when they are placed with familiar objects, thus forming a novelty learning set with great ease (Brown *et al.*, 1959). Also, in a maze-learning situation, baby monkeys showed consistent responsiveness to a new object when it replaced the original stimulus (Zimmermann, 1962).

One of the general observations we have made at our laboratory is that the low-protein subjects are highly reactive to changes in stimulus situations, such as minor changes in the test room, strangers in the laboratory, and the introduction of new discrimination objects in the learning situation. As was pointed out in the previous section, the low-protein animals were not inferior to the high-protein controls in the mean number of trials to reach criterion in simple learning situations. However, it was observed that at the end of the memory series when new sets of discrimination stimuli were introduced to initiate a simple learning set study, the low-protein animals were highly emotional. In addition, during the original learning in which each of the problems was learned to a criterion, often times the low-protein subjects would make a long sequence of incorrect responses. Thus, although they did not differ in the overall mean scores, the low-protein subjects were more variable than the high-protein subjects. For some of the low-protein animals the introduction of new stimuli in the discrimination learning situation was often accompanied by considerable emotional behavior. These animals often balked, producing a substantial but temporary increase in the total test time.

To test specifically for the presence of neophobic responses, the following experiment was designed. The 210-day high- and low-protein groups were tested on a set of fifty discrimination objects that had been presented repeatedly until there was significant recognition of the stimuli (Zimmermann *et al.*, 1970). Table VI diagrams the experimental paradigm to test for the presence of neophobic or neophilic responses. The subjects were presented with each of the fifty overlearned problems for six trials and then a seventh trial was introduced. On this trial, for half of the stimulus pairs, a new object replaced the formerly positive object, and on the other half of the problems, a new object replaced the formerly negative object. In this way responses away from the novel object when it replaced the positive stimulus would be the strongest indicator of neophobia. Responses to the new object when it replaced the negative object would be a measure of neophilia.

As might be expected from the report in the learning section, the low-protein subjects had slightly higher original learning scores. Figure 19 shows the performance of both groups in a six-trial learning set just prior to the introduction of the novel stimuli. Note that all animals were significantly above chance by the fourth trial in the six-trial sequence. However, on trial 7, when the novel stimulus was introduced, the groups behaved quite differently. Figure 20 shows the responses to the novel object when it replaced either the positive or negative stimulus. The low-protein subjects achieved over 90% correct on the sixth trial, but scored only 41% correct when the new object replaced the positive stimulus. These results indicated a significant neophobic reaction to the new objects by the malnourished monkeys. The high-protein subjects responded at a level of 82% correct on the sixth trial. But, when the novel stimulus replaced the negative stimulus, the high-protein subjects made 61% of their responses to it, although previously they had made only 15% of their responses to the negative stimulus on trial 6. These findings indicate a significant

TABLE VI

NEOPHOBIA EXPERIMENTAL PARADIGM

Original learning	Familiar object (reinforced)		Familiar object (nonreinforced)	
Test trial	Familiar object (reinforced)	Novel object (nonreinforced)	Novel object (reinforced)	Familiar object (nonreinforced)
Response interpretation	Responding to learned reinforcement contingencies	Neophilia	Responding to learned reinforcement contingencies	Neophobia

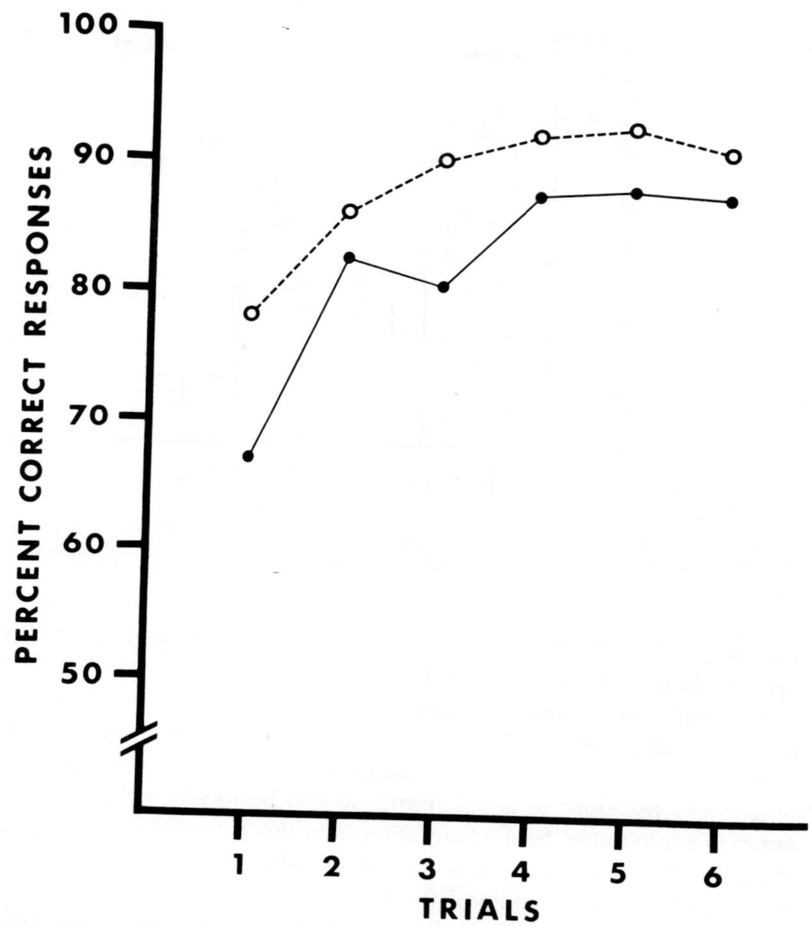

FIG. 19. Six-trial learning set performance immediately preceding the introduction of novel stimuli. ○····○, Low protein; ●—●, high protein.

amount of neophilic responses by the controls. Low-protein subjects made only 16% of their responses to the novel stimulus when it replaced the negative stimulus.

The qualitive observations made in the laboratory were, indeed, supported by this experiment. It also gave substance to the interpretation of the chain-pulling experiment in which it was found that the introduction of objects onto the chains resulted in a decrease in manipulatory activity

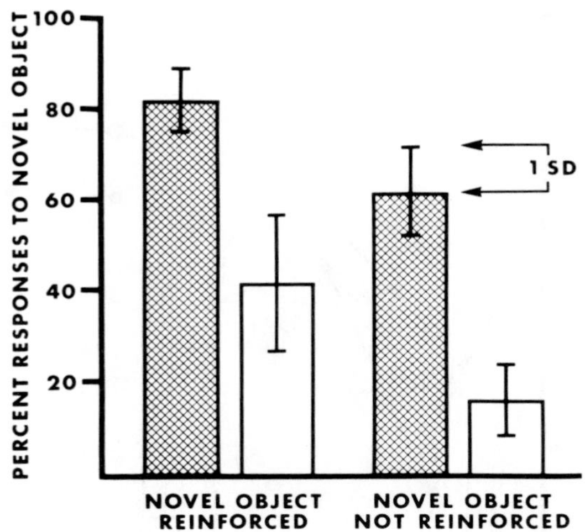

FIG. 20. Comparison of responses to familiar and novel stimuli. The low-protein diet is represented by the open bars; the high-protein diet is represented by the hatched bars.

on the part of the low-protein animals, whereas it increased this free operant in the high-protein subjects.

The general reaction of the low-protein monkeys to new or novel stimulation prompted the further investigation of this phenomenon by the development of a new test. It was thought that a more powerful determinant of aversive reactions to novel stimuli would be to place the animal in an approach–avoidance conflict situation in which it would have to approach the new stimuli in order to secure the reinforcement. An apparatus was constructed based on some of the known ecological characteristics of the monkey. The low-protein monkeys were observed to respond to frightening situations by freezing, self-clutching, or aimless bizarre movements. It did not seem that a traditional approach–avoidance alley or tunnel would provide the active and directed responses for which we were looking. However, climbing, when allowed, appeared to be the natural reaction of a monkey to fear-provoking stimuli. It was logical to take advantage of this response tendency in the design of the test apparatus. A 243.84-cm tall and 30.48-cm diameter cylinder was fashioned of wire mesh. This vertical tunnel had a door in the center and the bottom to permit entrance to the apparatus. An artist's conception of the apparatus is shown in Fig. 21. The cylinder stood vertically in the center of a 182.88 × 243.84 × 243.84 cm room. At the bottom of the cylinder, a small tube led to an automatic food dispenser from which reinforcements could be given. The tunnel was divided into 30.48-cm sections by external markers. The monkey was placed into the vertical tunnel and observed for a 10-minute session. The animals were all given twelve sessions in which general mobility within the apparatus was measured by recording the highest and lowest markers crossed in 15-second intervals.

Following this initial adaptation period, the animals were trained to shuttle in the tunnel. Whenever the monkey reached the top of the apparatus, the experimenter activated the feeder, which dispensed a piece of sugar-coated cereal into a tray at the bottom of the tunnel. Both high- and low-protein groups received twelve sessions with food available after each shuttle response. The session was terminated either when the animal earned thirty reinforcements or when the 10-minute test period was concluded.

Thirty additional sessions were given in the tunnel to test for the possible disruptive effects of novel stimuli on the previously established free operant response. The novel stimuli consisted of six inanimate junk objects. A single object was attached to a chain and suspended at the top, middle, or bottom of the apparatus. The stimulus order and the position of the object were randomized for each subject. Total number of reinforcements were recorded and observations of the animal's reaction to the objects were made at 15-second intervals. The reactions were placed into four categories: approach, avoidance, contact, or fear.

It might be anticipated from the learning studies that the low-protein monkey would have a higher rate of shuttling, as the result of a greater motivation for food, than the controls. The former continued to increase their shuttling rates with practice, whereas the latter showed a decline in final pre-object performance. The introduction of the objects resulted in a general reduction of shuttling behavior by the low-protein monkeys. The net change in shuttling behavior as indexed by mean differences in reinforcement rates before and after the introduction of objects is shown in Table VII. Total recovery of shuttling behavior by these animals occurred within twelve sessions, and terminal performance after thirty sessions with the objects was 35% higher than during original learning. The high-protein monkeys, on the other hand, continued to maintain low levels of shuttling behavior for food in the presence of the objects and showed only a slight, but insignificant, increase in performance following the introduction of the novel stimuli. Observational data taken from the animals during exposure to the objects are summarized in Fig. 22. The low-protein subjects showed significantly more avoidance and fear reactions than the controls. In fact, 2 of the low-protein animals refused to enter the tunnel with the objects present and had to be encouraged to ingress. The objects were also contacted a significantly greater number of times by the high-protein animals.

A final thirty-six sessions were presented with either one of a second

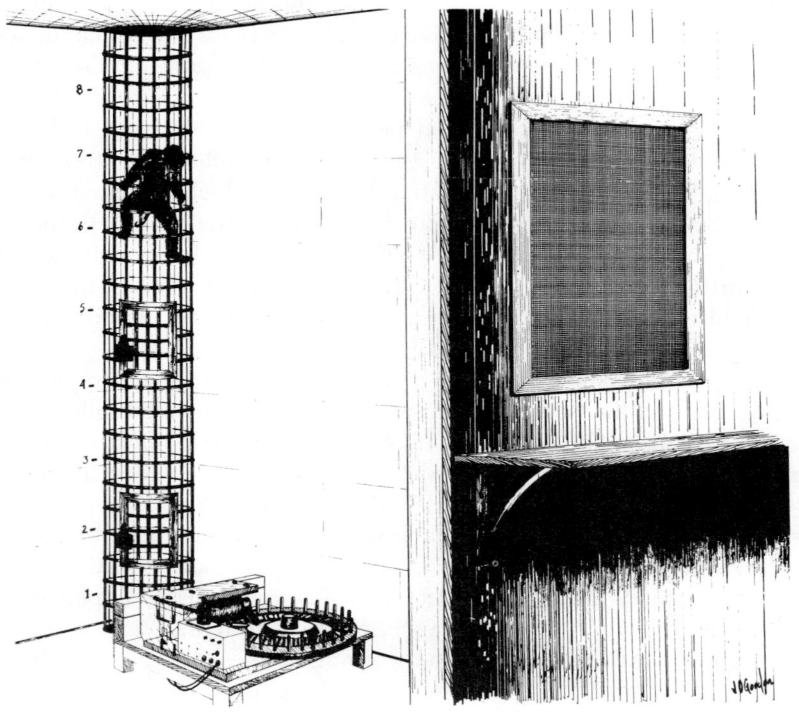

Fig. 21. Artist's impression of vertical tunnel.

set of three new objects or one of the three previously experienced objects. A significant decline in shuttling again occurred in the low-protein animals, but only to the new stimuli. The high-protein monkeys, in comparison, showed a significant increase in shuttling behavior with the familiar objects and a slight increase to the second set of novel objects. The low-protein animals again were able to recover their shuttling behavior as the objects became more familiar through increased exposure.

This is further evidence that monkeys deprived of protein may have a negative reaction to novel stimuli. There was a great increase in fear reactions, as shown in vocalizations, grimacing, and defecation to the initial presentation of the objects. By contrast, the high-protein monkeys exhibited approach and contact with the objects. These two experiments indicate strongly that the low-protein subjects react quite differently to a change in their environment than high-protein controls. A decrease in responsiveness to manipulatory stimuli or decrements in responses to

TABLE VII

NET CHANGE IN REINFORCEMENTS OBTAINED BEFORE AND AFTER INTRODUCTION OF DIFFERENT STIMULI INTO VERTICAL TUNNELS

	Novel objects 1	Novel objects 2	Familiar objects 2
	Initial 6-trial blocks		
Original learning			
210 Low protein	−4.91	—	—
210 High protein	+0.54	—	—
	Terminal 6-trial blocks		
Familiar objects I			
210 Low protein	—	−11.30 ($p < 0.05$)	−2.03
210 High protein	—	+4.29	+6.75 ($p < 0.025$)

food-getting behavior as the result of neophobic reactions would put an organism that is trying to survive in its normal ecological environment at a competitive disadvantage. In particular, if every new or strange environment is fearful and the reaction of the malnourished animal is to avoid or withdraw, then the stage is set for the production of organisms that may appear to be apathetic or lethargic. It is also a behavioral mechanism for reducing the amount of stimulation that an animal is willing to accept before withdrawing. Finally, retarded learning ability would be expected from these organisms, particularly if the learning situations produced novel stimulation and if reinforcement was predominantly from investigatory or manipulatory sources.

E. SOCIAL BEHAVIOR

Social interaction in the rhesus macaque is a highly complex form of behavior that requires considerable integration of innate and learned behavior patterns. The list of fifty-nine discernible elements and sixty-four compound patterns of behavior described by Altman (1962) would suggest that monkeys are required to make numerous and accurate social judgments. It is also apparent from group differences and from studies of monkeys raised in captivity that much of the social intercourse is acquired through experience with adults and peers. Furthermore, Altman

(1962) has suggested that concept formation and memory are essential to understanding the complex didactic interactions of the rhesus macaque. That is, it appears that the mature rhesus makes behavioral adjustment to other animals on the basis of predictable sequences of behavior and has the capacity for learning a multitude of fine discriminations between members of a group with long-term retention.

If the basic properties of learning, motivation, and perception are in any way affected by the variable of malnutrition, differences would be expected in the social behavior between malnourished animals and adequately fed controls. Malnutrition did, in fact, produce dramatic effects on the social behavior of the malnourished monkeys, and this discovery was in part the result of serendipity. In the initial experiments, our efforts were addressed primarily to measures of learning and motivation and were more concerned with the problems associated with these variables than social behavior. On the other hand, it was necessary to prevent the animals from

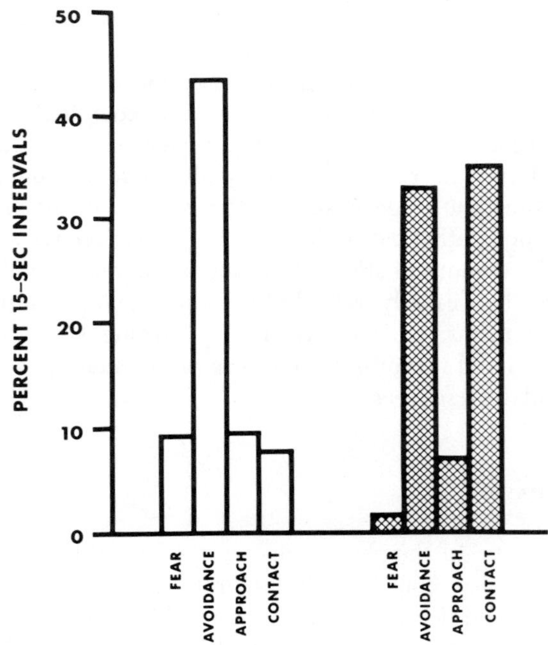

FIG. 22. Observations of monkey's reaction to novel stimuli placed in the vertical tunnel. The 210 low-protein diet is represented by the open bars; the 210 high-protein diet is represented by the hatched bars.

developing abnormal behavior patterns described to be the result of partial or total social isolation (Sackett, 1968). To guard against this eventuality, the monkeys were placed in a 91.44 × 152.40 × 274.32 cm cage every day for an hour play period, 5 days a week. Initially, systematic observations were not acquired during socialization intervals. After several weeks of this procedure, it was obvious that the groups of animals behaved differently in the social cage. At this time a systematic series of observations were made of the animals using three classes of behavior: (*1*) behavior directed toward others—action toward or initiation or contact with another animal in the cage; (*2*) behavior toward self, e.g., self-clutching, self-grooming, masturbation, and the sucking and mouthing of body members; (*3*) behavior directed toward the environment—action toward or initiating manual or oral contact with things such as cage parts, wood chips, and visually directed threats toward the observer. The behaviors were considered mutually exclusive and only one behavior was recorded at a time. The response durations were recorded by pressing one of three buttons that activated a clock. At the end of the observation period, which lasted 10 minutes, each activity was recorded and observations of the next animal commenced.

The 210-day diet groups differed in social behavior. The low-protein animals spent less than 6% of their time directing activity toward others, whereas the high-protein group spent 30% of their time in this activity. On the other hand, the low-protein subjects spent over 60% of their time in self-stimulating activities, whereas this score in the high-protein subjects did not exceed 45%. With continued deprivation the time spent in self-directed activity reached over 75% for the low-protein animals, as they spent less and less time in activities directed toward the environment.

With the completion of our new laboratory, we were able to construct a more spacious and sophisticated social room for testing. The new social testing was conducted in a room that measured 243.84 × 228.60 × 203.20 cm. Vertical and horizontal exercise bars spanned the room, two chains varying in length hung from the ceiling, and two shelves were attached to the walls at different levels from the floor. The floor of the room was covered with a layer of sawdust. An artist's impression of the social room is shown in Fig. 23.

The observer sat behind a wire screen and recorded his observations by pressing a set of keys mounted on a hand-held box. The keys activated microswitches that were electrically connected with an Esterline-Angus recorder. The animals were approximately 300 days of age at the beginning of this 2-year experiment. However, the animals had been given experience at least 5 days a week for 1 hour sessions from 120 days of age. Five categories of behavior were selected for observation in the first experiment. These were (*1*) approach–play behavior in which one of the animals approached another and initiated social contact resulting in continued reactions in the form of play, running, jumping, following the leader, or rough and tumble play; (*2*) avoidance–submission, running from an approach, not reciprocating when another animal attempted to elicit play, lying passively while another animal attempted to initiate play (not to be confused with grooming behavior); (*3*) self-clutching and self-stimulation, for example, masturbation or grasping another animal, an attempt to maintain ventral contact with another by wrapping himself around the animals, or clutching himself; (*4*) grooming and sexual behavior—active or passive grooming interactions, mounting or presenting; (*5*) non-social behavior, all behavior directed toward inanimate objects in the environment, the room, or the observer. Measures of frequency, duration, and rate of response were analyzed from the Esterline-Angus recordings. The mean totals for the various

FIG. 23. Artist's impression of the social room.

TABLE VIII

MEAN TOTALS, FREQUENCY, AND DURATION OF SOCIAL BEHAVIOR IN FIRST EXPERIMENT[a]

Behavior	Frequency		% Duration		Rates	
	210-HP	210-LP	210-HP	210-LP	210-HP	210-LP
Avoidance–submission	1.07	1.22	7	9	0.80	0.97
	(N.S.)		(N.S.)		(N.S.)	
Sexual	0.41	0.08	5	1	0.86	0.49
	($p < 0.005$)		($p < 0.005$)		($p < 0.057$)	
Approach–play	2.81	1.22	40	19	2.63	2.27
	($p < 0.005$)		($p < 0.019$)		(N.S.)	
Clutching	0.07	0.10	1	3	0.22	0.73
	(N.S.)		($p < 0.005$)		($p < 0.019$)	
Nonsocial	2.96	2.29	47	68	3.55	8.93
	($p < 0.005$)		($p < 0.005$)		(N.S.)	

[a] HP, high protein; LP, low protein; N.S., not significant.

response measures and categories of behavior between groups are summarized in Table VIII. The frequency measure indicated that the high-protein animals exhibited significantly greater approach–play, sexual behavior, and nonsocial behavior than the low-protein animals. Neither clutching behavior nor avoidance–submission behavior was significant.

Analysis of the second response measure, percent duration, disclosed that the high-protein animals spent significantly more time in approach–play and sexual behavior and significantly less time in clutching behavior and nonsocial forms of responses than the low-protein group. Again, avoidance–submission differences were not statistically significant.

In the analysis of rate (i.e., the frequency of behavior over time), the high-protein animals showed significantly higher rates only for the category of sexual behavior, and significantly lower rates for clutching and nonsocial behavior, when compared to the low-protein animals. Approach–play and avoidance–submission differences did not approach significance on this measure. It should be pointed out that although there were differences in the self-clutching response between the groups, there was a considerable reduction in this behavior during the testing period for the low-protein subjects.

Although it was observed that the avoidance–submission behavior constituted only a small portion of the animals' total activity, qualitative differences between the groups were observed. On numerous occasions, behavior initiated as approach–play in the low-protein group resulted in what might be characterized as aggression or brutality. Unlike the rough and tumble play behavior found in the high-protein animals, the low-protein subjects exhibited little or no reciprocity in mouthing, biting, or chewing. Inasmuch as this behavior was recorded as approach–play for the aggressor and as avoidance–submission for the submissive animals, the overall quanitification of differences in aggression all but disappeared when the data were summarized. It was thought that a specific category of aggression that included data from both the aggressor and the submissive animal would more accurately separate social differences between the high- and low-protein animals.

To summarize this second experiment on social behavior, the high-protein animals appeared to be more normal in the development of social behavior than the low-protein group. The high-protein animals engaged in approach–play and sexual behaviors for longer periods of time, but spent less time in clutching and nonsocial behaviors, than the low-protein animals. Conversely, the low-protein animals spent longer time intervals per event or bout in socially abnormal behaviors than did the high-protein monkeys. Specifically, the protein-deprived monkeys spend longer periods of time gazing blankly around the room, walking aimlessly, or sitting curled into a tight ball. In addition to these quantitative differences, qualitative differences between the groups in terms of the appearance of aggression as the result of a social interaction were apparent.

In order to evaluate this qualitative difference observed in the previous social behavior, testing was continued, but the categories of behavior observed were altered. At this time counters and clocks replaced the Esterline-Angus recorder in the measurement of duration and frequency of social behaviors. The categories recorded in the second phase of this experiment were (*1*) social aggressive interactions, including approach, contact, mounting, biting, chewing, grabbing, and pulling without reciprocity between the aggressor and the aggressee, in which the animals aggressed against appeared motivated by fear, as exemplified in brutality, escape, and submission; (*2*) social tactual contact behavior in which the animals mutually came into physical contact, stayed quietly in a group, and groomed or mounted one another; (*3*) social approach–play, characterized by chasing, running, jumping, rough and tumble play with mutual participation which did not appear motivated by fear and included active approach, mouthing, and biting with reciprocity; (*4*) object-oriented (nonsocial) behavior, for example, chain pulling and chewing, playing in the sawdust, swinging from the poles, and licking or chewing the wood, bars, and other cage parts; (*5*) undirected (nonsocial) behavior such as sittting, standing, or pacing without visible direction, self-clutching, and self-stimulation.

The results of the second phase of the experiment are shown in Table IX. As in the first phase, the high-protein monkeys continued to show more positive social response than the low-protein monkeys. For example, they engaged in approach–play behavior more frequently, for longer periods of time, and at higher rates than the low-protein animals. The high-protein subjects also showed significantly higher occurrences of tactual play responses. The rates of tactual contact and percent duration were higher for the high-protein group but were not statistically significant.

TABLE IX

MEAN TOTALS, FREQUENCY, AND DURATION OF SOCIAL BEHAVIOR IN FIRST EXPERIMENT [a]

	Frequency		% Duration		Rates	
Behavior	210-HP	210-LP	210-HP	210-LP	210-HP	210-LP
Aggression	0.39	0.56	3	3	0.75	1.00
	(N.S.)		(N.S.)		(N.S.)	
Tactual contact	1.56	0.78	20	12	3.96	3.53
	($p < 0.005$)		(N.S.)		(N.S.)	
Approach–play	2.91	1.57	35	17	3.62	2.40
	($p < 0.005$)		($p < 0.005$)		($p < 0.005$)	
Nonsocial object	2.05	2.09	23	43	2.96	5.96
	(N.S.)		($p < 0.005$)		($p < 0.005$)	
Nonsocial	1.89	1.82	19	25	2.58	3.67
	(N.S.)		($p < 0.005$)		($p < 0.057$)	

[a] HP, high protein; LP, low protein; N.S., not significant.

In comparison to the greater social responsiveness of the adequately nourished subjects, the low-protein animals engaged in nonsocial types of behavior nearly as often as the high-protein animals. But, when the low-protein group exhibited object-oriented or undirected nonsocial responses, they did so for longer periods of time.

Although the malnourished monkeys showed a higher frequency and a higher rate of aggressive behavior than did the high-protein monkeys, the differences did not reach statistical significance. However, the use of the absolute measure of analysis of aggression did not control for the initial differences in the amount of social behavior between the two groups. Along with approach–play behavior, aggression comprised what might be described as a composite measure of active social behavior. If the high-protein animals were to exhibit a higher active social responsiveness, then it could increase the likelihood of more aggressive behavior, and at the same time, constitute a smaller proportion of the animal's active social behavior relative to the low-protein group. In short, aggressive behavior would comprise an unequal amount of behavior between the two groups in the context of approach–play behavior. To control for this variable, an aggression ratio was derived by dividing aggression scores by the sum of aggression and approach–play scores. Table X shows the results of this analysis. As can be seen, the proportion of the aggressive responses in the total active social behavior was significantly higher for the low-protein monkeys in all measures.

TABLE X

MEAN TOTALS, FREQUENCY, AND DURATION OF AGGRESSION RATIO

	Frequency	% Duration	Rates
210-Day low protein	0.27	17	0.29
210-Day high protein	0.13	8	0.17
	($p < 0.033$)	($p < 0.033$)	($p < 0.033$)

Thus, not only were the low-protein animals less social in their behavior in the social tests, but when they did socialize, a great deal of their social responses resulted in aggressive acts that were basically aversive to the stimulus animals. With repeated testing, we found that there was a decrease in both social contacts and aggression by the low-protein monkey as the animals developed behavior patterns to keep them apart. This observation has not been quantified, but in the later tests the social contact fell almost to zero in the low-protein animals. The malnourished monkeys had a tendency to sit apart and showed a high degree of displacement when other monkeys approached. By keeping a distance between one another, the animals minimize the aversive reactions associated with social contact.

The low-protein monkeys were very aggressive toward one another when they did make social contact, but these relationships did not appear to be stable. In addition, the low-protein animals appeared more aggressive toward the experimenter than controls in food-testing situations. These observations led us to make an examination of the relative constancy of the dominance behavior in food and nonfood competition. Mason (1961) reported that animals raised in isolation for a year did not develop stable dominance relationships, whereas feral animals did. Since our low-protein animals were beginning to behave in a manner similar to social isolates in the social situation, it was believed that dominance relationships should be critically examined.

To test for dominance, two different social situations were developed. In a food competition test the animals were paired in all possible combinations within age groups in the social room. One section of the wall next to the experimenter was modified so that a tube projected into the social room through which reinforcements could be deposited. All of the animals were trained to find food at the reinforcement cup prior to the pairings. In the food competition studies, sugar-coated cereal was dropped down the tube into the room every 30 seconds until twenty reinforcements had been delivered. The animal that secured and consumed the reinforcement was recorded for each of the reinforcements. In the second, or free-play social dominance test, each animal of each cage group was placed in the social room with 4 animals of the opposite dietary group. For example, a 120-day low-protein animal was placed in the social situation with 4 120-day high-protein monkeys and vice versa. The animals were observed and dominance scores were recorded for 10-minute sessions of the odd monkey. The dominance scores included behaviors such as initiation, approach, displacement, aggressive behavior, and biting. The animals were tested in each of the conditions for ten sessions. The results of the experiments are summarized in Table XI. As can be seen, the age control animals and the high-protein 210-day monkeys were more dominant in the social dominance test, but there were no differences in the food competition tests between these two groups. The 120-day high- and low-protein animals do not differ on the social dominance test, but they were significantly different in the food competition tests. It should be pointed out that in the first two comparisons there was a weight difference of over 1 kg. Thus, the finding that the larger animal is more dominant is consistent with the results reported by Angermier *et al.* (1968). The two younger groups were more nearly equal in size, and under these conditions the low-protein subjects secured more of the reinforcements. In general, under the conditions of food reinforcement, the low-protein animals were not as submissive as they were in the free-play social situation. The general trait of dominance, which usually appears to be stable in the rhesus monkey (Warren and Maroney, 1958; Maroney *et al.*, 1959), did not remain stable across situations involving the low-protein animals. In the social dominance test, not one 380-day low-protein animal was dominant over a high-protein monkey in the free-play situation. However, in the food competition, 3 of the malnourished animals received more food

TABLE XI

t-TESTS BETWEEN AGE-PAIRED GROUPS ON SOCIAL AND FOOD DOMINANCE SCORES

Animal	Group	Social dominance	Food competition
3508	Age control	103.9	15.3
3509	Age control	53.2	8.45
4303	Age control	101.5	7.65
4310	Age control	145.1	4.75
1	380 Low protein	24.1	14.6
2	380 Low protein	16.8	14.2
3	380 Low protein	28.4	1.9
4	380 Low protein	1.6	11.85
		$t = 4.224$,[a] $df = 6$	$t = 0.43$, $df = 6$
5	210 Low protein	3.0	4.85
6	210 Low protein	2.2	11.1
5766	210 Low protein	2.2	12.3
5872	210 Low protein	10.5	13.1
5976	210 Low protein	16.4	5.4
5979	210 Low protein	18.8	9.2
5754	210 High protein	21.4	11.27
5755	210 High protein	154.5	11.03
5756	210 High protein	137.2	7.47
5758	210 High protein	158.9	12.97
		$t = 4.196$,[a] $df = 8$	$t = 0.675$, $df = 8$
8	120 Low protein	93.0	14.25
9	120 Low protein	166.6	14.8
10	120 Low protein	11.8	15.8
13	120 Low protein	65.9	11.45
15	120 High protein	50.5	2.85
16	120 High protein	58.6	3.5
17	120 High protein	24.9	3.8
18	120 High protein	47.4	13.5
		$t = 1.181$, $df = 6$	$t = 3.019$,[b] $df = 6$

[a] $p < 0.01$.
[b] $p < 0.05$.

than 3 of the age controls. In all cases, at least one-half of the low-protein subjects secured more reinforcements than the high-protein controls.

V. SUMMARY

The variable of protein malnutrition apparently produces changes in behavior of the developing rhesus monkey that are diffuse. Those behaviors that are not affected include general activity, as well as many forms of learning. These results are consistent with the data reported by Mitchell (1970) on the effects of social isolation on learning in the baby monkey. He found little or no learning differences after extensive adaptation and testing. Harlow and Harlow (1965) bring up one point that could be used to criticize our studies in learning. They suggest that many different tests should be given over a long period of time. They tested their animals on thousands of trials and hundreds of problems. It is possible that we did not test our animals long enough on any one set of problems, such as six-trial learning set. Differences might have arisen, but there are limits to the times we can test these animals. However, an extensive series of learning set problems is planned for the future.

The effects of malnutrition on behavior are more subtle than had first been anticipated and probably have a major effect on the development of motivational systems. Although the malnourished monkey is food oriented, it does not show greater resistance to extinction after food has been removed. There was no evidence of secondary reinforcement at work. Rather, the low-protein monkey appears to be a discriminating organism and sensitive to changes in food reward. Contrary to previous reports, a specific hunger for the high-protein diet was found in the low-protein subjects. The elevated drive for food additionally alters the animal socially in a food competition situation, although it remains submissive to a larger animal in normal social testing. From the studies of manipulation, learning, and social food competition, it might be hypothesized that the malnourished monkey has a channelization of drive; that is, the low-protein monkeys are so concerned with food-oriented behaviors that other behaviors, such as normal social reactions and manipulation of things in the environment, are delayed or do not appear at all. Thus, the dominance of food-seeking or food-oriented behaviors would prevent the development of normal non-food-oriented acts. Barnes *et al.* (1968) also described this increase in motivation for food in previously malnourished and protein-restricted pigs. In our studies of attention, we would expect the high drive animal to be less attentive to the subtle cues and more direct in his behavior. Therefore, on discrimination requiring observing or attention responses, it would be inferior.

This explanation of the behavior of the malnourished monkeys does not account for the novelty and neophobia studies. There is little hint of why a high food drive animal should show aversion to new stimuli. The neophobia data, the social behavior, and the curiosity findings, however, do find a counterpart in the literature. Our low-protein animals are strikingly similar to animals that have been raised in social isolation (Harlow, 1965a, b). Our animals did not appear deficient in learning capacities but were deficient in social behavior and normal curiosity. Harlow has not reported on studies involving the attention dimension. However, Fuller (1967) has suggested that the abnormal behavior of puppies reared in social isolation is the result of their inability to attend and respond to appropriate stimuli in the environment. Isolated animals appear stressed by a stimulus overload. It is quite possible that the condition of protein–calorie malnutrition may produce a similar set of inappropriate reactions in the baby monkey. It is interesting to note, however, that our quantitative studies support the qualitative observations made by Kerr and Waisman (1968) in which they reported that animals raised on low-protein diets show a lack of curiosity and retarded peer group interaction.

One effect of the low-protein diet mentioned is that the animals on such protein-restricted diets do not mature as early as adequately nourished animals. In the high-protein group, the testicles of the males descended at about 1230 days of age, and the monkeys began to engage in grooming, presenting, and mounting with the single female in the group who began to develop sexual folds and coloring at about 1170 days of age. Thus, one of the major effects of the low-protein diet may be retarded development. A major characteristic of the behavior of very young monkeys in the social situation is the lack of peer activity and a preoccupation with nonsocial stimuli in the environment. Low-protein monkeys may be behaving like very young monkeys in maintaining infantile types of responses that are not compatible with normal social development. The low-protein monkeys displayed infantile behaviors such as the fear grimace and a continual high pitched scream. By contrast, the high-protein control animals developed adultlike forms of threat postures and expressions as shown by van Hooff (1962). There was almost complete absence of sexual behavior in the low-protein monkeys. It is very possible that these animals could have developed quite normally intellectually. The lack of attention could also be a function of maturity and more closely allied to motivational variables than to association of learning variables. Fear of new objects is not uncommon in the young monkey. It is interesting to note that behavior differences attributable to growth and maturity have been reported in protein-deprived rats. Anderson and Smith (1932) postulated that the activity and maze-learning ability of protein-deprived rats were more like that of normal control animals of their own weight and size than like adequately fed rats their own age but of different weights.

The principal findings resulting from our investigation of the effects of malnutrition on the behavior development in the rhesus monkey is that motivational and social variables are dramatically affected. These are drastic and widespread abnormalities that develop and persist over a long period of time. The malnourished monkey is reactive to stimulus change, fearful, yet aggressive in the social situation. Its behaviors are, in a sense, very maladaptive to a changing environment. Moreover, no environment changes more rapidly than the social situation. It is possible that there are a series of mutually interacting patterns of neophobia, food preoccupation, and immaturity that contribute to the behavior abnormalities found in our low-protein animals. Further research with more animals is needed and may throw additional light on the source of the behavioral aberrations. Of particular importance is the manner in which the malnourished monkeys respond to dietary rehabilitation. With the return to normal growth, normal behavior patterns may emerge.

References

Aldrich-Blake, F. P. G., Bunn, T. K., Dunbar, R. I. M., and Headly, P. M. (1971). Observations on baboons, *Papio anubis,* in an arid region in Ethiopia. *Folia Primatol.* **15**, 1–35.

Altmann, S. A. (1962). A field study of the sociobiology of rhesus monkey, *Macaca mulatta. Ann. N.Y. Acad. Sci.* **102**, 338–435.

Anderson, J. E., and Smith, A. H. (1932). Relation of performance to age and nutritive condition in the white rat. *J. Comp. Psychol.* **13**, 409–446.

Angermier, W. F., Phelps, J. B., Murray, S., and Howansteine, J. (1968). Dominance in monkeys: Sex differences. *Psychon. Sci.* **12**, 344.

Barnes, R. H. (1971). Nutrition and man's intellect and behavior. *Fed. Proc., Fed. Amer. Soc. Exp. Biol.* **30**, 1429–1433.

Barnes, R. H., Cunnold, S. R., Zimmermann, R. R., Simmons, H., MacLeod, R. R., and Krook, L. (1966). Influence of nutritional deprivations in early life on learning behavior of rats as measured by performance in a water maze. *J. Nutr.* **89**, 399–410.

Barnes, R. H., Moore, A. U., Reid, I. M., and Pond, W. G. (1967). Learning behavior following nutritional deprivation in early life. *J. Amer. Diet. Ass.* **51**, 34–39.

Barnes, R. H., Reid, I. M., Pond, W. G., and Moore, A. U. (1968). The use of experimental animals in studying behavioral abnormalities following recovery from early malnutrition. *In* "Calorie Deficiencies and Protein Deficiencies" (R. A. McCance and E. M. Widdowson, eds.), pp. 277–285. Little, Brown, Boston, Massachusetts.

Barnes, R. H., Moore, A. U., and Pond, W. G. (1969). Behavioral abnormalities in young adult pigs caused by malnutrition in early life. *J. Nutr.* **100**, 149–155.

Behar, M., Ascoli, W., and Scrimshaw, N. S. (1958). An investigation into the causes of death in children in four rural communities in Guatemala. *Bull. W. H. O.* **19**, 1093.

Bingham, H. C. (1932). Gorillas in a native habitat. *Carnegie Inst. Wash. Publ.* **426**, 1–66.

Blomquist, A. J., and Harlow, H. F. (1961). The infant rhesus monkey program at the University of Wisconsin primate laboratory. *Proc. Anim. Care Panel* **11**, 57–64.

Booth, A. H. (1956). The distribution of primates in the Gold Coast. *J. West Afr. Sci. Ass.* **2**, 122–133.

Booth, A. H. (1957). Observations on the natural history of the Olive Colobus monkey, *Procolobus versus. Proc. Zool. Soc. London* **129**, 421–430.

Bradfield, R. B. (1968a). Colloquium on protein deficiencies and calorie deficiencies. *Amer. J. Clin. Nutr.* **21**, 130–133.

Bradfield, R. B. (1968b). Changes in hair associated with protein-calorie malnutrition. *In* "Calorie Deficiencies and Protein Deficiencies" (R. A. McCance and E. M. Widdowson, eds.), p. 213. Little, Brown, Boston, Massachusetts.

Brockman, L. M., and Ricciuti, H. N. (1971). Severe protein-calorie malnutrition and cognitive development in infancy and early childhood. *Develop. Psychobio.* **4**, 312–319.

Brown, W. L., Overall, J. E., and Blodgett, H. C. (1959). Novelty learning sets in rhesus monkeys. *J. Comp. Physiol. Psychol.* **52**, 330–335.

Burgess, A., and Dean, R. F. A. (1962). "Malnutrition and Food Habits." Macmillan, New York. Cited in Barnes (1971).

Butler, R. A. (1965). Investigative behavior. *In* "Behavior of Nonhuman Primates: Modern Research Trends" (A. M. Schrier, H. F. Harlow, and F. Stollnitz, eds.), Vol. 2, pp. 463–493. Academic Press, New York.

Cansdale, G. S. (1944). *Galago demidovii. J. Soc. Presrv. Fauna Emp.* **50**, 7.

Carpenter, C. R. (1934). A field study of the behavior and social relations of howling monkeys. *Comp. Psychol. Monogr.* **10**, 1–168.

Carpenter, C. R. (1935). Behavior of the red spider monkey (Ateles goeffroyi) in Panama. *J. Mammal.* **16**, 171–180.

Carpenter, C. R. (1940). A field study in Siam of the behavior and social relations of the gibbon (*Hylobates lar*). *Comp. Psychol. Monog.* **16**, 1-212.

Chalmers, N. R. (1968). Group composition, ecology, and daily activities of free living Magabeys in Uganda. *Folia Primatol.* **8**, 247–262.

Collias, N. E., and Southwick, C. H. (1952). A field study of population density and social organization in howling monkeys. *Proc. Amer. Phil. Soc.* **96**, 144–156.

Collier, G. H., and Squibb, R. L. (1967). Diet and activity. *J. Comp. Physiol. Psychol.* **64**, 409–413.

Collier, G. H., Squibb, R. L., and Jackson, F. (1965). Activity as a function of diet: 1. Spontaneous activity. *Psychon. Sci.* **3**, 173–174.

Coursin, D. B. (1965). Effect of undernutrition on CNS function. *Nutr. Rev.* **23**, 65.

Cowley, J. J., and Griesel, R. D. (1962). Pre- and post-natal effects of a low protein diet on the behavior of the white rat. *Psychol. Afr.* **9**, 216–225.

Cravioto, J., and Robles, B. (1965). Evolution of adaptive and motor behavior during rehabilitation from kwashiorkor. *Amer. J. Orthopsychiat.* **35**, 449–464.

Cravioto, J., DeLicardie, E. R., and Birch, H. G. (1966). Nutrition, growth, and neurointegrative development: an experimental and ecologic study. *Pediatrics* Suppl., Part 2, 319.

Crook, J. H. (1968). Gelada baboon herd structure and movement. *Symp. Zool. Soc. London* **18**, 237–258.

Crook, J. H., and Aldrich-Blake, F. G. (1968). Ecological and behavioral contrasts between sympatric ground dwelling primates in Ethiopia. *Folia Primatol.* **8**, 192–227.

Dart, R. (1963). The carnivorous propensity of baboons. *Symp. Zool. Soc. London* **10**, 49–56.

Davies, J. N. P. (1948). The essential pathology of kwashiorkor. *Lancet* **1**, 317–320.

Davis, D. D. (1962). Mammals of the lowland rain-forest of North Borneo. *Bull. Natu. Mus. St. Singapore* **31**, 1–129.

DeVore, I., and Hall, K. R. L. (1965). Baboon ecology. *In* "Primate Behavior: Field Studies of Monkeys and Apes."(I. DeVore, ed.), pp. 20–52. Holt, New York.

DeVore, I., and Washburn, S. L. (1963). Baboon ecology and human evolution. *In* "African Ecology and Human Evolution." (F. C. Howell and F. Bourliere, eds.), Viking Fund Publ. Anthropol. Wenner-Gren Found., New York.

Donisthorpe, J. (1958). A pilot study of the mountain gorilla (*G. g. beringei*) in S. W. Uganda February to September, 1957. *S. Afr. J. Sci.* **54**, 195–217.

Enders, R. K. (1930). Notes on some mammals from Barro Colorado Island, Canal Zone. *J. Mammal.* **11**, 280–292.

FAO WHO (1962). Joint Food and Agriculture Organization and the World Health Organization Expert Committee on Nutrition. *World Health Organ. Tech. Rep. Ser.* **245**.

Fooden, J. (1964). Stomach contents and gastro-intestinal proportions in wild shot Guianian monkeys. *Amer. J. Phys. Anthropol.* **22**, 227–232.

Frankova, S. (1968). Nutritional and psychological factors in the development of spontaneous behavior in the rat. *In* "Malnutrition, Learning, and Behavior" (N. S. Scrimshaw and J. E. Gordon, eds.), pp. 312–322. MIT Press, Cambridge, Massachusetts.

Frankova, S., and Barnes, R. H. (1968a). Influence of malnutrition in early life on exploratory behavior of rats. *J. Nutr.* **96**, 477–484.

Frankova, S., and Barnes, R. H. (1968b). Effects of malnutrition in early life on avoidance conditioning and behavior of adult rats. *J. Nutr.* **96**, 485–493.

Fuller, J. L. (1967). Experimental deprivation and later behavior. *Science* **158**, 1645–1652.

Gee, E. P. (1961). The distribution and feeding habits of the golden langur, *Presbytis geei. J. Bombay Natur. Soc.* **58**, 1-12.

Geist, C. R., Zimmermann, R. R., and Strobel, D. A. (1972). Effect of protein-calorie malnutrition on food consumption, weight gain, serus proteins, and activity in the developing rhesus monkey (*Macaca mulatta*). *Lab. Anim. Sci.* **22**, 369–377.

Gerber, M., and Dean, R. F. A. (1967). Gessell tests on African children. *Pediatrics* **20**, 1055.

Gongora, J., and McFie, J. (1959). Malaria, malnutrition and mortality. *Trans. Roy. Soc. Trop. Med. Hyg.* **53**, 238.

Goodall, J. (1963). Feeding behavior of wild chimpanzees. *Symp. Zool. Soc. London* **10**, 39–47.

Goodall, J. (1965). Chimpanzees of the Gombe Stream Reserve. *In* "Primate Behavior, Field Studies of Monkeys and Apes" (I. DeVore. ed.), pp. 425–473. Holt, New York.

Gopalin, C. (1968). Kwashiorkor and marasmus: evolution and distinguishing features. *In* "Calorie Deficiencies and Protein Deficiencies" (R. A. McCance and E. M. Widdowson, eds.), pp. 49–58. Little, Brown, Boston, Massachusetts.

Graham, G. G., Cordano, A., and Baertl, J. M. (1963). Studies in infantile nutrition. 11. Effect of protein and calorie intake on weight gain. *J. Nutr.* **81**, 249–254.

Hall, K. R. L. (1965). The behavior and ecology of the wild patas monkey, *Erythrocebus patas,* in Uganda. *J. Zool.* **148**, 15–87.

Hansen, J. D. L. (1968). Features and treatment of kwashiorkor at the cape. *In* "Calorie Deficiencies and Protein Deficiencies" (R. A. McCance and E. M. Widdowson, eds.), pp. 33–47. Little, Brown, Boston, Massachusetts.

Harlow, H. F. (1950). Learning and satiation of response in intrinsically motivated complex puzzle performance by monkeys. *J. Comp. Physio. Psychol.* **43**, 289–294.

Harlow, H. F. (1959). Learning set and error factor theory. *In* "Psychology: A Study of a Science" (S. Koch, ed.), Vol. 2, pp. 492–537. McGraw-Hill, New York.

Harlow, H. F. (1965a). The effects of early social deprivation on primates. *Tire Part, Symp. Bell Air II, Desafferentation Exp. Clin.,* pp. 66–77.

Harlow, H. F. (1965b). Total social isolation in monkeys. *Proc. Nat. Acad. Sci. U.S.* **54**, 90–97.

Harlow, H. F., and Harlow, M. K. (1965). The affectional systems. *In* "Behavior of Nonhuman Primates" (*A. M. Schrier, H. F. Harlow, and F. Stollnitz, eds.*), Vol. 2, pp. 287–344. Academic Press, New York.

Harlow, H. F., and Zimmermann, R. R. (1959). Affectional responses in the infant monkey. *Science* **130**, 421–432.

Harlow, H. F., Harlow, M. K., and Meyer, D. R. (1950). Learning motivated by a manipulation drive. *J. Exp. Psychol.* **40**, 228–234.

Harlow, H. F., Blazek, N. C., and McClearn, G. E. (1956). Manipulatory motivation in the infant rhesus monkey. *J. Comp. Physiol Psychol.* **49**, 444–448.

Harris, R. S., ed. (1971). "Feeding and Nutrition on Nonhuman Primates." Academic Press, New York.

Harrison, J. L. (1954). The natural food of some rats and other mammals. *Bull. Raffles Mus.* **25**, 157–165.

Harrison, J. L. (1962). The apes and monkeys of Malaya. *Malay. Mus. Pamphlets* **9**.

Hill, W. C. O. (1957). "Primates Comparative Anatomy and Taxanomy, Vol. 3: Hapalidae." Edinburgh Univ. Press, Edinburgh.

Hill, W. C. O. (1960). "Primates, Comparative Anatomy and Taxonomy, Vol. 4: *Cebidae.*" Edinburgh Univ. Press, Edinburgh.

Hill, W. C. O. (1962). Primates. "Comparative Anatomy and Taxonomy, Vol. 5: *Cebidae,*" Part B. Edinburgh Univ. Press, Edinburgh.

Hillman, N. M., and Riopelle, A. J. (1971). Acceptance and palatability of foods by protein-deprived monkeys. *Percep. Mot. Skills* **33**, 918.

Jay, P. (1965). The common langur of North India. *In* "Primate Behavior: Field Studies of Monkeys and Apes" (I. DeVore, ed.), pp. 197–249. Holt, New York.

Jelliffe, D. B. (1955). Infant Nutrition in the Subtropics and Tropics. *World Health Organ. Monogr. Ser.* **29**.

Jelliffe, D. B. (1959). Protein-calorie malnutrition in tropical pre-school children. *J. Pediat.* **54**, 227–256.

Jelliffe, D. B., and Welbourn, H. F. (1963). Clinical signs of mild-moderate protein-calorie malnutrition of early childhood. *In* "Mild-Moderate Forms of Protein-Calorie Malnutrition" (G. Blix, ed.), pp. 12–29. Almqvist & Wiksell, Stockholm.

Jones, C., and Pi, S. J. (1968). Comparative ecology of *Cercocebus albigena* (Gray) and *Cercocebus torguatus* (Kerr) in Rio Muni, West Africa. *Folia Primatol.* 9, 99–133.

Jones, T. S., and Cave, A. J. E. (1960). Diet, Longevity, and dental disease in the Sierra Leone chimpanzee. *Proc. Zool. Soc. London* **135**, 147–155.

Kawai, M., and Mizuhara, H. (1959). An ecological study of the wild mountain gorilla (*G. g. beringei*). *Primates* **2**, 1–42.

Kern, J. A. (1964). Observations on the habits of the proboscis monkeys, *Nasalis larvatus*, made in the Brunei Bay area, Borneo. *Zoologica* (New York) **49**, 183–192.

Kerr, G. R., and Waisman, H. A. (1968). A primate model for the quantitative study of malnutrition. *In* "Malnutrition, Learning, and Behavior" (N. S. Scrimshaw and J. E. Gordon, eds.), pp. 240–249. MIT Press, Cambridge, Massachusetts.

Keys, A., Brozek, J., Henschel, A., Mickelson, O., and Taylor, H. L. (1950). "The Biology of Starvation." Univ. of Minnesota Press, Minneapolis.

Klein, R. E., Gilbert, O., Canosa, C., and DeLeon, R. (1969). Performance of malnourished in comparison with adequately nourished children. Annu. Meet. Amer. Ass. Advan. Sci., Boston.

Kortlandt, A. (1962). Chimpanzees in the wild. *Sci. Amer.* **206**, 128–138.

Latham, M. C. (1968). Pre- and post-kwashiorkor and marasmus. *In* "Calorie Deficiencies and Protein Deficiencies" (R. A. McCance and E. M. Widdowson, eds.), pp. 23–30. Little, Brown, Boston, Massachusetts.

Latham, M. C., and Velez, H. (1966). *Proc. Int. Cong. Nutr., 7th, Hamburg.* (Abstr.) Cited in Latham (1968, pp. 28).

Lim, B. L. (1967). Note on the food habits of the *Ptilocercus lowii* and the *Echinosorex gymnurus* Raffles (Moonrat) in West Malaysia with reference to "ecological labeling" by parasite patterns. *J. Zool.* **152**, 373–380.

McClure, H. E. (1964). Some observations of primates in Climax Dipterocarp forest near Kuala Lumpur, Malaya. *Primates* **5**, 39–58.

McLaren, D. S. (1968). Vitamin deficiencies complicating the severer forms of protein-calorie malnutrition, with special reference to vitamin A. *In* "Calorie, Deficiencies and Protein Deficiencies" (R. R. McCance and E. M. Widdowson, eds.), pp. 191–199. Little, Brown, Boston, Massachusetts.

MacRoberts, M. H. (1970). The social organization of Barbary apes (*Macaca sylvana*) on Gibralter. *Amer. J. Phys. Anthropol.* 33, 83–100.

Maroney, R. J., Warren, J. M., and Sinha, M. M. (1959). Stability of social dominance hierarchies in monkeys (*Macaca mulatta*). *J. Social Psychol.* **50**, 285–293.

Mason, W. A. (1961). The effects of social restriction on the behavior of rhesus monkeys. III. Dominance tests. *J. Comp. Physiol. Psychol.* **54**, 694–699.

Mason, W. A. (1968). Early social deprivation in the nonhuman primates: Implications for human behavior. *In* "Environmental Influences" (D. C. Glass, ed.), pp. 70–100. Rockefeller Univ. and Russell Sage Found., New York.

Mason, W. A., and Green, P. C. (1962). The effects of social restriction on the behavior of rhesus monkeys. IV. Response to a novel environment and to an alien species. *J. Comp. Physiol. Psychol.* **55**, 363–368.

Melville, M. K. (1968). Ecology and activity of Himalayan foothill rhesus monkeys (*Macaca* mulatta). *Ecology* **49**, 110–123.

Menzel, E. W., Jr., Davenport, R. K., Jr., and Rogers, C. M. (1961). Some aspects of behavior toward novelty in young chimpanzees. *J. Comp. Physiol. Psychol.* **54**, 16–19.

Menzel, E. W., Jr., Davenport, R. K., Jr., and Rogers, C. M. (1963). The effects of environmental restriction upon the chimpanzees responsiveness to objects. *J. Comp. Physiol. Psychol.* **56**, 78–85.

Merfield, F. G., and Miller, H. (1956). "Gorilla Hunter." Farrar, Straus, New York.

Mitchell, G. (1970). Abnormal behavior in primates. *In* "Primate Behavior: Developments in Field and Laboratory Research" (L. A. Rosenblum, ed.), Vol. 1, pp. 195–249. Academic Press, New York.

Monckeberg, F. (1968). Adaptation to chronic calorie and protein restriction in infants. *In* "Calorie Deficiencies and Protein Deficiencies" (R. A. McCance and E. M. Widdowson, eds.), pp. 91–106. Little, Brown, Boston, Massachusetts.

Napier, J. R., and Napier, P. H. (1967). "A Handbook of Living Primates." Academic Press, New York.

Oates, J. (1966). Secrets of Spanish Guinea. *Animals* 3, 142–145.

Ochs, K. (1964). Pygmies in my drawing-room. *Animals* 3, 142–145.

Ordy, J. M., Samorajski, T., Zimmermann, R. R., and Rady, P. M. (1966). Effects of postnatal protein deficiency on weight gain, serum proteins, enzymes, cholesterol and liver ultrastructure in a subhuman primate (*Macaca mulatta*). *Amer. J. Pathol.* **48**, 769–791.

Petter, J. J. (1962). Ecological and behavioral studies of Madagascan lemurs in the field. *Ann. N.Y. Acad. Sci.* **102**, 267–281.

Petter, J. J. (1965). The lemurs of Madagascar. *In* "Primate Behavior: Field Studies of Monkeys and Apes" (I. DeVore, ed.), pp. 292–319. Holt, Rinehart, New York.

Pineda, O. (1968). Metabolic adaptation to nutritional stress. *In* "Calorie Deficiencies and Protein Deficiencies" (R. A. McCance and E. M. Widdowson, eds.), pp. 75–87. Little, Brown, Boston, Massachusetts.

Pocock, R. J. (1939). "The Fauna of British India, Mammalia." Taylor & Francis, London. 1939.

Portman, O. W. (1970). Nutritional requirements. *In* "Feeding and Nutrition of Nonhuman Primates" (R. S. Harris, ed.), pp. 27–49. Academic Press, New York.

Reynolds, V. (1967). "The Apes." Dutton, New York.

Reynolds, V., and Reynolds, F. (1965). Chimpanzees of the Budongo Forest. *In* "Primate Behavior: Field Studies of Monkeys and Apes" (I. DeVore, ed.), pp. 368–424. Holt, New York.

Rowell, T. E. (1966). Forest living baboons in Uganda. *J. Zool.* **149**, 344–363.

Sackett, G. P. (1968). Abnormal behavior in laboratory-reared rhesus monkeys. *In* "Abnormal Behavior in Animals" (M. W. Fox, ed.), pp. 293–331. Saunders, Philadelphia, Pennsylvania.

Sanderson, I. T. (1957). "The Monkey Kingdom." Hamish Hamilton, London.

Sauer, E. G. F., and Sauer, E. M. (1963). The South-West African Bush-baby of the Galago senegalensis group. *J. South-West Afr. Sci. Soc.* **16**, 5–35.

Schaller, G. B. (1961). The orang-utan in Sarawak. *Zoologica* (New York) **46**, 73–82.

Schaller, G. B. (1963). "The Mountain Gorilla: Ecology and Behavior." Univ. of Chicago Press, Chicago, Illinois.

Shaw, G. A. (1879). A few notes upon four species of lemurs, specimens of which were brought alive to England in 1878. *Proc. Zool. Soc. London* pp. 132–136.

Sidowski, J. B., and Lockard, R. B. (1966). Some preliminary considerations in research. *In* "Experimental Methods and Instrumentation in Psychology" (J. B. Sidowski, ed.), pp. 3–32. McGraw-Hill, New York.

Simonds, P. E. (1965). The bonnet macaque in South India. *In* "Primate Behavior: Studies of Monkeys and Apes" (I. DeVore, ed.), pp. 175–196. Holt, New York.

Srikantia, S. G. (1968). The cause of edema in protein-calorie malnutrition. *In*

"Calorie Deficiencies and Protein Deficiencies" (R. A. McCance and E. M. Widdowson, eds.), pp. 203–211. Little, Brown, Boston, Massachusetts.

Stanley, W. B. (1919). Carnivorous apes in Sierra Leone. *Sierra Leone Stud.* pp. 3–19.

Stoch, M. S., and Smythe, P. M. (1968). Undernutrition during infancy, and subsequent brain growth and development. *In* "Malnutrition, Learning, and Behavior" (N. S. Scrimshaw and J. E. Gordon, eds.), pp. 278–288. MIT Press, Cambridge, Massachusetts.

Stoffer, G. R., Zimmermann, R. R., and Strobel, D. A. (1973). Punishment of oral-genital self-stimulation in young rhesus monkeys. *Percept. Mot. Skills* **36**, 199–202.

Stollnitz, F. (1965). Spatial variables, observing response and discrimination learning sets. *Psychol. Rev.* **72**, 247–261.

Strobel, D. A., and Zimmermann, R. R. (1971). Manipulatory responsiveness in protein-malnourished monkeys. *Psychon. Sci.* **24**, 19–20.

Struhsaker, T. T. (1967). Ecology of vervet monkeys (Cercopithecus aethiops) in the Masai-Amboseli Game Reserve, Kenya. *Ecology* **48**, 891–904.

Trowell, H. C., Davies, J. N. P., and Dean, R. F. A. (1954). "Kwashiorkor." Arnold, London. Cited in Bradfield (1968b).

van Hooff, J. A. R. A. M. (1962). Facial expressions in higher primates. *Symp. Zool. Soc. London,* **8**, 97–125.

Vis, H. L. (1968). General and specific metabolic patterns of marasmic kwashiorkor in the Kivu area. *In* "Calorie Deficiencies and Protein Deficiencies" (R. A. McCance and E. M. Widdowson, eds.), pp. 119–132. Little, Brown, Boston, Massachusetts.

Viteri, F., Behar, M., Arroyave, G., and Scrimshaw, N. S. (1964). Clinical aspects of protein malnutrition. *In* "Mammalian Protein Metabolism" (H. N. Munro and J. B. Allison, eds.), Vol. 2, pp. 523–541. Academic Press, New York.

Warren, J. M., and Maroney, R. J. (1958). Competitive social interaction between monkeys. *J. Social Psychol.* **48**, 223–233.

Washburn, S. L., and DeVore, I. (1961). The social life of baboons. *Sci. Amer.* **204**, 62–71.

Waterlow, J. C., Cravioto, J., and Stephen, J. M. L. (1960). Protein malnutrition in man. *Advan. Protein Chem.* **15**, 131.

Wayburne, S. (1968). Malnutrition in Johannesburg. *In* "Calorie Deficiencies and Protein Deficiencies" (R. A. McCance and E. M. Widdowson, eds.), pp. 7–20. Little, Brown, Boston, Massachusetts.

Wells, A. M., Geist, C. R., and Zimmerman, R. R. (1972). The influence of environmental and nutritional factors on problem solving in the rat, *Percep. Mot. Skills* **35**, 235–244.

Wharton, C. H. (1950). Notes on the Phillipine treeshrew *Urogale everetti* (Thomas, 1892). **31**, 352–354.

Whitehead, R. G., and Dean, R. F. A. (1964a). Serum amino acids in kwashiorkor. I. Relationship to clinical condition. *Amer. J. Clin. Nutr.* **14**, 313–319.

Whitehead, R. G., and Dean, R. F. A. (1964b). Serum amino acids in kwashiorkor. II. An abbreviated method of estimation and its application. *Amer. J. Clin. Nutr.* **14**, 320–330.

Young, P. T. (1955). The role of hedonic processes in motivation. *In* "Nebraska symposium on motivation" (M. R. Jones, ed.), pp. 193–237. Univ. of Nebraska Press, Lincoln.

Zimmermann, R. R. (1962). Form generalization in the infant monkey. *J. Comp. Physiol. Psychol.* **55**, 918–923.

Zimmermann, R. R. (1969). Early weaning and weight gain in infant rhesus monkeys. *Lab. Anim. Care* **19**, 644–647.

Zimmermann, R. R. (1972). Unpublished research.

Zimmermann, R. R., and Geist, C. R. (1972). A highly palatable and easy to make diet for producing protein-calorie malnutrition in the rhesus monkey. *Lab. Primate Newslett.* **11**, 1–3.

Zimmermann, R. R., and Strobel, D. A. (1969). Effects of protein malnutrition on visual curiosity, manipulation, and social behavior in the infant rhesus monkey. *Proc. 77th Annu. Conv. Amer. Psychol. Ass.* pp. 241–242.

Zimmerman, R. R., and Wells, A. M. (1971). Performance of malnourished rats on the Hebb-Williams closed-field maze learning task. *Percept. Mot. Skills* **33**, 1043–1050.

Zimmermann, R. R., Strobel, D. A., and Maguire, D. (1970). Neophobic reactions in protein malnourished infant monkeys. *Proc. 78th Annu. Conv. Amer. Psychol. Ass.* pp. 197–198.

The Borneo Orang-Utan: Population Structure and Dynamics in Relationship to Ecology and Reproductive Strategy*

DAVID AGEE HORR

Department of Anthropology
Brandeis University
Waltham, Massachusetts

THE STUDY

This chapter is primarily based on field data gathered between September, 1967, and November, 1969, in Sabah, Malaysian Borneo. Discounting brief encounters with other individuals, over 1200 hours of direct observation were made on 27 orang-utans in an area of some 8 square

* Accepted for publication March, 1972. Preliminary versions of this paper have been presented at Yale University, March, 1970, the American Anthropological Association meetings, November, 1971, and the American Association of Physical Anthropologists, April, 1972.

miles along both banks of the Lokan River, approximately 60 miles inland from the east coast of Borneo. Observations were concentrated on 8 individuals within a 1½-square mile study area.

An additional 250 hours of observation on 7 individuals were collected between July and mid-September, 1971, in a 3-km study site on the Sengatta River in the Kutai Nature Reserve, some 15 miles from the east coast of Kalimantan, Indonesian Borneo. The Kutai study area is approximately 350 miles directly south of the Lokan (Fig. 1). A brief report of this study has appeared in the *Borneo Research Bulletin* (Horr, 1972).

Data from the Kutai reserve confirm the findings made previously at the Lokan site and help validate the applicability of these findings to Bornean orang-utans living in lowland primary rainforest.

Both study areas are located along rivers and are largely composed of typical, primary, lowland dipterocarp rainforest.* Elevations range from 50 up to 1000 ft above sea level. The Sabah site included areas of riverine seasonal swamp and some true permanent swamp and, thus, represents most of the normal orang-utan habitat types with the exception of coastal Nipa-Mangrove and mountainous terrain such as Mount Kinabalu.

Since no reliable long-term data on free-ranging orang-utans were available in 1967, and conflicting versions of their population structure were found in the literature, the first year in Sabah was devoted to a survey of a large area of jungle, and the second year to more detailed observations of habituated animals in the 1½-square-mile study area. Orang-utans were located by searching the jungle along paths or transects using two-man teams. Once contacted, every attempt was made to remain with orang-utans—day and night—since relocation was often difficult, especially with nonhabituated animals.

II. POPULATION STRUCTURE AND ECOLOGY

A. Population Composition and Structure

1. Population Structure

The Sabah data confirm the short-term observations of Carpenter (1938), Harrisson (1962), Schaller (1961), Yoshiba (1964), Davenport (1967), and others that orang-utans live in small population units, usually incomplete breeding units, which forage independently in the jungle

* Sabah forest identifications by J. E. D. Fox, Forest Ecologist, Sabah Forest Department.

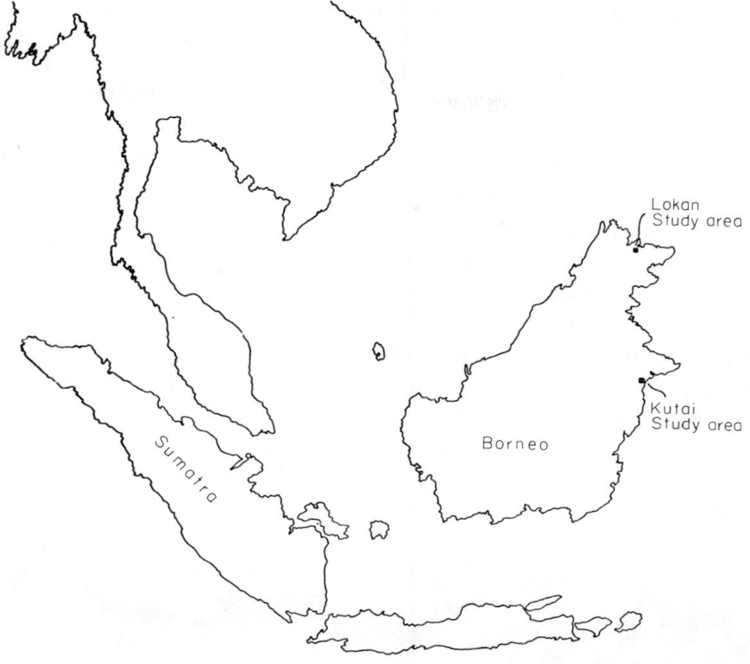

FIG. 1. Orang-utan study sites 1967–1971.

(Horr, 1972). Beyond this, the Sabah study conclusively established that this dispersed population structure is normal for orang-utans living in their primary forest habitat and is not simply the artifact of heavy human predation for food or museum collections. This is true in Sabah and was again confirmed at the Indonesian site. Reports by MacKinnon (1971) that orang-utans are migratory in large numbers may reflect abnormal conditions caused by crowding.

This paper describes the nature of normal orang-utan population structure in lowland Borneo rain forest as established in two typical areas and presents some speculations as to why this organizational strategy may be adaptive for orang-utans.

2. *Population Units*

There are three categories of orang-utan basic population units that forage more-or-less independently in the jungle.

The largest of these and the only persisting social unit is *the adult female with her still-dependent offspring* (Fig. 2). An adult female often has as many as 2 such offspring foraging with her, and a third may be relatively nearby, as will be explained later. Female–offspring units forage in confined home ranges amounting to ¼ square mile or less, depending on the type of jungle. These ranges are relatively permanent; at least for an 18-month period no basic change in range was observed. A female's home range may overlap that of at least one other adult male.

Adult males (Fig. 3) forage independently as solitary individuals. The exact size of the male range is not precisely known as we were not able to follow adult males for more than a few days at a time. The distances they moved and the relative infrequency with which we reencountered given males in a given area implies that their ranges are significantly larger than those of female–offspring units, perhaps 2 square miles.* For example, in one 8-day period we followed an adult male for a straight-line distance of nearly 2 miles (although his circuitous route was considerably longer). In the course of a year, several adult males may be

FIG. 2. Adult female with her young infant. Copyright by David Horr.

* Since the time this paper was accepted for publication, a computer simulation investigation of the model proposed herein strongly indicates that, to maintain adequate population levels, adult male range is probably between 2 and 3 square miles (Horr and Ester, 1975).

FIG. 3. A fully-mature adult male. Copyright by David Horr.

seen at different times in a given place, but long periods of time elapse between sightings of these individual males when compared with the frequency of sightings of individual female–offspring units in a given area.

Adult male ranges overlap the ranges of several adult females.

The third category of population unit is the *developing juvenile* (Fig. 4), which shows increasing separation from its mother to the point of foraging independently. This is, of course, only a transition stage; at some point the juveniles form independent, individually foraging, nonbreeding population units.*

3. Development Cycle

A brief look at the development of orang-utans will help in understanding their population dynamics.

The normal gestation period is 275 days (Napier and Napier, 1967). For the first year of life infant orang-utans cling almost continuously to

* Although reports have been made of small groups of juvenile males foraging together (Harrisson, 1962), no such groups were detected during this study.

Fig. 4. Young, semi-independent juvenile female feeds on fruit. Copyright by David Horr.

their mother's bodies during the daily feeding cycle. During the second year, they take larger amounts of solid food and spend greater amounts of time off the mother's body. During the third year, young orang-utans are becoming rather independent; not only do they feed away from the mother, but they can construct sleeping nests of their own—although orang-utans may still sleep in their mother's nest even when they are young juveniles.

Juvenile females remain in close association with their mothers for the next 2–4 years. During this time they establish their own conservative home range and ultimately breed, starting at age 7 or 8 years (Fig. 5)* to set up female–offspring units of their own.

The development of juvenile males is less clearly known (Fig. 6). Since we encounter apparently solitary juvenile males in the forest and do not normally find juvenile males in close association with adult females, the data indicate that at this stage they are already establishing the larger ranges described earlier for adult males. The age of onset of breeding is not yet known for free-ranging male orang-utans, although it might be as early as the eighth year.

Adult males and adult females assume the ranging patterns described earlier.

* Schultz (1941) reported instances in which fetuses are found in females younger than 9 years of age based on dental irruption.

FIG. 5. A seven-year-old female. Copyright by David Horr.

Old adult males assume more restricted ranges and ultimately spend a proportionally greater amount of their time foraging near or on the ground.* No equivalent old adult females were encountered in the study.

Although the survival rate of males as a class may be less than that of females due to differential mortality in early life, the life expectancy of adult females may be shorter than that of adult males due to child rearing or some other factor. However, the apparent lack of old, semiterrestrial females may simply be because the size dimorphism of males makes old males incapable of full arborealism, whereas the smaller females might function adequately in the trees even after age reduces their agility.

4. *Encounters*

a. *Encounter Patterns.* Encounters between basic population units are infrequent. In Sabah, observed or inferred encounters between adult males and adult females numbered 5 times—although avoidance of meeting was observed more frequently. Intensive observations of a mother–offspring unit over a 16-month period yielded only one direct encounter with an adult male, although observations were not entirely continuous

* It should be mentioned that male and female orang-utans of all ages spend time on the ground, feeding, drinking, walking, and, if the canopy is inadequate, escaping from observers; in one instance, a juvenile female was observed playing in a mudhole.

FIG. 6. A subadult male. Copyright by David Horr.

during the period and other encounters were, of course, possible. It is clear, however, that males and females are infrequently in contact. Encounters between mother–offspring units are correspondingly infrequent, although contact may, of course, be greater if there is closer overlap of home ranges than observed here. Only one encounter, violent and brief, was observed between adult males.

Accompanying this low frequency of association between adult orang-utans, the level of social interaction—when compared with other apes and higher primates—is quite low, even within mother–offspring units.

b. *Female Breeding Pattern.* The data indicate that orang-utan females with offspring do not normally breed again until their infant's third year. Since the estrous cycle of females is not detectable by outward changes in the perineal region, female breeding cycles must be inferred by behavioral data, that is, by their reaction to the presence of adult males. Several kinds of data point to a minimum 2½–3 year breeding cycle: (1) young offspring accompanying females are estimated to be spaced at least 2½ years apart; (2) observed male–female consort pairs never included females with young clinging infants, and females with clinging infants have been observed on many occasions to avoid or directly threaten adult males; (3) two females whose activity was well known over a long period of time began actively approaching adult males only when their young infants began to enter the more independent stage including nest building; (4) finally, in two instances reported by another observer, when adult males attempted to mate with females carrying clinging infants, the interaction was violent with the female attempting to avoid the male and both episodes were described as "rape" (MacKinnon, 1971).

B. Important Features of Habitat

1. *Diet*

Orang-utans in their natural habitat are primarily vegetarian. The nutritionally crucial parts of their diets are fruits, but orang-utans consume great quantities of young or fleshy leaves, the inner bark of trees, and terminal bamboo shoots, as well as an occasional orchid, some species of insects such as termites, and very infrequently dirt from termite mounds. Although orang-utans will accept eggs and ground meat in captivity, no evidence of meat eating or egg stealing was observed in the wild.

2. *Habitat Structure*

As opposed to temperate environments where one finds large numbers of a few species, lowland rainforest in the tropics is characterized by having a large number of plant species in a given area, but normally no large concentration of any particular species in any given spot. This limit on quantity for any given food species is further affected by the irregular nature of fruiting cycles within plant species; that is, some trees skip a year or more between fruition. At the same time, however, fruiting cycles of different species are not rigidly synchronized with each other by strong climatic seasonality, so that in most months something will be in fruit.

Other diet items, leaves, inner bark, bamboo, etc., are available throughout the year, but are not sufficient in themselves to nourish orang-utans. For the orang-utans this means that food of some sort including fruit is nearly everywhere available in its habitat, but never are there large concentrations of fruit, the critical diet item, either in time or space.

In any given day, orang-utans attempt to eat some item of fruit, however scanty the supply. Orang-utan appetites are large. With the exception of some plant species, a single adult male can strip a tree of most of its fruit in a day or two. Therefore, availability of fruit must be considered a major limiting factor.

3. *Predation*

The other aspect of orang-utan habitat to consider here is the lack of effective natural predators for adult Bornean orang-utans. Although a medium-sized cat (clouded leopard, *Felis nebulosa*) is found in Borneo and probably preys occasionally on isolated juveniles, there are no predators (except man) that present a major threat to adult orang-utans.*

III. INTERPRETATION

We can now try to derive a model for the adaptive nature of orang-utan population organization in terms of feeding strategy, predator pressure, and reproductive success.

A. Feeding–Predator Strategy

To be optimally successful, an animal population should not unduly overload the normal food supply of its habitat. This, of course, involves limiting the absolute numbers of individuals of that species in a given area. However, the way in which those individuals are distributed through the habitat also relates closely to the population's optimum utilization of the available food supply. Part of the explanation of orang-utan social structure, then, is related to its optimum habitat utilization strategy.

Female–offspring units move in small home ranges. By definition, these must provide sufficient food throughout the year for the adult female and her still-dependent offspring. The rather even distribution of mother-offspring unit home ranges, with some overlap among them, reflects the evenly although thinly dispersed nature of the food supply described earlier.

A small range is also important to adult females since movement in the jungle canopy is laborious, and the adult female is normally carrying a dependent infant and may also be accompanied by a young juvenile who would have correspondingly greater difficulty in moving in the canopy. Individual orang-utan females tend to follow the same routes in any given part of their range, so that optimum pathways are well known and safety of movement is enhanced.

Adult males are vigorous animals unencumbered by young. Thus, they can range over a wider area. This is important as it means that the adult male does not overload the available food supply for any given female–offspring unit since he is in her home range for reasonably short periods of time during any one visit, and visits are infrequent, only amounting to a few times a year. Adult males and older juvenile males could also respond to seasonal variations in hill and riverine fruiting patterns.

It should be noted here that in the absence of serious natural preda-

* In Sumatra, tigers are terrestrial, not arboreal, predators.

tion, adult males do not serve any protective function for adult females and offspring. There are, therefore, no constraints upon orang-utans to group for defense as in the case of the smaller monkey species in the same forests. In the case of human predation, orang-utans have no effective aggressive defense. In fact, their highly dispersed organization and low level of activity and vocalization are the best mechanism for avoiding detection and, therefore, predation by man.

In relationship to the structure of the food supply, then, the fragmented orang-utan social organization represents an optimum feeding strategy. Large cohesive groups of orang-utans would have to range over wider areas, as the food supply in a given spot at any given time would be rapidly exhausted. Movement over a large area would also be somewhat less adaptive for the reasons cited above. MacKinnon (1971) has reported large concentrations of orang-utans from the Segama River system, some 30 miles southeast of the Lokan study area. This was most probably the result of crowding due to extensive logging activities, and MacKinnon reports migratory behavior, increased contact, and large increase in aggressive encounters as a result.

B. Reproductive Strategy

To understand more fully the nature of orang-utan population structure, it is important to look at the data from the point of view or reproductive strategy and relative parental investment by each sex.

If we assume that male and female reproductive success depend on the degree to which each contributes its own genes to successive generations (Hamilton, 1964; Trivers, 1972), then we can see how orang-utan population structure in its normal habitat contributes to maximum reproductive success for both males and females.

1. *Differential Parental Investment*

Adult female orang-utans have a very large parental investment in their offspring. Not only is energy invested during pregnancy, but the long maturation time of young orang-utans and the low rate of reproduction means that females ensure their maximum reproductive success by a large investment in individual offspring. Although twinning has been reported from zoo populations, the helplessness of young orang-utans and the difficulties of the arboreal habitat make it unadaptive for orang-utan females to carry and care for more than one highly dependent infant at a time. Females, therefore, appear not to breed more than once every 2–3 years as a maximum rate. That is, they avoid breeding until each offspring has reached a level of independence sufficient to allow them to care for anew, highly dependent offspring.

The adult male investment in individual offspring is quite negligible. In fact, it is largely limited to insemination of the female since he is not around long enough to provide any parental care for offspring or serve any protective function for females or young. In a certain sense, males aid their reproductive success by not overloading the food supply of female–offspring units and thereby not competing with offspring and their caretakers for food. In the absence of any positive function for his offspring, thet male is free to maximize his contribution to the next generation by inseminating as many females as possible.

Given the dispersed distribution of adult females and their long breeding cycle, the adult male best enhances his reproductive success by ranging over an area sufficiently large to cover the ranges of several adult females. A preliminary computer simulation study of the model of orang-utan social structure presented here and previously (Horr, 1972) supports this model as functional, while rejecting a model limiting male range to one square mile (Horr and Ester, 1975).

2. *Male Reproductive Tactics*

a. *Contact Vocalizations.* Despite the low level of orang-utan vocalization, males, and to some extent females, have developed a loud trumpeting vocalization which, in the case of adult males, can carry for up to a mile in the jungle under ideal circumstances. This is a segmented vocalization, each segment commencing with a low throaty sound, rising rapidly to a high pitched squeal, then dropping slightly in pitch before being repeated. These segments may be repeated up to 11 or 12 times in one trumpeting sequence. Such a vocalization is structurally well designed for locating the sender in space (Horr, 1969).

In the few instances when males were observed while making this vocalization, the large laryngeal throat sac was inflated. This may assist in providing more air and, thus, a louder volume or may serve in part as a resonating chamber for the vocalization, or both. The throat sac was occasionally, although by no means always, inflated during threat, which did not include the trumpeting vocalization.

This vocalization may be given several times in a single day or by a young adult male, but very infrequently by old males. Normally, a male vocalizes once a day or less. This call is often given when the animal is at rest either in the early morning or late in the afternoon. Occasionally, it is heard at night.

It is postulated that this vocalization is a classic locator call which, by announcing the presence and location of adult males, serves to space adult males in male–male competition while attracting females who are sexually receptive. The *general* location of females is relatively predictable in the jungle, whereas that of males is highly unpredictable due to their large range. Therefore, this mechanism is important in increasing male reproductive success by announcing his presence and location to females.

My explanation of the function of this call is derived from observed reactions to it by other orang-utans: in the case of females, those which were not sexually receptive tended to ignore or even move away from the call. In one instance a female with a clinging infant responded to the call of an adult by giving a threat vocalization even though the male was not yet physically near her. Adult females who were apparently receptive, as described earlier, moved toward the call, in one case with considerable enthusiasm.

The usual observed reaction of adult males to this call is to ignore it if the calling animal is distant. Calling by one adult male did not normally elicit calls from others; however, in one case an adult male gave the trumpeting vocalization near, although out of sight of, a large adult male who was feeding in a fruit tree with an adult female. In this instance, the male in the fruit tree responded with an extremely violent threat display involving vocalizing, breaking and throwing branches, and even charging in the direction of the still-invisible calling male. The other male left the scene with no further vocalization and our search efforts to locate him were in vain.

To summarize, the trumpeting vocalization is given by adult males as they move through their home ranges, thereby identifying their location. Calls are given at varying frequencies, depending on the age and, presumably, the breeding level of the caller. Adult females avoid or are attracted, depending on their state of sexual receptivity, whereas adult males ignore, avoid, or threaten the caller. It is not known whether orang-utans are identifiable as individuals to other orang-utans by their trumpeting vocalizations.

b. *Male–Male Competition.* This model of orang-utan strategy implies strong male–male competition for females as a scarce resource.

Although direct evidence for male competition in the form of dominance or aggressive encounters is limited, certain indirect evidence also points in the direction of such male competition.

In addition to the nonpositive reaction of males to the contact vocalization of other males, the nearly nonexistent incidence of observed contacts between adult males would imply purposeful male avoidance of each other rather than the product of pure chance.

Orang-utan sexual dimorphism, in particular the great difference in size between males and females, could perhaps be best explained in terms of male–male competition for females. Extreme dimorphism in primates such as baboons has often been explained in terms of predator pressure with selection for large, robust adult males that can drive off predators and thereby protect the troop and the young (DeVore and Washburn, 1963). The orang-utan clearly does not conform to this model; not only is its environment nearly predator-free, but males do not remain with females or young individuals to protect them. In any case, group selection arguments are perhaps in themselves fallacious (Williams, 1966). If, however, one postulates that, in competition for females, increased individual body size contributes to individual reproductive success, one may analyze size dimorphism in terms of male–male competition and sexual selection. Baboon size dimorphism could also be explained under such a model. The lack of such dimorphism in gibbons, which have virtually no male–male competition for females, would also support this explanation.

Male–male competition could also account for the extreme features of the male orang-utan face—elaborate facial structure, heavy cheek flanges, beard, large inflating throat sac, etc.—all of which could function in male–male competition for females even though there are no predators on which to use these "weapons." In this context, it is interesting to note that male "size" vis-a-vis females is not a matter of double linear dimensions, rather much of the additional size of the male is in terms of weight and trunk bulk (Schultz, 1941).

IV. CONCOMITANT BEHAVIOR PATTERNS

The orang-utan has developed an unusual social organization for a higher primate species. It has responded to the problem of predation by remaining highly arboreal and by outgrowing its potential arboreal predators in sheer size. In relationship to the structure of the habitat, however, the orang-utan has become so large that a highly dispersed social organization is its optimum feeding strategy to avoid overloading the food supply. The long female breeding cycle contributes to the dispersion of adult males.

This has had certain repercussions on orang-utan behavior. One is a much lessened amount of social interaction and social behavior. Although orang-utan adults may go for days or weeks without contacting other adults, in some instances an encounter in the forest may result in no observable interaction at all. Male and female orang-utans have been observed to move and feed through the same tree crown like ships that pass in the night. Although they were clearly aware of each other's presence, each remained as aloof as the other.

Interaction between young orang-utans in the wild is also limited. Even young orang-utans in semicaptivity with many other young individuals around play far less with each other than might be expected from most higher primates. The percentage of parallel play to total play when young orang-utans are together is very high.

Grooming, often taken as the *sine qua non* of primate positive social behavior, is virtually never observed between orang-utans.

Not only has "troop structure" become diffuse, but the mechanisms for reinforcing social bonds have apparently been somewhat diminished.

At the same time, the importance of the mother–infant bond—critical for all higher primates—is reemphasized in the orang-utan. The meaningful social units in orang-utan society are uterine kin lines. This is not only true of the mother–offspring units, but it is probable that most wider contacts among orang-utans follow uterine kin lines. If females do tend to establish ranges near those of their mother, the wider ranging of the male offspring would also be adaptive in that it would help ensure that his bond to the uterine kin group would not result in a tendency toward exclusively sibling mating. The virtual absence of adult male participation in the socialization of young orang-utans also raised interesting questions about the learning of sex-differentiated behaviors in young males (Horr, 1973).

V. SUMMARY

Under normal conditions, Bornean orang-utans live in small, incomplete breeding units which forage independently in the jungle. Female–offspring units forage in small conservative home ranges of about ¼ square mile, whereas adult male home ranges cover a wide area overlapping the ranges of several adult females. Contacts between adults are rare, and breeding for any given adult female normally occurs no more than once every 2–3 years.

This social organization, which is unusual for a higher primate, reflects the feeding and reproductive strategy of orang-utans. Orang-utan food is evenly but thinly distributed in space and time. Since males serve no protective function against predators for females or young, orang-utans have organized their habitat space into contiguous, semioverlapping female ranges that can support female–offspring units indefinitely with males impinging upon the food supply of any given female for a relatively small percentage of the year. In terms of reproductive strategy, males have virtually no parental investment in young and maximize their reproductive success by competing with other males to inseminate as many females as possible, leaving females to maximize their reproductive success by investing large amounts of time and energy in the care of individual offspring.

Male–male competition is probably maintained through spacing calls, rare but violent threat behavior (and presumably fighting in some instances), large male body size, and elaborate facial structures.

The fragmented nature of orang-utan society is accompanied by a much reduced level of social behavior and interaction, quite apart from the reduced incidence of social encounters.

A final consequence is the nearly absolute role in interaction and socialization assumed by the uterine kin group and in particular the adult female—extreme emphasis even for a higher primate species.

Acknowledgments

The field work on which the foregoing is based was supported by NIMH grant MH–13156. Additional support was obtained from the William F. Milton Fund, Harvard University.

Particular credit must go to: Dr. Richard K. Davenport, Jr., who first surveyed the Lokan area in 1964 and who gave the author invaluable advice and assistance; Professor Irven DeVore, who sponsored the initial research; Mr. J. E. D. Fox, who provided invaluable botanical insights; Mr. G. S. de Silva, who enabled us to work in Sabah; and Mr. Walman Sinaga, who enabled us to work in Kalimantan.

The staff of the Sabah Forest Department and the Lamag District Office were crucial in enabling us to work in the Lokan area, and the field stations of Pertamina and C.G.C. served the same crucial role in the Kutai. Peter and Carol Rodman, who joined the study to investigate the ecological relationships of the sympathic primate species, kindly assisted me in the Kutai Nature Reserve where they had relocated the study.

Finally, my wife Sabine, and my son, Bill, were really the ones who made it all possible.

References

Carpenter, C. R. (1938). A survey of wildlife conditions in Atjeh, North Sumatra. *Neth. Comm. Int. Prot.* No. 12, pp. 1–33.

Davenport, R. K. (1967). The orang-utan in Sabah. *Folia Primatol.* **5**, 247–263.

DeVore, I., and Washburn, S. L. (1963). Baboon ecology and human evolution. "African Ecology and Human Evolution," Viking Fund Publications in Anthropology, No. 36, pp. 335–367. Wenner-Gren, New York.

Hamilton, W. D. (1964). The genetical evolution of social behavior. *J. Theor. Biol.* **7**, 1–16.

Harrisson, B. (1962). "Orang-Utan." Doubleday, Garden City, New York.

Horr, D. A. (1969). Communication and behavior of the slow loris. Doctoral dissertation, Harvard University, Cambridge, Massachusetts.

Horr, D. A. (1972). The Borneo orang-utan. *Borneo Res. Bull.* **4** (2), pp. 46–50.

Horr, D. A. (1973). Orang-utan primary socialization: Mothers and siblings. Paper presented at the AAPA meeting, April 13, 1973; (1974). *Amer. J. Phys. Anthropol.* **40**(1), 140 (Abstr.).

Horr, D. A., and Ester, M. (1975). Orang-utan social structure and habitat: A computor simulation of a free-ranging population. *In* "The Measures of Man: Methodologies in Human Biology" (E. Giles and J. Friedlaender, eds.). Schenkman Publ. Co., Cambridge, Massachusetts.

MacKinnon, J. (1971). The orang-utan in Sabah today. *Oryx* **11**, (2/3), 140–191.

Napier, J. R., and Napier, P. H. (1967). "A Handbook of Living Primates" Academic Press, New York.

Richards, P. W. (1964). "The Tropical Rain Forest." Harvard Univ. Press, Cambridge, Massachusetts.

Schaller, G. B. (1961). The orang-utan in Sarawak. *Zoologica (New York)* **46** (2), 73–82.

Schutz, A. (1941). Growth and development of the orang-utan. *Carnegie Inst., Washington, Contrib. Embryol.* No. 29, pp. 58–111, (*Publ.* No. 525).

Trivers, R. L. (1972). Parental Investment and Sexual Selection. *In* "Sexual Selection and the Descent of Man, 1871–1971." (B. Campbell, ed.), pp. 136–179. Aldine, Chicago.

Williams, G. C. (1966). "Adaptation and Natural Selection: A Critique of Some Current Evolutionary Thought." Princeton Univ. Press, Princeton, New Jersey.

Yoshiba, K. (1964). Report of the preliminary survey on the orang-utan in North Borneo. *Primates* **5**, (1/2), 11–26.

Basic Data and Concepts on the Social Organization of *Macaca fascicularis*

WALTER ANGST

Zoological Institute of Basel University
Basel, Switzerland

I. INTRODUCTION

This paper is a preliminary attempt to extract a general concept of the social organization of *Macaca fascicularis* (Raffles, 1821) from observa-

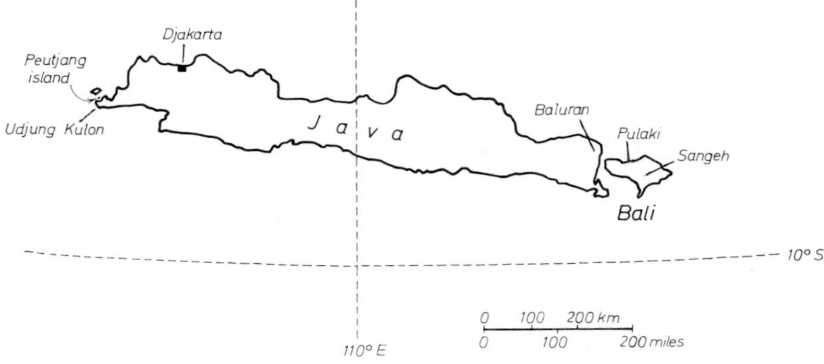

FIG. 1. Areas in which the author studied wild *Macaca fascicularis*. Djakarta is only shown for orientation.

tions in captivity and in the wild. The chapter also presents a considerable portion of the more detailed data collected in the wild.

I lived in Indonesia from September 1969 until January 1971, and studied the long-tailed macaque mainly at six localities (Fig. 1): 20 days in Sangeh (Bali), 13 days in Pulaki (Bali) (both temple areas with protected monkeys); 14 days in the Baluran game reserve (east Java), and the rest of the time in the Udjung Kulon nature reserve in west Java, especially on Peutjang island. Much time was lost by traveling, waiting, and other duties. The great difficulties in observing the monkeys in the forest led me to a more ecological study (Figs. 2 and 3). Hence the data on social organization of the crab-eater in Udjung Kulon are very few. Occasionally I also saw groups of wild *M. fascicularis* at other localities in Java, Bali, and Singapore.

In addition, since 1958, I have observed the crab-eater group in the Basel Zoo. In 1961 I started to make notes and since 1962 I have identified all monkeys individually. Scientific observation began in 1965; I was abroad from January 1968 until April 1969, and again, as mentioned above, from September 1969 until January 1971. Fortunately, reidentification of the known individuals and new identification of offspring was possible both times upon my return. Sixty-seven monkeys have been removed from this group since 1960. For discussion of the social organization of the Basel Zoo group, I chose two dates; the number of monkeys, their age, and kinship are presented in Tables XXV and XXVI, respectively, for each date. (Kinship here always refers to matrilinear genealogies.)

Many aspects of social organization have to be neglected here. The most important one among these is the sexual aspect (Fig. 4), which will be

FIG. 2. (Top) Grooming pair (females) in the forest of Peutjang island. This picture was taken by use of a 500-mm telelens under optimal conditions. Usually the monkeys were hidden. (Bottom) A subgroup, probably comprising exclusively female members of the same genealogy, belonging to the eastern group of Sangeh (Bali). This picture was taken by use of a normal 55-mm lens.

FIG. 3. Pulaki temple on Bali with monkeys. Here observation was no problem.

treated elsewhere. Preliminary conclusions from my field data indicate that the species has a birth peak during the rainy season, from about December until May, in both Java and Bali. Accordingly, they have a mating season in summer, but it is not well delineated. In the Basel Zoo, the same is true. Consort relationships (Fig. 5) can be observed throughout the year, both in the wild and in the zoo. At least in the zoo, they do not affect the female's status according to the criteria used in this paper.

II. GROUP COMPOSITION

A. GROUP COHESION

Recognition of individual monkeys, partially enhanced by trapping and marking, showed that the crab-eating macaque in the wild lives in stable groups and that each group has a home range. In the mainland part of

FIG. 4. Ketut, the central alpha male of Pulaki group, copulating with an adult female. He shows the copulation grimace. A juvenile female is presenting under the adult one.

Udjung Kulon, the groups were found clearly isolated from each other. But this does not mean, that their home ranges did not overlap. On Peutjang island, where the density of monkeys was exceedingly high (preliminary calculations indicate over 400 individuals per square kilometer), the monkeys seemed to be scattered all over, and I often was not able to delimit a group reliably. Figure 6 shows, on the one hand, the scattering of a group and, on the other hand, the use of the same area by four groups. When sleeping the groups were concentrated markedly in all areas that I have visited, but only in Sangeh (Bali) was I able to measure this feature (Table I). The group dispersion was plotted on a map and then measured with an accuracy of 5 meters. The ratios of day-to-night group diameters turned out surprisingly similar. Since the same result was obtained by Vessey (1973) for a group of rhesus monkeys, these figures seem quite reliable. Kurland (1973) also reports a concentration of Bornean groups on the trees for sleep. Probably due to the smallness of these groups, the monkeys were only "spread over 1 to 4 trees, up to 15 m apart" at night, whereas the Balinese groups were more widely

FIG. 5. (Top) Hidji, the present alpha male of the Basel Zoo group and his consort Sumba, grooming each other simultaneously. (Bottom) Pogog, the central alpha male of Bukit group (Bali), with his consort, watching other individuals of the group.

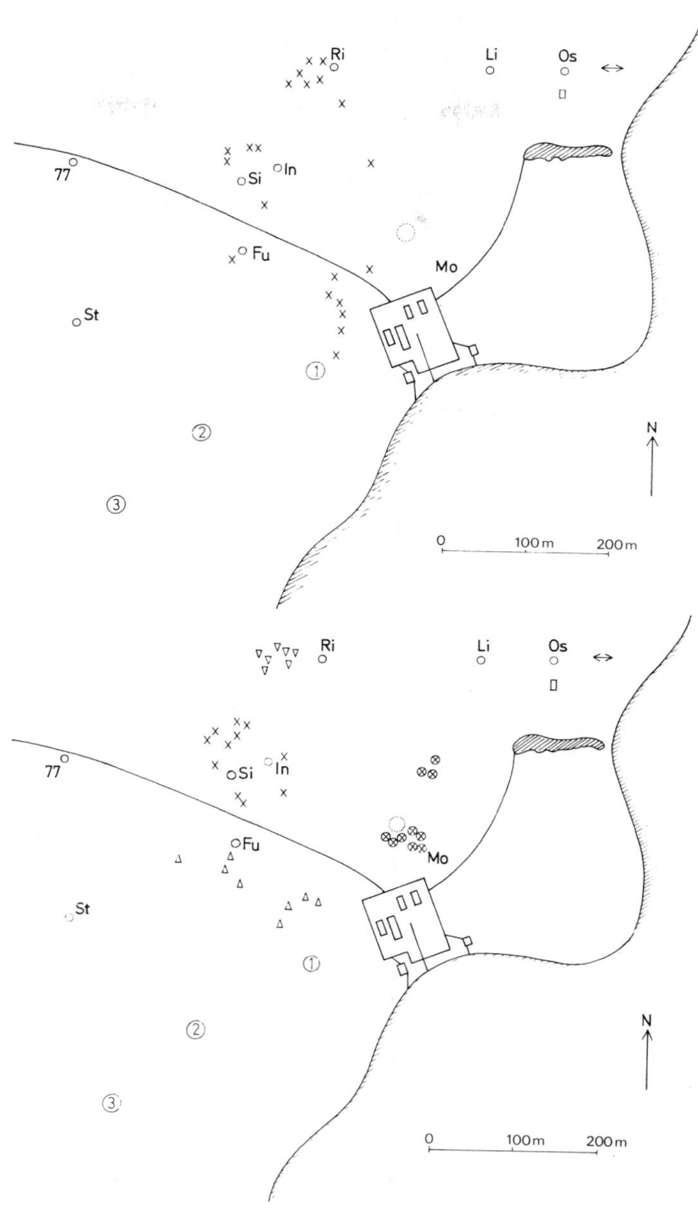

FIG. 6. Peutjang island, area around the field station. (Top) One group (×), extremely scattered, occupies the whole area considered (morning of May 28, 1970). (Bottom) Four groups (×, ▽, ⊗, △) within the same area (morning of October 29, 1970). In both cases it was not certain for all individuals (especially for the ones near "Ri") to which group they belonged.

TABLE I
AVERAGE OF MAXIMUM GROUP DIAMETERS (SANGEH)

Group (No. of individuals)	Daytime (meters)	Night (meters)	Proportion
Eastern group (ca. 59)	110	55	2:1
Central group (47)	90	45	2:1
Western group (18)	75	30	2.5:1
Macaca mulatta (73)[a]	70	35	2:1

[a] La Parguera Island, Puerto Rico (Vessey, 1973).

spread (Table I). On Peutjang island the diameter even averaged about 60 meters.

B. Composition of Wild Groups

The data in the literature on group size are quite divergent. Lord Medway (1969) indicated a range of 8 to 40 or more individuals for Malaysia and Singapore. Bernstein (1967) reported numbers of 14, 15, 20, and 70 in a forest of Malaysia, and 18, including 2 hybrids (with *Macaca nemestrina*) for a group in Kuala Lumpur (Bernstein, 1968). Fooden (1971) estimated nine groups in Thailand, ranging from 7 to 100 individuals with a mean of 35.3. Kurt and Sinaga (1970) counted fourteen groups in Sumatra, ranging from 6 to 20 individuals with an average of 16. Kurland (1973) counted twelve groups in east Kalimantan (Indonesian Borneo), and found a range of 10 to 30 individuals with a mean of 18.2. Lord Medway and Fooden both included urban and protected groups in their figures, whereas the data from Sumatra and Borneo, providing low numbers, consider entirely wild, forest-dwelling groups. It is known mainly from Japanese macaques that provisionized groups increase their sizes to a level much higher than ever found in groups independent of humans. On the other hand, I must add, that for my part I found it impossible to count a forest group completely, except when a group crossed a large open area. It is conceivable that some of the low numbers quoted are affected by missing a part of the group. Tables II and III summarize extensive counts of human-dependent groups in Singapore and Malaysia. I assume that the low numbers in the recent count of the Penang groups are due to a recent splitting. My own data on human-dependent groups in Bali (Table IV) fit well the figures for Singapore and Penang 1965 (Table II). The largest group observed (Bali; 85 individuals) was one

TABLE II
Composition of Wild, but Human-Dependent Groups in Malaysia and Singapore

Age of monkeys	Penang A-troop 1960 (Furuya, 1965)	Singapore C-troop 1960 (Furuya, 1965)	Singapore C-troop 1965 (Shirek-Ellefson, 1967)	Singapore A-troop 1960 (Furuya, 1965)	Singapore B-troop 1960 (Furuya, 1965)	Singapore E-troop 1960 (Furuya, 1965)
Adult males (over 6 years)	7	6	4	4	3	2
Subadult males (4–6 years)	3	4	2	1	1	—
Adult females (over 4 years)	20	18	22	22	14	5
Subadult females (4 years)		2	—	6	2	—
Juveniles (3 years)	3	—	2	3	2	-
Juveniles (2 years)	6	5	3	1	7	2
Juveniles (1 year)	9	9	11	4	6	4
Infants (under 1 year)	14	3	15	4	3	-
Age unknown	10	—	—	-	-	-
Total	72	47	59	45	38	13

Mean $\overline{X} = 18.2$; socionomic sex ratio = 1:1.70.

TABLE III
URBAN GROUPS IN MALAYSIA [a]

Monkeys	Penang Waterfall a	Penang Waterfall b	Penang Waterfall c	Penang Waterfall d	Royal Selangor Golf Club, Kuala Lumpur	Ramanchandran's House, Kenny Hills, Kuala Lumpur
Adult males	3	4	7	6	1	2
Adult females	4	4	13	12	4	2
Juveniles	1	6	6	8	5	1
Infants	—	1	11	3	3	2
Total	8	15	37	29	13	7

Mean $\overline{X} = 18.2$; socionomic sex ratio = 1:1.70.

[a] Data from C. H. Southwick and F. C. Cadigan, Jr., 1972.

TABLE IV

COMPOSITION OF WILD, BUT HUMAN-DEPENDENT GROUPS IN BALI (1970)

Age of monkeys	Sangeh, eastern group	Sangeh, central group	Sangeh, western group	Sangeh, River group [a]	Pulaki, Pulaki group [a]	Pulaki, Bukit group [a]	Pulaki, Tirte group
Adult males (older than 6 years)	4	1	1	2(+?)	7+2	6+3	2
Subadult males (4–6 years)	2	—	2	1 +?	6	3+?	–
Adult females (older than 4 years)	20+?	14	5	13 +?	15+?	22+?	6
Subadult females (4 years)	4 ?	1	–	2 +?	1+?	1+?	–
Juveniles (3 years)	3 ?	2	1	2 +?	1+?	6+?	–
Juveniles (2 years)	9 ?	8	3	7 +?	9+?	8+?	3
Juveniles (1 year)	10 ?	7	4	8 +?	14+?	23+?	2
Infants (under 1 year)	5 ?	14	2	9 +?	6+?	13+?	4
Total	57–61	47	18	44 +?	61+?	85+?	17

Mean of total $\overline{X} = 47.3$; socionomic sex ratio = 1:2.48.

[a] Plus and number (on top line) indicate the number of semisolitary males following the group.

that mainly lived in the hills and thus was not often fed by humans. The mean group size in the three areas was 37.6 individuals.

Forest groups seem to have at least an equal size. Several accounts give lower figures, but the more reliable the count, the higher is the number obtained. Table V represents two counts from Malaysia, and Table VI my own data from Java. The mean size in both areas together is 42.8 individuals per group. I was able to get the complete count of only one single group, which often crossed a large meadow and of which I could recognize individually all adult males, including 3 semisolitary followers. Also I was able to count the Baluran group twice when it crossed a bay, but it is easily possible that some semisolitary males were not with it at that time. The numbers of the Peutjang groups listed were calculated on the basis of trapped and marked animals and the frequencies of spotting them in relation to unmarked individuals.

TABLE V
FOREST GROUPS IN MALAYSIA [a]

Monkeys	Cape Rachado a	Cape Rachado b
Adult males	2	3
Adult females	6	16
Juveniles	12	14
Infants	2	8
Total	22	41

Mean $\overline{X} = 31.5$; socionomic sex ratio = 1:4.40.

[a] C. H. Southwick and F. C. Cadigan, Jr., 1972

Crab-eating macaques usually have multimale groups. All forest groups that I could observe closely but was not able to count completely, contained more than 1 adult male. One-male groups in this species are an exception and so far are only observed under human-dependent conditions. Two hypotheses can be offered for the occurrence of one-male groups in this species: (1) a small part of a multimale group breaks off and forms a split group, clustering around a single male (such a group eventually would grow into a multimale group); (2) by human interference, all semisolitary and peripheral males are trapped or killed and only a single, intolerant central male survives. This would be a more stable situation and probably is true for the two one-male groups observed in Sangeh (Table IV; see also Section III, E). I think in undisturbed forest groups the socionomic sex ratio does not surpass 1:2.5. The ratio of

TABLE VI
COMPOSITION OF FOREST GROUPS IN JAVA (1970)

Age of monkeys	Udjung Kulon, Tjidaun group (count)[a]	Baluran, Kelor group (count)[a]	P. Peutjang, Kobak group (estimation)	P. Peutjang, Biwak group (estimation)
Adult males (over 6 years)	3+3	3(+?)	6	9
Subadult males (4–6 years)	4	5	3	9
Adult females (over 4 years)	11	13	12	18
Subadult females (4 years)	1	7(?)	3	7
Juveniles (3 years)	5	5	6	6
Juveniles (2 years)	6	9	3	4
Juveniles (1 year)	7	12	4	2
Infants (under 1 year)	1	2	2	3
Total	41	56(+?)	39	58

Mean of total $\overline{X} = 48.5$; socionomic sex ratio = 1:1.60.

[a] Plus and number (on top line) indicate the number of semisolitary males following the group.

1:4.4, given for Cape Rachado (Table V), probably does not include peripheral and semisolitary males, which usually escape counts.

C. Composition of Captive Groups

Captive groups used for behavorial studies, which are known to the author, are listed in Tables VII and VIII. With the exception perhaps of the group observed by Goustard (1961) (Table VIII), all compositions may be found in the wild and, thus, are representative for a type of natural group. Best, of course, are groups that approach the average composition of wild groups, as do the first three groups listed in Table VII.

III. SOCIAL ORGANIZATION

A. The Concept of Personal Relationships

In a natural group of macaques, all individuals, with the exception of the youngest infants, know each other personally. Embedded in this knowledge is a specific disposition of how to react to each other in a given situation. Each of these individual dispositions is the result of a genetic base, on the one hand, and of social experience, on the other hand. Social experience here means all previous social interactions in which the individual has taken part, and all observations of other interactions. In addition to genetically based changes resulting from aging and seasonality, every new stimulus is a new input able to modify the reactive system. Thus, the response disposition of an individual is continuously developing from birth to death, and includes sudden changes from time to time. The dyadic response disposition of A toward B, I call the "personal relationship" of A toward B. Of course, the dyadic, unidirectional relationships in a group influence each other in a complex manner. The effects range from "mediation" to "suppression" (Kummer, 1971). The whole network of personal relationships, including all interdependencies, is what is called the "social structure" or "social organization."

We cannot see the social structure, but we can observe the behavior that is determined by the existing social structure and which affects the future structure. Ideally, from the observer's point of view, we should describe the personal relationship of A toward B as the set of probabilities of all responses by A toward B in any situation, so far as they are codetermined by social experience. To illustrate this concept, let me give a simplified example, involving three interacting monkeys: male A grabs

TABLE VII
COMPOSITION OF CAPTIVE GROUPS UNDER INVESTIGATION 1

Age of monkeys	Monkey Jungle (semiwild) (F. Du Mond, inside group, Aug. 1972)[a]	Yerkes Field Station (I. S. Bernstein[b] Dec. 1 1972)	Basel Zoo (W. Angst, July 1, 1967)	Basel Zoo (same group) (W. Angst, July 1, 1972)	Univ. of California Berkeley (T. de Benedictis,[b] Aug. 1972)
Adult males (over 6 years)	6(+?)	4	4	1	1
Subadult males (4–6 years)	Minimum 3	10	7	4	1
Adult females (over 4 years)	Minimum 9	10	26	25	4
Subadult females (4 years)	?	1	3	–	2
Juveniles (3 years)	?	2	5	9	3
Juveniles (2 years)	?	8	9	17	2
Juveniles (1 year)	?	7	12	3	1
Infants (under 1 year)	Minimum 2	9	11	8	2
Total	Minimum 30	51	77	67	16

[a] Data were obtained by the author of this paper.

[b] Personal communication.

TABLE VIII
COMPOSITION OF CAPTIVE GROUPS UNDER INVESTIGATION 2

Age of monkeys	Group studied by M. Goustard (Paris 1961)	Mulhouse Zoo (J.-C. Fady, 1967)	Univ. of Utrecht (J.A.R.A.M. van Hooff,[a] Group Akim, Nov. 1972)	Univ. of Utrecht (J.A.R.A.M. van Hooff,[a] Group Gleuf, Nov. 1972)
Adult males (over 6 years)	—	2	1	2
Subadult males (4–6 years)	2	–	2	1
Adult females (over 4 years)	1	16	2	3
Subadult females (4 years)	1	1(?)	1	–
Juveniles (3 years)	4	18 (Juveniles 3 years to Infants)	2	2
Juveniles (2 years)	2		1	1
Juveniles (1 year)	—		1	2
Infants (under 1 year)	—		2	3
Total	10	37	12	14

[a] Personal communication.

and bites juvenile B; B screams and soon his mother C approaches and starts to scream too; then male A lets B run away and chases C. Let us just consider the behavior of C. Being the mother of screaming B, she shows a high probability of protecting her offspring. In other words, her relationship toward B includes a high probability of protective behavior. The strong stimulus now elicits her approach for protection, which includes a certain probability of aggression in favor of B. Her relationship toward A includes high probabilities both for submissive behavior and flight. This inhibits a full aggression in the above sequence, but instead enhances the likelihood of screaming. Although much simplified, this example also shows that for purposes of analysis the personal relationship may be considered a component of motivation, i.e., motivation that is mainly the outcome of social experience.

In macaques, I propose, there are five basic components of the personal relationship: (1) familiarity; (2) mother–infant attraction; (3) infant–mother attraction; (4) sexual attraction; (5) dominance–inferiority relation.

1. Familiarity. Experiments with macaques, in which artificial groups are formed or strangers are added to groups, show that more aggression occurs among strange than among familiar individuals (Kawai, 1960; Bernstein, 1964b, 1969; Bernstein and Mason, 1963; Furuya, 1965; Hansen *et al.*, 1966). Rosenblum and Lowe (1971) demonstrated that, in an artificially composed group of young squirrel monkeys, the individuals preferred those partners for interaction with which they had been reared together. In some cases this was still evident after a year.

Also within a natural group of macaques, we find different degrees of familiarity. My hypothesis is that the higher the familiarity between two individuals, the more "positive" is their mutual relationship, i.e., the more they stay together, groom each other, cooperate in agonistic episodes, and the less they are aggressive toward each other. Familiarity is closely related to ontogeny and has a long-term effect. An infant becomes first and most familiar with its mother, then with the older siblings and other individuals most often near to its mother, then more and more with age—sex mates, and so on.

What is familiarity? The base of it is an acquaintance, a greater or lesser experience with each other. There is some evidence that acquaintance by itself has a positive effect on attractions. Thus, Zajonc (1971) concluded from experiments with adult humans and from data in the literature mainly on animals that "The mere repeated exposure of an individual to a given stimulus object is a sufficient condition for the enhancement of his attraction toward it, . . ." (pp. 147–148). This hypothesis has not as yet been tested, but I expect that it also holds true for social acquaintance in *Macaca fascicularis*. Superimposed on this important base of all personal relationships is its reinforcement by the behavior of the partners. The result (in terms of response disposition) of basic acquaintance and modifying reinforcement is what I call familiarity. Positive reinforcement (e.g., playing, grooming, hugging) increases the familiarity, negative reinforcement (e.g., avoidance and aggression) decreases it. Familiarity has the most general impact on personal relationship.

2. Infant–mother attraction. Until weaning, the infant has a specific attraction toward its mother, related to physical attachment and nursing. But from the beginning the familiarity affect is superimposed. After reaching a maximum level, the total familiarity dwindles somewhat, due to negative reinforcement by the mother, but, especially in females, effective for the whole life of the offspring.

3. Mother–infant attraction. By giving birth a mother immediately becomes strongly and specifically attracted to her infant. This specific attraction, influenced by hormone status, is transitional; it decreases, as the time of weaning approaches, and when a new baby is born, the newborn becomes the focus of the mother–infant attraction. The remaining attraction toward the weaned infant is probably again mainly the effect of familiarity.

4. Sexual attraction. Sexual attraction only becomes important in mature individuals, and it fluctuates very much. During the estrus of a female, it can be so effective that it overrides nearly everything else. Furthermore, a consort relation can have a positive long-term effect on the familiarity between male and female.

5. Dominance–inferiority relation. This topic is treated in some more detail below (Section III, B).

The relative efficiency of these five fundamental components of the personal relationship can differ according to sex, age, physiological state of the individual, and the situation, but familiarity has the greatest effect. The same is true for the interdependence of these aspects. Of course, the personal relationship is comprised of many other components than the five discussed above. But these other elements are to a great extent derived from the basic ones and often are of limited influence on the behavior observed. Relevant examples of these would be attention relations ("attention structure"; Chance, 1967), fascilitation relations, leader–follower relations, and imitation relations. The age–sex mate preference in play (Fady, 1969) is, I think, the result of trial and error learning with positive reinforcement by the most similarly disposed playmates (Fig. 7). Also play increases familiarity and thus has a long-lasting positive effect on the personal relationship.

Among different species the variables determining the relationships

FIG. 7. Rough-and-tumble play between Kreuz and Asru (Basel Zoo).

and their functioning have different genetic bases, and, thus, we find a range of species-specific social structures. Some species-specific modifications by the environment cannot be excluded, but otherwise the environmental factors mainly cause the variability of social structures within the species.

B. DOMINANCE HIERARCHY

When feeding forest-living *Macaca fascicularis* for the first time, by offering them bananas in trees, the dominance hierarchy among the eating individuals became immediately evident. There was never more than one individual at a time with the bananas, and, sometimes an eating individual would retreat from an approaching one. There was no harrassing and fighting, as can be observed under comparable conditions in baboons. Further observation of wild *M. fascicularis* revealed the higher complexity of the phenomenon, but the results corresponded completely with the ones from captive groups. A consistent dominance hierarchy was also reported by Shirek-Ellefson (1967) for the groups living in Singapore Botanical Gardens. The dominance hierarchy is treated in the following for wild and captive groups together.

1. Criteria of Dominance

There has always been much discussion about how to measure dominance in primates (e.g., see the review by Bernstein, 1970). The concept of personal relationships, however, helps to provide valid criteria for the crab-eater; but one cannot generalize from this to other species. Rowell (1966), for example, using more-or-less the same concept, failed to find such criteria in a group of captive *Papio anubis.*

As already mentioned, dominance or inferiority has to be considered as a component of the personal relationship. Furthermore, it was stated that, for analysis, the personal relationship can be treated like motivation. Moreover, the long-tailed macaque has compound expressions in its repertoire that seem to be directly motivated (in the sense mentioned) by dominance and inferiority, respectively (Table IX). As indicators for dominance, we can use the spreading of the ears and the opening of the mouth during a threat (Fig. 8); as indicators for inferiority, we can use teeth-baring (grimace) and more complex expressions comprising this (Fig. 8). The dominance indicators are not fully reliable, because sometimes, especially young individuals, behave dominantly without really being so. The inferiority indicators, except for the teeth-bare by males during copulation (Fig. 4), on the other hand, have a "direction consistency index" (Rowell, 1966) of 100%. This may be true because each individual has a general tendency for dominance rather than for being submissive and only bares teeth if the motivation is clear and sufficient. Shirek-Ellefson (1967) writes that "With only a few exceptions, scream threats are given by less dominant troop members toward more dominant troop members" (p. 91). Unfortunately, the exceptions are mentioned nowhere else.

A recent paper by Missakian (1972) suggests that the same criteria as used here could also be used for rhesus monkeys. For crab-eaters, approach–retreat and attack–flee interactions, especially in captivity, can be misleading, because the influence of protectors, who sometimes are not even visible, may make a particular inferior individual regularly "win" over a certain dominant one. Such "dependent dominance" (Kawai, 1965) is often a transition stage leading to reversal of "basic rank." In the following, we are concerned with the "basic rank," for which the criteria were set above.

Besides the "open-mouth threat," *Macaca fascicularis* has another form of threat, the "white-pout threat" (Shirek-Ellefson, 1967) or *Hetzen* (Angst, 1974), (Fig. 9), which is invalid as a dominance indicator. Shirek-Ellefson came to the same conclusion. Similar to the criteria introduced by Sade (1967) for rhesus monkeys, some types of agonistic encounters, together with the expressions performed, can also serve as

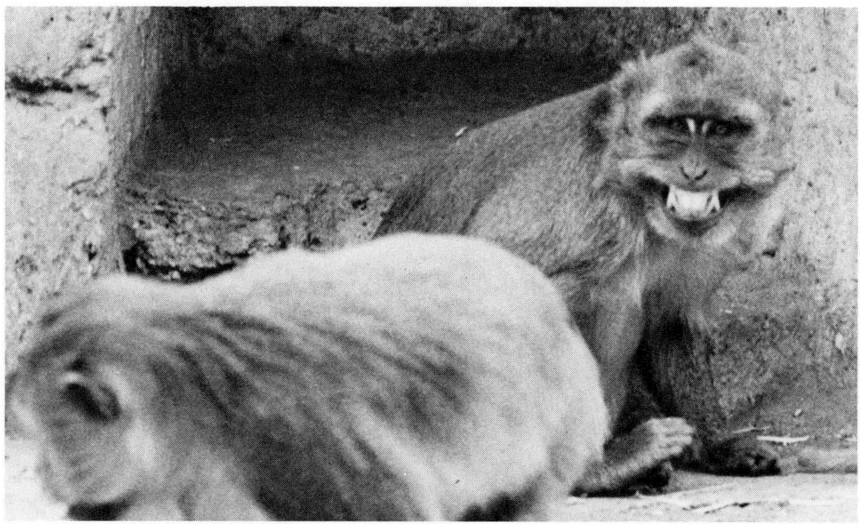

FIG. 8. (Top) Open-mouth threat, an indicator of dominance, shown by Molu (left). In the background sits Nila, an adult daughter of Molu, and on the right, Ratu, the top-ranking female in the Basel Zoo group. (Bottom) Grimace, a reliable indicator of inferiority (excluding the copulation context), shown by Tilu toward Hidji (Basel Zoo).

TABLE IX

CRITERIA FOR DOMINANCE AND INFERIORITY IN *Macaca fascicularis*

Gross motivation:	Name of compound expression by different authors: Van Hooff (1967)	Shirek-Ellefson (1967)	Angst (1974)
Dominant–aggressive	Staring open-mouth face	Open-mouth threat	*Stummes Drohen* + *Drohen mit Grunzen* [a]
Inferior–submissive	Silent bared-teeth face	Grimace	*Zähneblinken* [b]
Inferior–aid-claiming			*Kreischen*
Inferior—aggressive—aid claiming	Staring, bared-teeth, scream face	Scream threat	*Verkreischen*

[a] Not fully reliable.

[b] With exclusion of the copulation context.

dominance indicators in the crab-eater. For instance, if animal A challenges B with a white-pout threat as the facial expression, and B holds ground or even drives A back, while showing an open-mouth threat, B is certainly dominant. But otherwise, also in agonistic episodes, the facial expressions and the vocalizations listed in Table IX are crucial, rather than the spatial outcome of the encounter. Under natural conditions the two kinds of criteria are better correlated than in captivity. The dominance hierarchies given in the following section are based only on the criteria presented above. In determining the rank position of each individual at the chosen day, I used my records of the nearest date. If there were not sufficient records within a year before the day chosen and within 3 months after, the particular dominance relation was left open.

2. *Dominance Hierarchy of Adult and Subadult Males*

Table X presents a dominance hierarchy matrix from Bali, and Table XI a matrix from the Basel Zoo (Fig. 10). All dominance relations obtained

FIG. 9. Giet, of Bukit group (Bali), performing the "white-pout threat." This type of threat does not permit conclusions on dominance.

TABLE X
DOMINANCE HIERARCHY AMONG ADULT MALES OF BUKIT GROUP (PULAKI, BALI)

	Sondra	Item	Oko	Pogog	Wajan	Giet	Purna	Taman	Uwuk
Sondra		+	+	+			+		
Item			?		+	+		+	
Oko				?		?	+	+	+
Pogog					+	+	+	+	+
Wajan						+	+	+	
Giet							+		
Purna								+	+
Taman									+
Uwuk									

FIG. 10. Willy, the alpha male of the Basel Zoo group, and Ceram with baby (1966).

TABLE XI
DOMINANCE HIERARCHY AMONG ADULT AND SUBADULT MALES (JULY 1, 1967) IN BASEL ZOO

	>	A	C	D	U	AA	R	V	AB	Z	X	Y	
Willy	(A)		+	+	+	+	+	+	+	+	+	+	Willy
Fredy	(C)			+	−	+	+	+		+		+	< Fredy <
Fipps	(D)				+	+	+	+	+	+		+	Fipps > Ulrich
Ulrich	(U)		+			+	+	+	+	+		+	Urs
Urs	(AA)						+	+	+	+		+	Roland
Roland	(R)							+	+	+		+	Victor
Victor	(V)								+	+	+	+	Hugo
Hugo	(AB)									+			Kaspar
Kaspar	(Z)										+	+	Xaver
Xaver	(X)											+	Suso
Suso	(Y)												

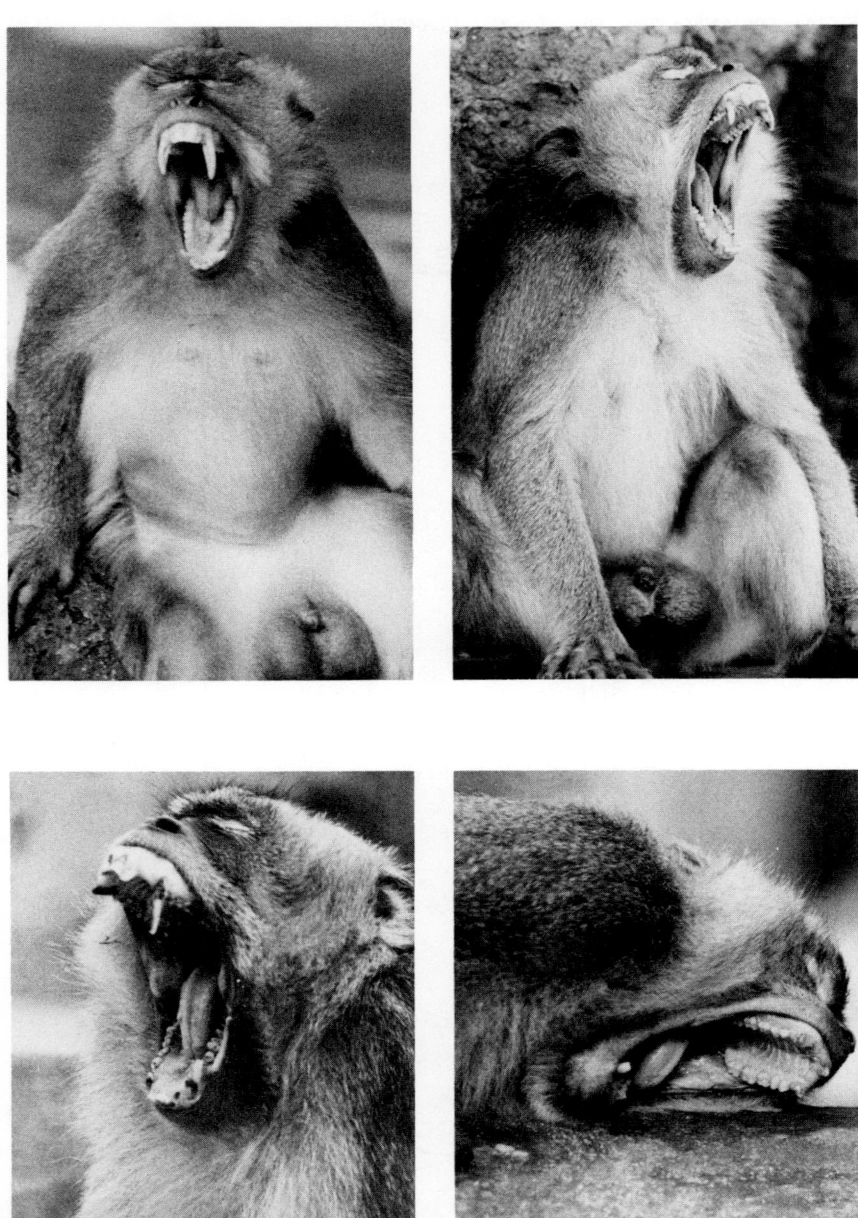

Fig. 11. Canine condition in males as criterion for age estimation (Bali). (Top left) Pendek, showing fully intact canines. Estimated age, 9 years. (Top right) Tegeg, showing considerably worn canines. Estimated age, 18 years. (Bottom left) Njoman, showing very worn canines. Estimated age, 23 years. (Bottom right) Timur, showing almost completely worn canines. Estimated age, 28 years.

in the wild fitted a linear hierarchy in the respective groups. But the day in 1967 chosen for the zoo group happened to be one during a period where there was an exception to linearity. As a transitional stage, this deviation from linearity may also occur in the wild.

I estimated the age of all adult males observed in Bali (Fig. 11) and tried to find a correlation between age and dominance rank (Table XII). I considered exclusively the proven relations according to our criteria. The following results were obtained:

younger (11–20 years) dominating older	13
older (21–30 years) dominating younger	2
$N =$	15

($p < 0.01$)
(Binomial test)

indicating that males in the prime of age dominate old males (Figs. 12, 13, 14). Probably when approaching 20 years the males start to get senile and thus drop in the dominance hierarchy. Males under 10 years seem just to begin rising in rank and place themselves within the hierarchy of the males over 10. In all twenty-two dominance relations between adult and subadult males, the adult males were dominant. Among the subadult males there also seemed to be a linear hierarchy.

For reasons discussed later, the number of adult males in Basel Zoo was always small. But we have the advantage of knowing the exact age and kinship of the individuals born after 1960, and a few before. Table XIII gives the age–rank relations among the adult and subadult males at the two dates selected for this paper (for 1967 I chose a later date than I did for the other tables, in order to demonstrate the stabilized linear hierarchy). Here we find a subadult male, Ulrich, dominant over 3 adult males. Ulrich is the oldest son of the highest-ranking female, Tusnelda (compare Table XXVI). Also her second son, Urs, dominates the youngest adult male and all subadult males except his older brother. Cooperation and protection, based on familiarity mainly, are the reasons for such raised status, but rank dynamics, a voluminous topic by itself, will be treated in a subsequent report. I believe, however, that confinement enhances such processes to a certain extent and that in the wild they are less common. Finally, in the 1967 situation (Table XIII) Hugo is also noteworthy: from other data, it is very likely that he is a younger brother of Willy (Table XXVI)). In the 1972 hierarchy we see the decline of the subadult male Dua. The basic cause of this is that Dua has by far the lowest-ranking mother among the 5 males.

By generalizing from the individual dominance relations among adult and subadult males, we can state two correlations. First, in males from subadulthood up to an age just before senility, older individuals generally

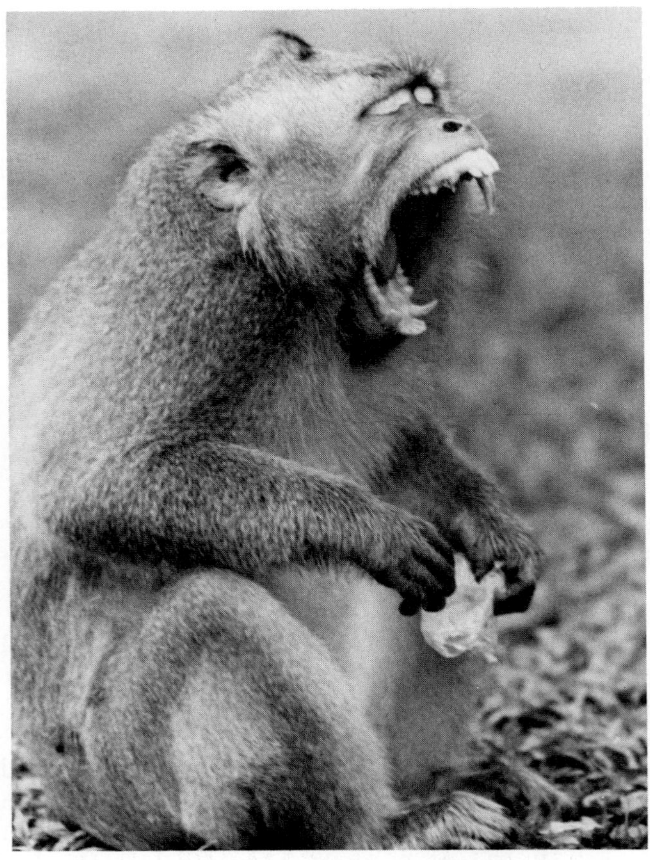

FIG. 12. Barat, the alpha male of the western group in Sangeh, yawning. His right upper canine is broken off.

tend to dominate younger ones. Senile males drop in rank. Second, a factor modifying the first one, high-ranking kinship is correlated with a rise in rank over older males and vice versa. A short look at the crab-eater group living inside Gould's Monkey Jungle provided further support for the rank-age correlation statement.

3. *Dominance Hierarchy of Adult and Subadult Females*

Because during the 5 weeks of observation in Bali, I could identify all adult females in only two of the seven groups, my data on female status

TABLE XII

DOMINANCE HIERARCHIES OF ADULT MALES IN RELATION TO AGE (BALI, 1970)[a]

Bukit group		Pulaki group		Sangeh East	
Age order (years estimated)	Dominance order	Age order (years estimated)	Dominance order	Age order (years estimated)	Dominance order
Pogog (25)	Sondra*	Putu (26)	Oklan*	Timur (28)	*Radja*
Wajan (23)	Item?*	Made (24)	*Ketut*	*Radja* (15)	Timur
Uwuk (23)	Oko?*	Njoman (23)	Deger?*	Pendek (9)	Pendek
Purna (23)	*Pogog*	*Ketut* (19)	Gusti	Legit (7)	Legit
Item (17)	Wajan	Raden (18)	Raden		
Giet (14)	Giet	Oklan (15)	Njoman?	Tirte group	
Oko (13)	Purna	Deger (13)	Putu	Tegeg (18)	*Taka*
Sondra (11)	Taman	Gusti (11)	Made	*Taka* (14)	Tegeg
Taman (8)	Uwuk	Garus (7)	Garus		

[a] Italics indicate alpha males, asterisks indicate semisolitary males.

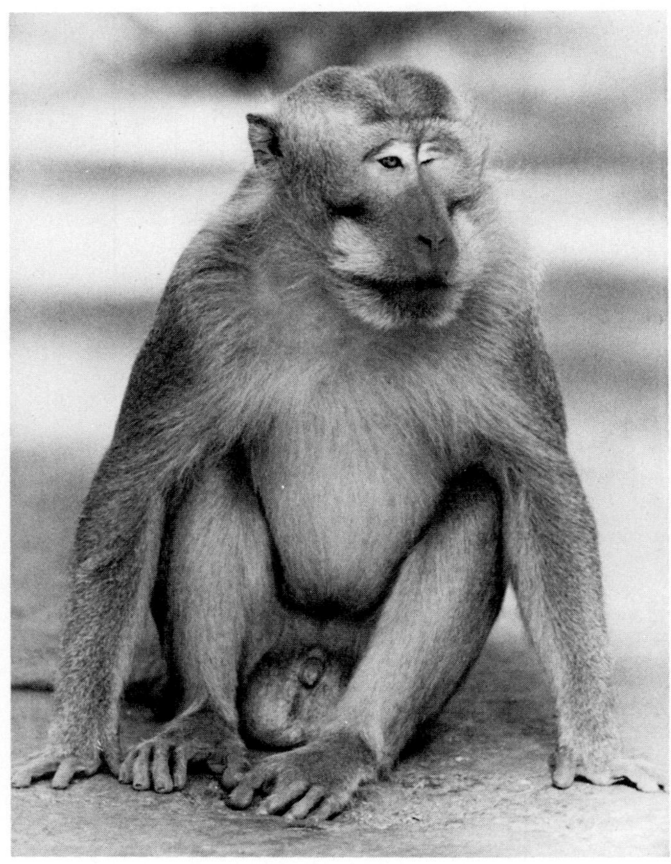

Fig. 13. Taka, the alpha male of Tirte group (Bali), who lacks his left eye.

are scarce and so I give both matrices (Tables XIV and XV). All dominance relations obtained fit a linear hierarchy.

In the zoo groups (Table XVI), we see some inconsistencies, almost exclusively concerning subadult and young adult females. The age–rank correlation in females is less pronounced than in males. Subadult females are already integrated into the adult hierarchy. Only very old females, approaching 30 years, drop in the hierarchy (3 cases so far). On the other hand, the influence of kinship on status is much stronger in females than in males. All adult females of one geneology tend to have the same status in relation to other genealogies (Table XVII). The alpha female in 1972,

FIG. 14. Opat is grooming Hidji, who is yawning (Basel Zoo).

TABLE XIII
DOMINANCE HIERARCHY OF ADULT AND SUBADULT MALES IN RELATION TO AGE (BASEL ZOO)

October 9, 1967		July 1, 1972	
Age order (years)	Dominance order	Age order (years)	Dominance order
Willy (15?)	Willy	Hidji (7)	Hidji
Fredy (9?)	Ulrich	Dua (6)	Tilu
Fipps (9?)	Fredy	Tilu (6)	Opat
Roland (7)	Fipps	Opat (6)	Lima
Ulrich (5)	Urs	Lima (4)	Dua
Victor (5)	Roland		
Xaver (5)	Victor		
Suso (5)	Hugo		
Kaspar (5)	Kaspar		
Urs (4)	Xaver		
Hugo (4)	Suso		

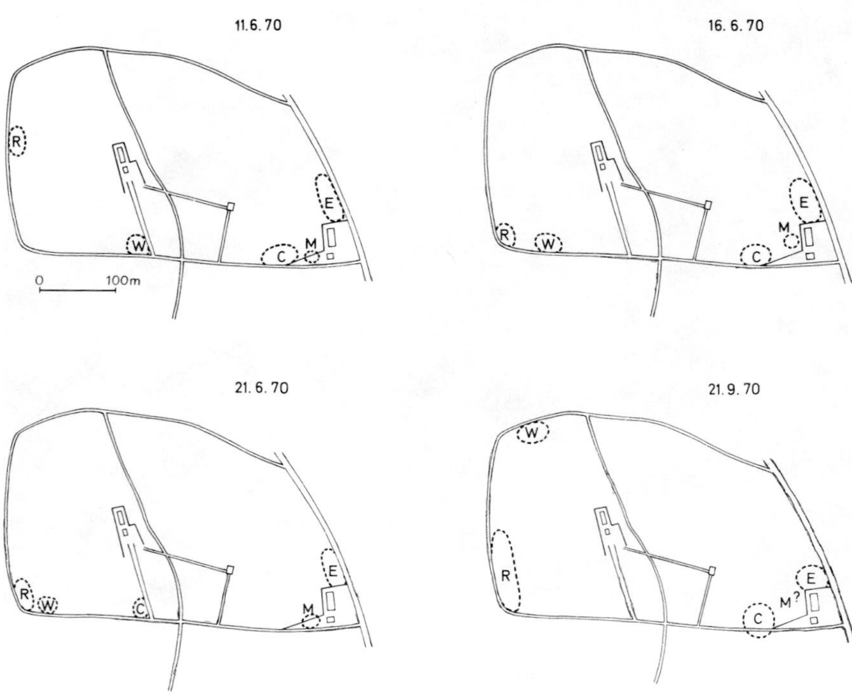

FIG. 15. Four examples of sleeping sites of the groups spending the night in the Holy Monkey Forest in Sangeh, Bali (map after Jahn, modified). E—eastern group, M—Subadult male subgroup of eastern group, C—central group, W—western group, R—river group. Note that C on 21.6.70, uses the same sleeping site as W on 11.6.70.

TABLE XIV
DOMINANCE HIERARCHY AMONG ADULT FEMALES IN SANGEH (WESTERN GROUP)

	Kumis	Wedji	Konek	Brati	Bulan
Kumis		+			+
Wedji			+	+	
Konek				+	+
Brati					
Bulan					

TABLE XV
DOMINANCE HIERARCHY AMONG ADULT FEMALES IN SANGEH (CENTRAL GROUP)

	Putri	Rasni	Mangku	Rines	Darni	Rateh	Niti	Komang	Njaling	Blojok	Sepir	Rai	Alit	Laba
Putri		+	+											
Rasni			+	+				+			+			+
Mangku													+	
Rines							+							
Darni						+								
Rateh							+							
Niti								+	+					
Komang									+					
Njaling												+		
Blojok														
Sepir												+		
Rai														
Alit														
Laba														

TABLE XVI

Dominance Hierarchy among Adult and Subadult Females in Basel Zoo (July 1, 1972)

	>	r	ai	am	at	p	ar	s	ag	o	ak	w	z	e	ae	g	al	ad	b	y	ac	ag	l	g	as	ah	an	af	ap	ao
Tusnelda	r		+	+	+	+	+	+	+	+	+	+	+	+	+	+		+	+	+	+	+			+					+
Ottilie	ai			+	+	+		+	+	+	+		+	+		+		+	+	+		+	+	+	+			+		+
Ambon	am					+	+	+	+			+				+													+	
Timor	at					+	+	+			+	−	+			+		−		+					+			+	+	
Eulalia	p							+	+	+	+		+	+		+	+	+	+	+	+	+		+				+	+	+
Riau	ar							+		+	+	+				+		+			+			+				+		
Flores	s								+	+	+	+	+	+	+	+		+	+	+						+		+		+
Kosi	ag									+	+	+	+	+	+		+	+		+		+		+	+					
Molu	o											+	+	+	+	+							+	+					+	
Klothilde	ak												+	+	+	+		+				+	+		+				+	+
Waltraut	w				+								+		+		+			+	+	+		+	+					
Dana	z													+	+	+	+	+	+	+	+	+		+			+	+	+	+
Esmeralda	e															+	+												+	
Ceram	ae																+	+		+				+	+		+	+		
Susi	g																+	+	+			+	+	+					+	
Liselotte	al																	+		+	+	+			−				+	
Adi	ad				+															+		+		−				+	+	+
Sieglinde	b																										+			
Regina	y																					+	+		+			+	−	
Teun	ac																					+			+	+		+		+
Leti	ag																						+		+			+	+	+
Lombok	l																										+			
Kunigunde	g																	+												
Sabina	as																+											+		
Ewab	ah																										+			
Anastasia	an																													
Sumba	af																												+	
Pini	ap																			+										
Sangi	ao																													

(July 1, 1972)

	az	Ratu
	be	Nila
	ax	Banda
	at	Timor
	s	Flores
	am	Ambon
	o	Molu
	ar	Riau
	bd	Bintan
	ag	Kosi
	g	Susi
	af	Sumba
	z	Dana
	? av	Rindja
	ah	Ewab
Sumba <	bb	Bali
	? ay	Sula
	? ao	Sangi
		Mausi
	ae	Ceram
	? ag	Leti
Mausi <	? au	Moa
	? bc	Buru
	l	Lombok
	ap	Pini

TABLE XVII

INTRAGENEALOGICAL DOMINANCE HIERARCHY OF ADULT FEMALES (NUMBERING 2–4 INDIVIDUALS, 1967 AND 1972)

All genealogy-members rank next to each other	Members of one or two other genealogies are in between	Members of three to seven other genealogies are in between	N
9	2	3	14

Mother dominates daughter:	8	Daughter 1 dominates daughter 2:	2
Daughter dominates mother:	6	Daughter 2 dominates daughter 1:	3
	$N = 14$		$N = 5$

TABLE XVIII

DOMINANCE HIERARCHY AMONG JUVENILES AND INFANTS IN BASEL ZOO (JULY 1, 1972)

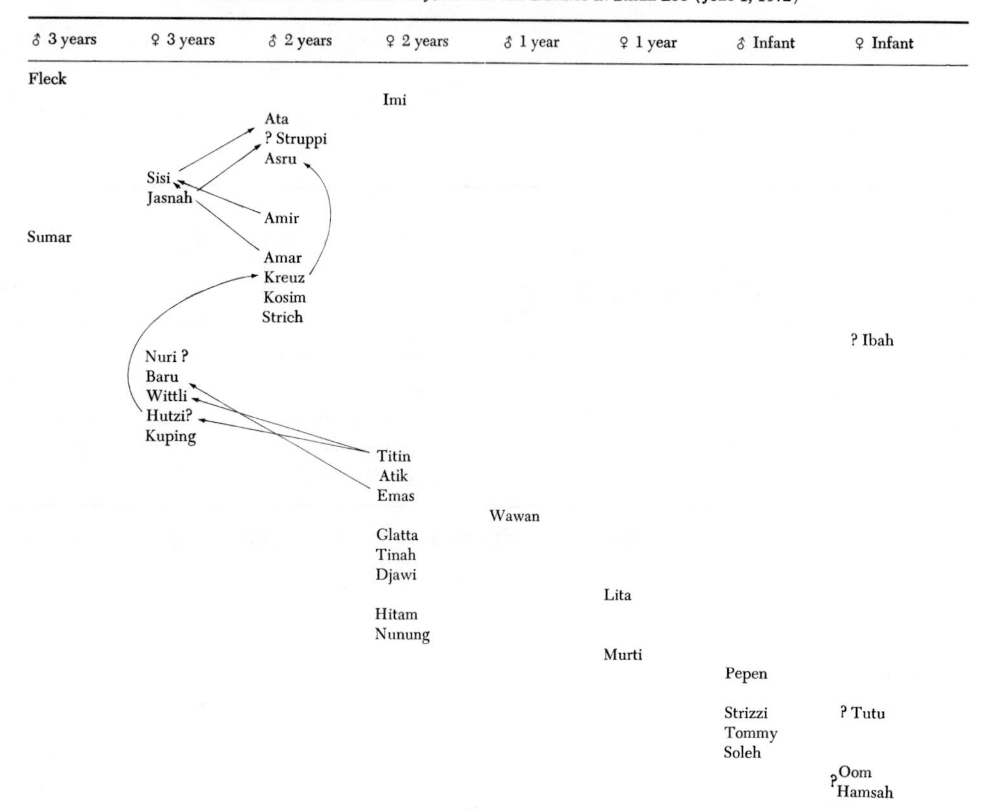

♂ 3 years	♀ 3 years	♂ 2 years	♀ 2 years	♂ 1 year	♀ 1 year	♂ Infant	♀ Infant
Fleck							
			Imi				
		Ata					
		? Struppi					
		Asru					
	Sisi						
	Jasnah						
		Amir					
Sumar							
		Amar					
		Kreuz					
		Kosim					
		Strich					
							? Ibah
	Nuri ?						
	Baru						
	Wittli						
	Hutzi?						
	Kuping						
			Titin				
			Atik				
			Emas				
				Wawan			
			Glatta				
			Tinah				
			Djawi				
					Lita		
			Hitam				
			Nunung				
					Murti		
						Pepen	
						Strizzi	? Tutu
						Tommy	
						Soleh	
							? Oom
							Hamsah

Ratu (Table XVI), is the second daughter of the 1967 alpha female, Tusnelda (Table XXVI); the beta female in 1967, Ottilie, is her first daughter. For other kinship relations see Tables XXV and XXVI. Within the genealogy, dominance criteria are rarely visible. In the Basel Zoo, the hierarchy among the adult females of one geneology seems to be rather randomly established and not even the relations between mothers and daughters are uniform (Table XVII). This is an indication that the geneology members are nearly equals and that dominance among them is functionally less important than among members of different geneologies. Mutual attraction is the prevailing feature among the members of the same genealogy; however, the data are not sufficient to allow firm conclusions at this time.

4. Dominance Hierarchy of Male Juveniles and Male Infants

The statements on juveniles and infants in general are based on zoo observations only. The dominance hierarchy in male juveniles and infants is generally linear and age-correlated. Out of 51 relations obtained among 18 juveniles for July 1, 1967, only 3 were not linear, and 7 were age-inverted.

I do not know the birth dates of the individuals born between 1968–1971. But these males, too, have an almost linear rank order, and, in the younger ones, whose exact age is known, the hierarchy is also age-correlated (Table XVIII). In the entire hierarchy of these juvenile males, there is no significant correlation with the hierarchy of their mothers. Nevertheless, it is reasonable to expect that ultimately male juveniles of high descent are more likely to overtake older ones than juvenile males of low descent.

5. Dominance Hierarchy of Female Juveniles and Female Infants

The data on dominance relations among female juveniles and infants in 1967 are not sufficient for any conclusions. In 1972 we find an almost linear hierarchy as in the males. Out of 44 relations (18 individuals) obtained for July 1, 1972, only 3 were not linear. Not knowing the exact age of these juveniles, I only can compare age classes with each other: older age classes generally dominate younger ones (Table XVIII). The two exceptions in this group, Imi and Ibah, both have high-ranking mothers. Within an age class the dominance hierarchy is correlated with the one among the mothers. For the 6 3-year-old females, the mothers' hierarchy, with one unknown relation left open, corresponds exactly with the one found among the juveniles. For the 2-year-olds, it is somewhat mixed up, and the correlation at this age is not yet significant. For the 2 yearlings, it is inverted but corresponds to age. Among the 4 female infants the hierarchy seems to be correlated with age and mothers' hierarchy at the same time.

6. Dominance Hierarchy among Different Age–Sex Classes

Shirek-Ellefson (1967) indicated a linear dominance hierarchy for the age–sex classes of her study group in Singapore Botanical Gardens. Of top rank were the adult males, next the subadult males, then the adult females, and finally the juveniles (of both sexes plus the subadult females). Moreover, she suggests (p. 32) that "Small infants are extensions of their mothers and cannot be fitted independently into the hierarchy." Our current data confirm this view. In the wild, I never observed a female dominating an adult male, although there were several quite old males. In the zoo, however, there were cases where a young adult male was inferior to one or several high-ranking females. But, also in the zoo, I never saw a male older than 10 years being inferior to a female. I do not have sufficient data to make any statements concerning the dominance relations between subadult males and adult plus subadult females in the wild. In the zoo, younger subadult males are often inferior to high-ranking females. The older they are, the more they are likely to dominate the females. Female juveniles are inferior to all older males and generally also to the males of the same age. A striking exception is the 2-year-old female Imi (Table XVIII). She has a rather high-ranking mother, Molu (this is an unproved conclusion from observation, since Imi was born during my absence). But this cannot be the sole reason for her extraordinary status. The basic cause for this situation most probably is the permanent protection of Imi by the highest-ranking female Ratu. This protection is probably crucially enhanced by the loss of Ratu's own newborn in 1971 and perhaps also in 1970.

As already indicated in discussing the intergenealogical hierarchy among adult and subadult females, the female offspring tend to reach the mother's status. This is already visible in young juveniles. In males it is similar, but they, when approaching adulthood, tend to overtake all females. Table XIX gives all dominance relations of juveniles toward adult females, which do not correspond to their mothers', as obtained for July 1, 1972. There are quite a number of relations in which a female dominates the offspring of a dominant mother, but there are only few relations in which a female is inferior to the offspring of an inferior mother. (The offspring are listed at the same level as their mothers and immediately below.) Although there are more data on some individuals than on others, the evidence of individual differences is reliable. The very low-ranking female Buru, for example, is certainly dominant over more youngsters than the medium-ranking female Rindja. Amar, son of the

TABLE XIX

Dominance Relations of Juveniles toward Adult Females Differing from Those of Their Mothers (Basel Zoo, July 1, 1972)

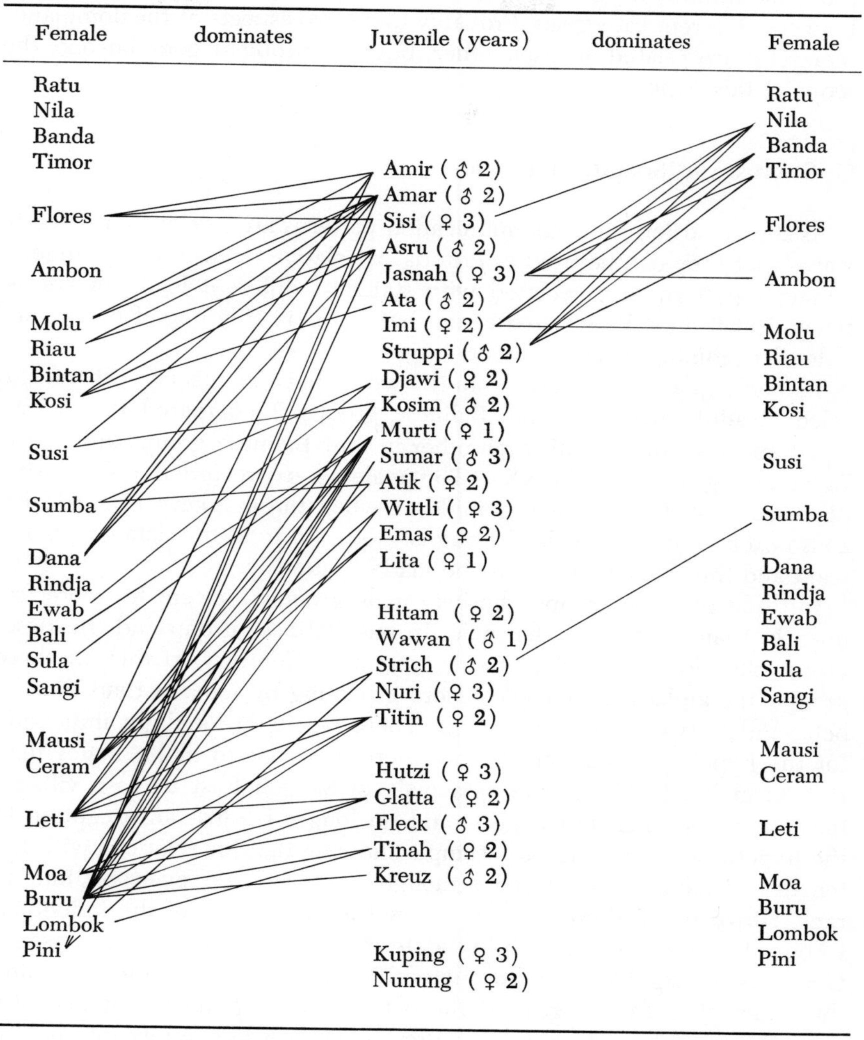

Female	dominates	Juvenile (years)	dominates	Female
Ratu				Ratu
Nila				Nila
Banda				Banda
Timor		Amir (♂ 2)		Timor
		Amar (♂ 2)		
Flores		Sisi (♀ 3)		Flores
		Asru (♂ 2)		
Ambon		Jasnah (♀ 3)		Ambon
		Ata (♂ 2)		
Molu		Imi (♀ 2)		Molu
Riau		Struppi (♂ 2)		Riau
Bintan		Djawi (♀ 2)		Bintan
Kosi		Kosim (♂ 2)		Kosi
		Murti (♀ 1)		
Susi		Sumar (♂ 3)		Susi
		Atik (♀ 2)		
Sumba		Wittli (♀ 3)		Sumba
		Emas (♀ 2)		
Dana		Lita (♀ 1)		Dana
Rindja				Rindja
Ewab		Hitam (♀ 2)		Ewab
Bali		Wawan (♂ 1)		Bali
Sula		Strich (♂ 2)		Sula
Sangi		Nuri (♀ 3)		Sangi
		Titin (♀ 2)		
Mausi				Mausi
Ceram		Hutzi (♀ 3)		Ceram
		Glatta (♀ 2)		
Leti		Fleck (♂ 3)		Leti
		Tinah (♀ 2)		
Moa		Kreuz (♂ 2)		Moa
Buru				Buru
Lombok				Lombok
Pini		Kuping (♀ 3)		Pini
		Nunung (♀ 2)		

fourth-ranking female, is inferior to many more females than Struppi, son of the eighth-ranking female.

There is not much of a dominance component in the relations of black infants toward other categories (and also among themselves). But, when aged about one-half year, they start to establish rank relations and at the end of the first year they have quite a definable status, which develops according to the rules described for juveniles. Corresponding to linearity, the rank stability is generally greater within an age–sex category than between different categories. Probably these two aspects of the dominance hierarchy are related to each other, but this problem goes beyond the scope of this paper.

C. Status and Social Behavior

As stated earlier, some communication patterns are so closely related to status that they can be used as criteria of relative rank. A great range of behavior patterns is correlated with status at different levels of consistency. But it must be stressed again that status is only one of many variables determining social behavior.

The only quantitative data from the wild I have in this context refer to selected adult males in some Balinese groups. I compared the central alpha and beta males with each other in four Balinese groups and in the Basel Zoo group (Table XX). The males were recorded only during phases of rest, but not during midday, when the monkeys often dozed. I also exclude data of males having consorts. In total the data are rather scant and thus conclusions must be made cautiously.

In three Balinese groups, the beta male grooms the females more frequently than does the alpha male. In one Balinese group and the Basel group, the alpha male does more grooming. Contrary to this, in three groups, the alpha male received more grooming by females than did the beta, and in two groups vice versa. These results are a slight indication for the hypothesis that females are more attracted to alpha males than to beta males. If Timur, the beta male of Sangeh East is, as a villager told me, the former alpha, he would no longer be just an exception to the hypothesis, but rather an example showing that his relations with the females remained unchanged although another male overtook him in rank. However, all these data also show the importance of the individual whose behavior is not fully determined by the social system. Generally (7 significant cases out of 10) males receive more grooming than they give. In two groups, the beta male self-groomed significantly more than the alpha. In two other groups this was not significant, and, in the fifth group, the alpha did somewhat more self-grooming. I suggest then that beta males, because they have less grooming contacts on the whole than alpha males, tend to groom themselves more frequently than alpha males. In all five groups, the alpha more often mounts a female than the beta does, but only 2 cases were significant. I am sure that more data would make every case significant. Table XXI gives the mounting

TABLE XX

A COMPARISON OF FREQUENCIES (IN HALF-MINUTE UNITS) OF SOME BEHAVIOR PATTERNS IN THE TWO TOP-RANKING CENTRAL MALES OF FIVE GROUPS

	Basel Zoo		Sangeh East		Bukit	
	α	β	α	β	α	β
Male grooms female	49/300 = 16.3% [a]	7/300 = 2.3%	5/507 = 1.0%	24/347 = 6.9% [a]	1/135 = 0.7%	24/153 = 15.7% [a]
Female grooms male	55/300 = 18.3%	82/300 = 27.3% [c]	68/507 = 13.4%	122/347 = 35.2% [a]	115/135 = 85.2% [a]	14/153 = 9.2%
Self-grooming	31/300 = 10.3%	25/300 = 8.3%	36/507 = 7.1%	26/347 = 7.6%	3/135 = 2.2%	6/153 = 3.9%
Male mounts female	7/300 = 2.3%	3/300 = 1.0%	5/507 = 1.0%	2/347 = 0.6%	5/135 = 3.7%	1/153 = 0.6%
Dominance displays	6/300 = 2.0% [b]	0/300 = 0.0%	3/507 = 0.6%	4/347 = 1.1%	1/135 = 0.7%	1/153 = 0.6%

	Pulaki		Tirte	
	α	β	α	β
Male grooms female	89/275 = 32.4% [a]	40/263 = 15.2%	1/176 = 0.6%	17/190 = 8.9% [a]
Female grooms male	153/275 = 56.7% [a]	56/263 = 21.3%	79/176 = 44.9% [a]	11/190 = 5.8%
Self-grooming	2/275 = 0.7%	28/263 = 10.6% [a]	1/176 = 0.6%	15/190 = 7.9% [a]
Male mounts female	12/275 = 4.4% [c]	2/263 = 0.8%	8/176 = 4.5% [b]	1/190 = 0.5%
Dominance displays	5/275 = 1.8%	0/263 = 0.0%	7/176 = 4.0%	1/190 = 0.5%

[a] $p < 0.001$ The chi-square test used for all p. Low values with Yates' correction. [b] $p < 0.05$. [c] $p < 0.01$.

scores of the 4 oldest males in Basel Zoo collected over a year. They are ranged according to their dominance hierarchy, and it is evident that the alpha male mounts females much more frequently than does any other male. The second- and third-ranking males had equal scores, but the former male more often mounted adult females than younger ones. The fourth male again had a lower score than the second and third male, and he almost exclusively mounted nonadult females. The only adult female that he was observed to mount once, was his mother, and this act was very short and without intromission.

TABLE XXI
MOUNTING FREQUENCY IN RELATION TO DOMINANCE RANK [a]

	Hidji (1)	Tilu (2)	Opat (3)	Dua (4)	
Male mounts adult female	126	15	11	(1)	(his mother)
Male mounts younger female	14	4	11	4	
Total	140	19	22	5	

[a] General observation (112 hours) from July 4, 1971 to June 6, 1972.

"Dominance displays," i.e., strutting with tail up and shaking of branches and trunks, are performed more frequently by the alpha male in three groups (only in one of which it is significant). In one group the scores are equal, and in the Sangeh East group, the "retired" alpha showed some, although not significantly, more dominance displays than the actual alpha.

D. Role Differentiation

In discussing the dominance hierarchy, we generalized the personal relationships referring to age–sex classes. Another possibility of categorization, basing on social interaction and "labor division," is achieved by describing roles, as shown by Bernstein (1964a, 1966), Bernstein and Sharpe (1966), Gartlan (1968), Crook (1970), and by Reynolds (1970). In accordance with Rowell (1972) the term "role" should be defined functionally. Therefore, I consider as a role the functional part of a social system, which is fulfilled by the specific behavior of particular individuals.

1. Group Protection Role

In crab-eaters, and probably in all macaques, we find the highest concentration of roles in the central alpha male, e.g., the social control role, leader role, and group-protection role. During 33 days in Bali, I observed 12 attacks by central alpha males (4 individuals) toward dogs (5) and humans (7). In 8 of these instances the alpha male supported the aggression of other group members, including 2 cases in which an endangered juvenile was screaming. In another case an individual was uttering alarm calls, and in 3 instances I noticed no other individuals reacting to the danger except by retreat. The group-protection role is not an exclusive attribute of the central alpha male. He is the most frequent and most effective individual in exerting this role, but, as shown above, there is a high amount of cooperation in group protection.

Let me illustrate this group-protection role by one example:

June 7, 1970, central group Sangeh, Bali

A 3-year-old female, feeding on a coconut palm, is cut off from her group by two boys, who walk in between and start to throw stones at the juvenile. The monkey jumps to the only other palm in reach and starts to descend. But one of the boys approaches quickly and, therefore, the juvenile stops about 4 meters above ground and starts to scream. An adult female (probably her mother, as indicated by other observations), together with the alpha male immediately rush to the place and pant-threaten the boys. The alpha male attacks the boys. One of them flees, the other counterattacks with a stick, threby leaving the palm with the juvenile on it. The juvenile descends and runs behind the alpha, who retreats. The juvenile pant-threatens the remaining boy.

One day I made a rather peculiar observation. Several females took turns in vigorously biting into a stick which lay in a field. Other females and young watched and screamed. The alpha male came and also bit into the stick. Then they all left it. When I collected the stick, several individuals uttered alarm calls. The stick turned out to be one that sometimes had been used by boys for chasing off the monkeys and now had been left behind. The stick thus served as a substitute for or association with the hostile boys, toward which aggression could be directed.

The important aspect of group protection, namely protection against other groups, will be described in Section IV.

2. *Alarm Role*

In the zoo, the alpha male utters alarm calls more frequently than any other individual. Imanishi (1965) hypothesized that in Japanese macaques the highest-ranking male present warns the group by giving the male-specific "kwan" call. In the Basel Zoo, the male alarm call is triggered by the presence of the keeper, who sometimes captures some individuals. In wild Japanese macaques, the kwan call is a response to the human observer. In this case also, the groups concerned have probably had some negative experience with humans. These cases suggest that the dominant male's alarm call is linked with the group-protection role. However, in Udjung Kulon, I rarely found evidence of this sort. Since I usually avoided disturbing the groups as much as possible, I could seldom identify the individuals calling. The cases in which the alarmer was clearly identifiable as a male were relatively few. Once an adult male, continuously alarm-calling, approached me and finally settled about 12 meters vertically above my head.

TABLE XXII
ALARM-CALLING WITHIN THE FIRST 3 MINUTES OF AN ENCOUNTER

No. of voices heard	No. of individuals seen			
	0	1	2–5	More than 5
0	∞	70	70	29
1	33	9	41	12
2	—	—	12	5
More than 2	—	—	3	7
?	—	—	14	2

Table XXII gives the numbers of individuals alarm-calling during short encounters. It shows clearly that the monkeys tend to avoid calling and, if alarm calls are uttered, it is usually only by one individual. I sometimes missed a single soft call that induced flight, but I also missed groups sneaking silently away on many occasions. Furthermore, I have evidence of social facilitation of flight without any calling.

During observations of a group where some alarm-calling was audible, the role character of alarm-calling became apparent. Generally, a single hidden individual called, and when that animal stopped, another one took over the role (Table XXIII). If, for a while, 2 or 3 individuals vocalized, then only 1 did so intensively.

To obtain more objective data, I started to listen to groups over prolonged periods of time and to record the number of voices per 0.5-minute unit (Table XXIV, p. 376). The amount of overlap in alarm-calling was in fact less than half of the amount expected for random distribution (of the recorded voices). This suggests that an alarm-calling individual inhibits others from doing the same. The alarm role is economized to its minimum and is fulfilled by a single individual. The determination of

TABLE XXIII
ALARM-CALLING AFTER THE INITIAL 3 MINUTES OF AN ENCOUNTER

No. of voices heard	No. of individuals seen			
	2–5		More than 5	
1	5	} 7	6	} 8
More than 1, consecutively	2		2	
More than 1, simultaneously	2		1	

the role performer seemed to vary from case to case, but the actor always was a rather distant individual, usually hidden in a tree.

TABLE XXIV
ALARM-CALLING DURING PROLONGED PERIODS OF OBSERVATION

Total number of 0.5-minute units recorded (11 periods)	2190
Number of 0.5-minute units with one voice alarm-calling	810
Number of 0.5-minute units with more than one voice alarm-calling	147
Number of 0.5-minute units with more than one alarm-calling voice as expected assuming random distribution of the calling by different individuals within each period	317.17

E. PERIPHERAL MALES

Looking at Table XII we see that several of the most dominant males are marked as semisolitary ones. There were indications that among the 3 semisolitary males following the Tjidaun group (Table VI), at least one was dominant over all central males. What then is a semisolitary male and how is it possible that a central male can be inferior to such an individual?

Itani and Kawamura (quoted in Imanishi, 1957) have established three categories of spatial–social positions which they call central, peripheral, and solitary. This classification was based on the spatial arrangement of the monkeys at artificial feeding grounds. In multimale groups of *Macaca fascicularis*, such as the ones described in Bali, we find a continuous range of positions from central through peripheral to semisolitary. I have no evidence for completely solitary individuals. The central males are well integrated and feed, move, rest, and sleep within the group, whose core consists of the adult females with their youngest offspring. These males are frequently and extensively groomed by these females and the alpha male often copulates with them. The less a male fulfills the criteria mentioned, the more he is considered to be peripheral. Semisolitary males are characterized by being apart from the group most of the time, not merely at its periphery, and by having even less contact with its members than the peripheral males. But semisolitary males may copulate with females that come to them.

I do not know why one adult male is central and another of the same age is semisolitary. But, by combining the ontogenetic findings from captivity and the situational results from the field, we at least can gain some idea of possible mechanisms involved. Let us compare the son of a high-ranking mother with the son of a low-ranking mother. When they become subadult, the central alpha male, and probably other adult central males, too, generally grow less tolerant of the young males and drive them away from the adult females. At the same time, the son of the low-status mother successively overcomes in rank the females dominating his mother. The male of high-rank descent, on the other hand, is already dominating almost all females. Overcoming the females in rank involves considerable struggle, in which the central males tend to support the females. Through this process, and even after some time, because the overtaken females remain aggressive for a while, the subadult male of low descent is pushed much more to the periphery of the group than the one of high descent. Additionally, due to more contacts with high-ranking females than with low-ranking ones (except his own kinship and special cases), the central alpha male is more familiar with the sons of the former. He more frequently interferes against the male of low descent than against the one of high descent, including those cases in which the two young males fight each other.

At this age the subadult males tend to form temporary exclusive subgroups, as I have observed in two Balinese groups. They are often in contact with the main part of the group, but sometimes move away from it and usually sleep separately nearby.

As they approach adulthood, the young males start to overtake other low-ranking adult males. In this process, individuality and cooperation are probably equally crucial. But the son of the high-status mother, due to his more positive relationship to the central alpha male especially, has a better chance to move into the central hierarchy than the son of a low-status mother. The latter is tolerated only when being fully submissive. The former is able to replace other males in the central hierarchy with the tolerance or even help of the central alpha male. If the young adult male of low descent is a tough individual, he will, nevertheless, overcome the central males in dyadic rank, but this he achieves only at the cost of being repulsed from the group, i.e., he becomes semisolitary. In the Pulaki and Bukit groups, the central males regularly, by cooperative aggression, drove the semisolitary males away from the females.

F. Social Organization in the Wild and in Captivity

In comparing the social organization of captive groups of *Macaca fascicularis* with those of wild groups, I came more or less to the same conclusions as Rowell (1967) in her general outline of the problem. The main handicap of captive groups is generally the fact that its members are put together rather than being born into the group. Therefore in captive groups the personal relationships are less developed and sometimes are atypical. However, the groups in the Monkey Jungle, Yerkes Field Station (being a split group from Monkey Jungle), and Basel Zoo are in this respect comparable to wild ones. Often captive groups have an unnatural composition. But, as mentioned earlier, the groups listed in Tables VII and VIII, with one exception, could all be found in the wild.

A deficiency, which probably applies to all captive groups of crab-eaters, with the exception of the ones in Monkey Jungle, is the spatial confinement, making it impossible to develop the natural adult male relations as described above. Adult males inferior to females, terribly suppressed, restricted to little corners of the cage, and so on, are all distortions of the natural structure due to confinement. In the usual enclosure situation, it is impossible for a male to rise in rank as a peripheral, not to speak of becoming a dominant semisolitary. In Basel Zoo, many males are therefore removed.

Because the animals are always near to each other and are fed concentratedly in time and space, as well as having little else to do, aggression and various other kinds of social behavior become hypertrophied. This is primarily a quantitative difference compared to natural conditions, but it also has consequences for the social structure. Social protection is more efficient in captivity than in the wild, and the protected young therefore rise considerably earlier in rank than they do in the wild. Qualitatively, however, the behavior observed in well-kept captive groups at least yields a valid model of many aspects of natural behavior.

G. Function of Social Organization in the Crab-Eating Macaque

Macaca fascicularis is a widespread pioneer species, living even on islands where no other primate species can live, and adapting to the most varied habitats. Periods of unfavorable conditions are common for this species. But, like any other species, it tends toward an optimal utilization of any given habitat. Nevertheless, the basic social organization is, as far as we know, the same everywhere. Thus, one may assume, this organization has some selective value.

Let me now pick out one single fragment of the whole problem: the peripheralization of males. This phenomenon is well established. Another fact is the total sex ratio of adults being skewed in favor of the females, although at birth it is almost equal. I suggest that the first phenomenon is the cause of the second. I saw many wounded adult males, both central and peripheral, as well as semisolitary ones. The groups occupy the safest places for day and night, so the expulsed individuals probably suffer more from predation and human killing. The groups also occupy the best and, during harsh times, probably almost all food localities. Needless to say, the frequent harrassments by the group and fights ending in the sea or other unsuitable places are an additional danger. Wounds hampering mobility can be fatal to a semisolitary male, whereas it hardly endangers the life of a central male. As a result of all these factors the number of males is reduced. Because a few males are sufficient to fertilize many females and because a high reproductive potential, represented by many females, has a high selective value, we can consider the other males as surplus individuals. Thus, accepting Wynne-Edwards' (1972) ideas of "conventional competition," we conclude as a hypothesis that the peripheralization of males in *M. fascicularis* is a social mechanism enhancing the optimal utilization of a habitat. If expulsed males can penetrate other groups, this could further the gene flow, but I have no data on this.

It is interesting to note that in the Basel Zoo group, very old, sterile females get somewhat peripheral, too, whereas low-ranking females with black babies become more central than those without babies. These differencies are due to the high-ranking females and their attraction toward mothers with small infants, and not to the central males. But it must be added that the carrying of black babies seems to inhibit to some extent aggression by the males, and this may also play a role.

Hypothetically we could establish an order of decreasing pressure by the environment on different sections of a crab-eater group: (*1*) semi-solitary males; (*2*) peripheral males; (*3*) exploring juveniles; (*4*) old peripheral females; (*5*) low-ranking females without babies and experienced juveniles; (*6*) central males, high-ranking females, and low-ranking females with babies. Of course, these categories would be somewhat interwoven. But the important result of this speculation is that the core of reproduction is the safest element.

IV. INTERGROUP BEHAVIOR

On Peujang island the groups were not territorial and even tolerated experimentally introduced strangers in their home ranges (Angst, 1973). There were indications, however, that between some groups there existed a dominance–inferiority relationship.

A. Sangeh

There were three groups of *Macaca fascicularis* living permanently in the Holy Monkey Forest of Sangeh (Table IV), a dipterocarp forest. They were fed daily by local people and tourists. The food was not distributed evenly, but along a path in such a way that the eastern group got most of the food. The central group got part of the food, and the western group only rarely got something. All three groups ranged throughout the entire forest, but their short excursions usually led to different surrounding areas. Among the three groups there was a linear rank order: the eastern group dominated the other two, the central group dominated the western one, and the western group was inferior to the other two. The rank order, as indicated above, was correlated with the size of the groups and was the reason for the different success in obtaining food from man. In the evening, a fourth group (river group; Table IV) often entered the forest from the west to sleep. It clearly dominated the western and central groups. I never observed contacts between the river group and the eastern group.

The inferior group used to retreat immediately when it encountered a dominant group while moving in the forest. This retreat was not dependent on location. The inferior group retreated on the approach of the alpha male of the dominant group, and even the beta male of the eastern group was able to displace an inferior group. I never saw an alpha male or the eastern beta male ("retired alpha") acting aggressively toward adult members of an inferior group. Nonetheless, adult members of an inferior group especially avoided those males.

By contrast, conflicts were rather frequent among young males and adult males of different groups. In such cases the alpha male of the inferior group sometimes powerfully helped his group members. Once, after a short conflict between the eastern and central groups, I saw an adult female (Mangku) of the retreating central group with a fresh wound on her head. However, I also observed nonagonistic contacts such as copulation, between an eastern subadult male (E1) and an adult female (Njaling) of the central group, for instance, and juveniles of the eastern and central groups playing together.

For sleeping, the eastern group always occupied the southeastern corner, next to the first feeding in the morning from bypassing villagers. The other three groups changed their sleeping sites, although showing some preferences. Especially noteworthy is that the same sleeping site may be used by different groups on different nights (see Fig. 15, p. 356).

B. Pulaki

Three groups lived in and around the Pulaki temple (see Table IV). The Pulaki group never went far away from the temple; the Tirte group entered the area when no other group was there, but otherwise ranged around the temple up to several hundred meters away; and the Bukit group ranged mainly in the hills, but made a visit to the temple area nearly every day.

The small Tirte group was inferior to both the other groups and avoided any contact if possible. Two times I saw the Pulaki group chase away the Tirte group. On one of these occasions, where only low-ranking males (3 adults and 2 or 3 subadults) attacked the Tirte group, the alpha male of Tirte group, Taka (Fig. 13), covered the flight of the Tirte group by holding ground against all 5 or 6 aggressors until he followed his group as the last individual.

The two big groups usually respected each other's presence and kept apart. But still, quite a range of retreats and agonistic contacts occurred between these two groups. Once I fed both groups continuously and led them together in front of the temple. Again peripheral low-ranking males and subadults started to quarrel a bit. But the two central alpha males finally sat and fed within 3 meters from each other and avoided any kind of aggression and even direct visual contact. On other occasions, juveniles of the Pulaki group had positive contacts (genital touching, play) with juveniles of both the Tirte and Bukit groups. Once an approximately 3½-year-old male of the Pulaki group mounted Tegeg, the beta male of the Tirte group. Similar contacts (with lip-smacking, presenting by the older one) occurred betweeen adult peripherals of the Pulaki and Bukit groups, on the one hand, and juveniles of the two groups, on the other. The semi-solitary males of Pulaki and Bukit groups seemed to have a dominance hierarchy among themselves. There was some evidence that also in this respect Sondra (Bukit) was all-dominant in that he also dominated Oklan (Pulaki). Oklan, in turn, most probably dominated Oko (Bukit).

C. Summary of Intergroup Behavior

The home ranges of the groups living in and around Pulaki overlapped extensively, and of the three groups in Sangeh almost completely. Neighboring groups had relationships with each other considered to be based ultimately on the relationships of the respective central alpha males, and well comparable to the personal relationships within a group. Seven of eight group relationships that were analyzed included a dominance–inferiority relation, and one (Pulaki–Bukit) was ranked neutral.

In forest-living groups, where the home ranges overlap as well, rank-neutral relations, due to less competition, are probably more frequent. Among the Balinese temple groups, the emergence of dominance is certainly much enhanced by competition over food offered by humans.

Larger groups consistently dominate smaller ones, or they are equal. One may assume that the number of adult males in a group has a strong impact on its status, just in terms of fighting potential. In the cases observed the differences were always very great, and then quantity may override quality. However, the dominance relation between the respective central alpha males is probably more fundamental.

Most intergroup contacts occur among juveniles, subadult, and peripheral adult males. They comprise agonistic and "friendly" components as well.

V. COMPARISON OF *Macaca fascicularis* WITH OTHER MACAQUES

The principles of social organization seem to be the same in all macaques. But for detailed comparison, data are available only for two other species, namely *Macaca mulatta* and *Macaca fuscata*. Continuing the main topic of the preceding sections, let us focus on the dominance hierarchy of the three species.

The interclass hierarchy, including much individual variation, seems to be the same in the three species. A general rank decline in old males, however, as described in the preceding, is reported neither for rhesus nor for Japanese macaques. But single cases have been described, e.g., by Koyama, (1967) for *M. fuscata* (the old male Bartollo) and by Southwick *et al.* (1965) for *M mulatta* (Shifty dominating the two old central leaders). Differences among the three species in this respect appear to be only minimal. A greater but still gradated difference may exist in the intergenealogical hierarchy of juvenile males. Both in rhesus (Sade, 1967; Missakian, 1972) and Japanese macaques (Koyama, 1967), the mother's status seems to affect the son's status toward age mates and even older males more strongly than in crab-eaters. But the age priority within the genealogy is well documented for all three species.

In adult females, the lineage hierarchy and Kawamura's (1965) hypothesis on intragenealogical hierarchy are rigidly fulfilled in *M. fuscata* and *M. mulatta* as well (Koyama, 1967; Sade, 1969; Missakian, 1972). In my zoo group there were more exceptions to the lineage hierarchy (Tables XXV and XXVI) than in the other two species. Furthermore, the mother, in nearly half of the cases, was inferior to her daughters, and the "youngest ascendency" (Koyama, 1967) was not evident either. For reliable conclusions, more data, if possible from the wild, are needed.

There is no evidence of dominant semisolitary males in rhesus monkeys (see Kaufmann, 1967). In Japanese macaques they occur (S. Mito, personal communication, and personal observation of the semisolitary male Ika on Koshima) but probably are less common than in long-tailed macaques.

My observations on intergroup behavior in the crab-eater fit very well the reports on rhesus (Altmann, 1962; Southwick *et al.*, 1965; Vandenbergh, 1967; Vessey, 1968; Hausfater, 1972). Unfortunately, I have no data concerning the shifting of males from one group to another, as reported of rhesus monkeys (Boelkins and Wilson, 1972).

Finally, I want to present a striking similarity between rhesus and crab-eater. In order to obtain information on the general frequency of aggression in forest-living groups, I listened to them over prolonged per-

TABLE XXV

MATRILINEAR GENEALOGIES OF THE 77 INDIVIDUALS LIVING IN THE BASEL ZOO ON JULY 1, 1967

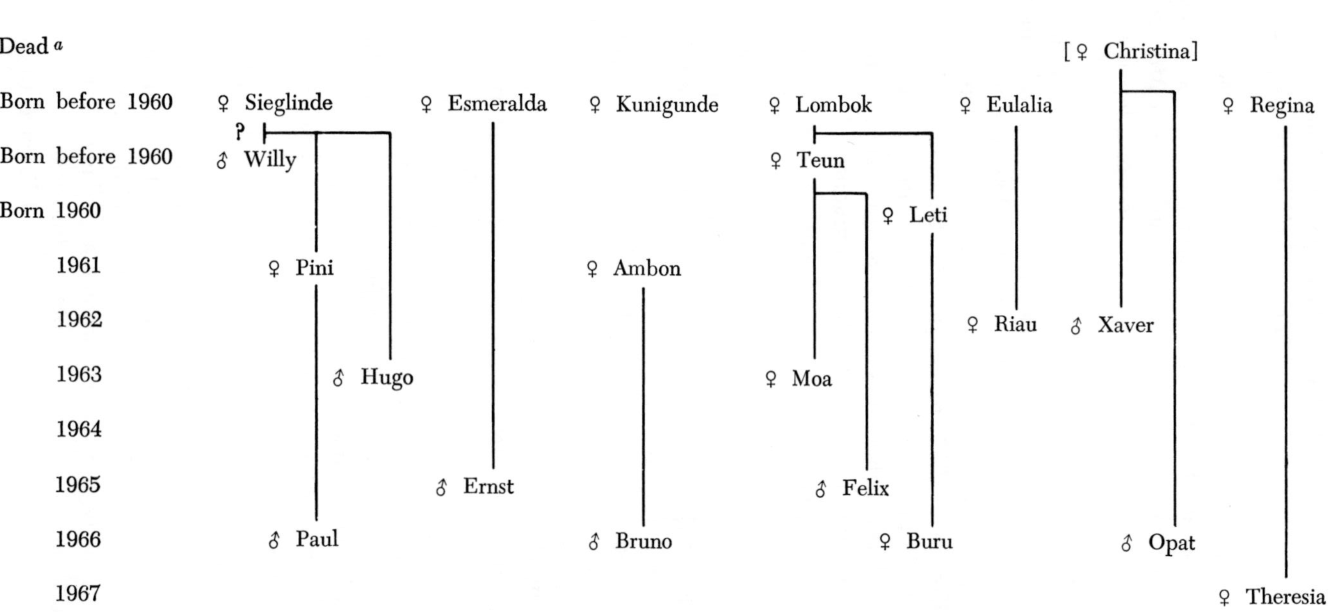

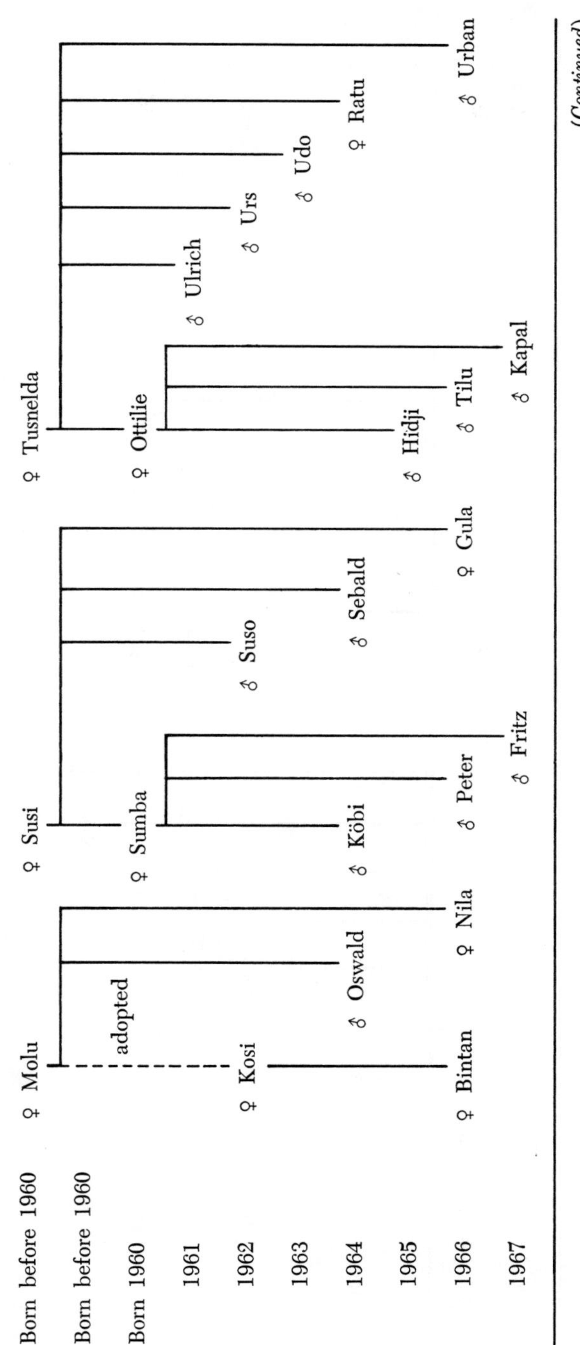
Dead
Born before 1960
Born before 1960
Born 1960
1961
1962
1963
1964
1965
1966
1967
♀ Molu
adopted
♀ Kosi
♂ Oswald
♀ Bintan
♀ Nila
♀ Susi
♀ Sumba
♂ Köbi
♂ Peter
♂ Fritz
♂ Suso
♂ Sebald
♀ Gula
♀ Tusnelda
♀ Ottilie
♂ Hidji
♂ Tilu
♂ Kapal
♂ Ulrich
♂ Urs
♂ Udo
♀ Ratu
♂ Urban

(Continued)

TABLE XXV (Continued)

MATRILINEAR GENEALOGIES OF THE 77 INDIVIDUALS LIVING IN THE BASEL ZOO ON JULY 1, 1967

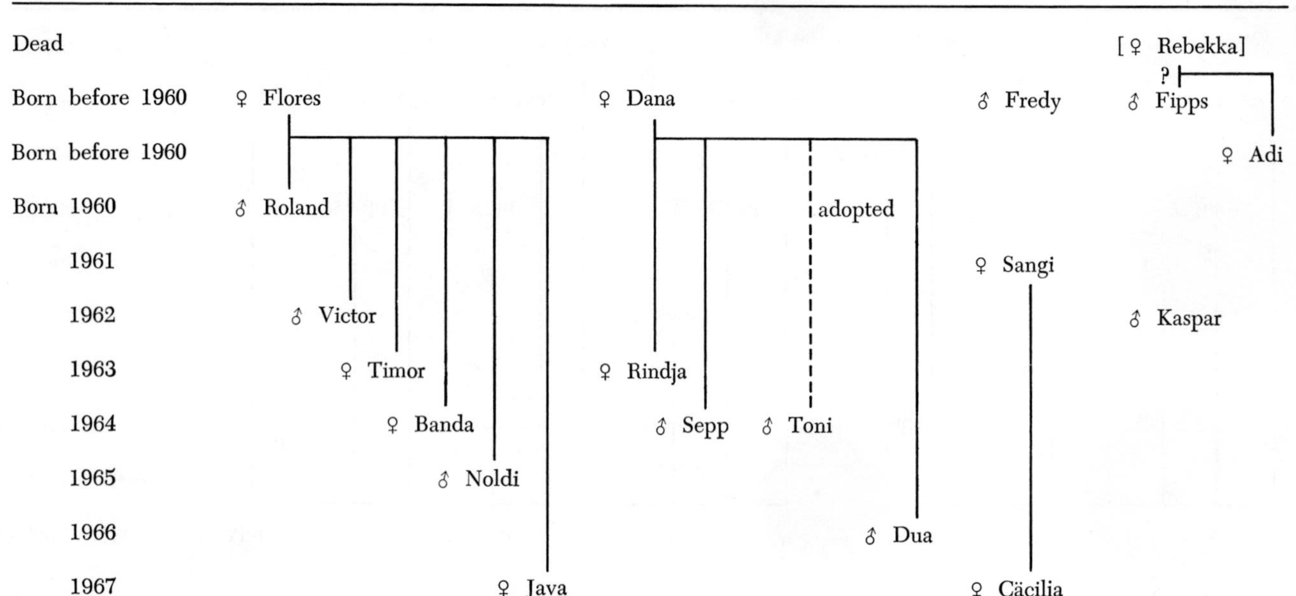

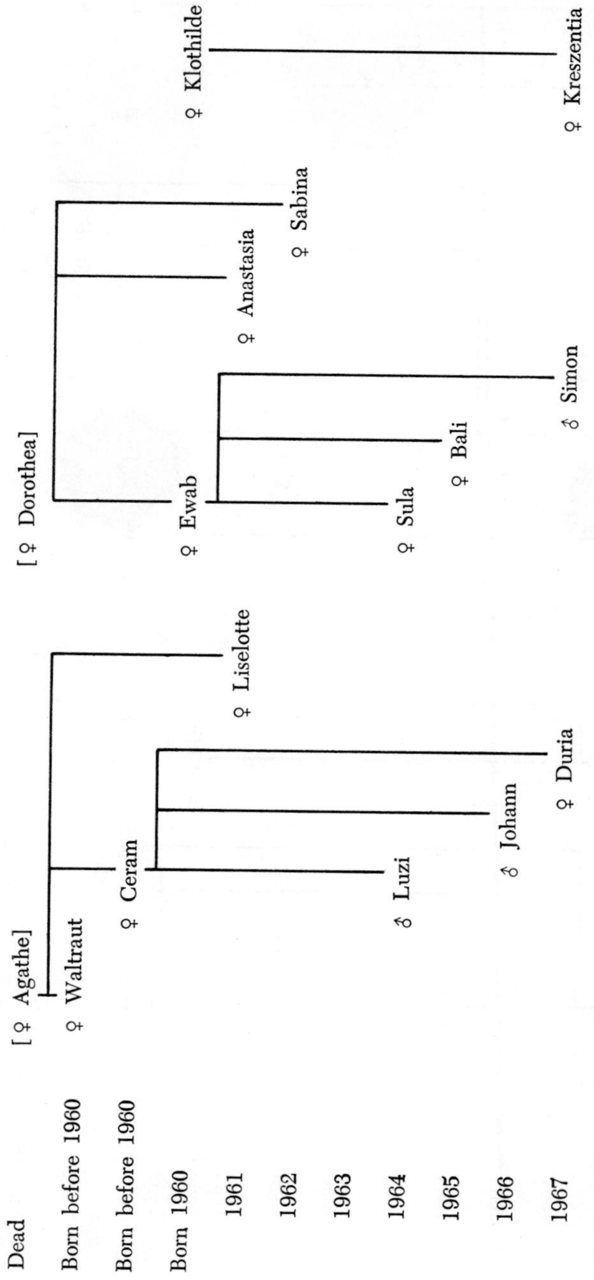

[a] Dead individuals are listed only when necessary for tracing the kinship among living individuals.

TABLE XXVI

MATRILINEAR GENEALOGIES OF THE 67 INDIVIDUALS LIVING IN THE BASEL ZOO ON JULY 1, 1972

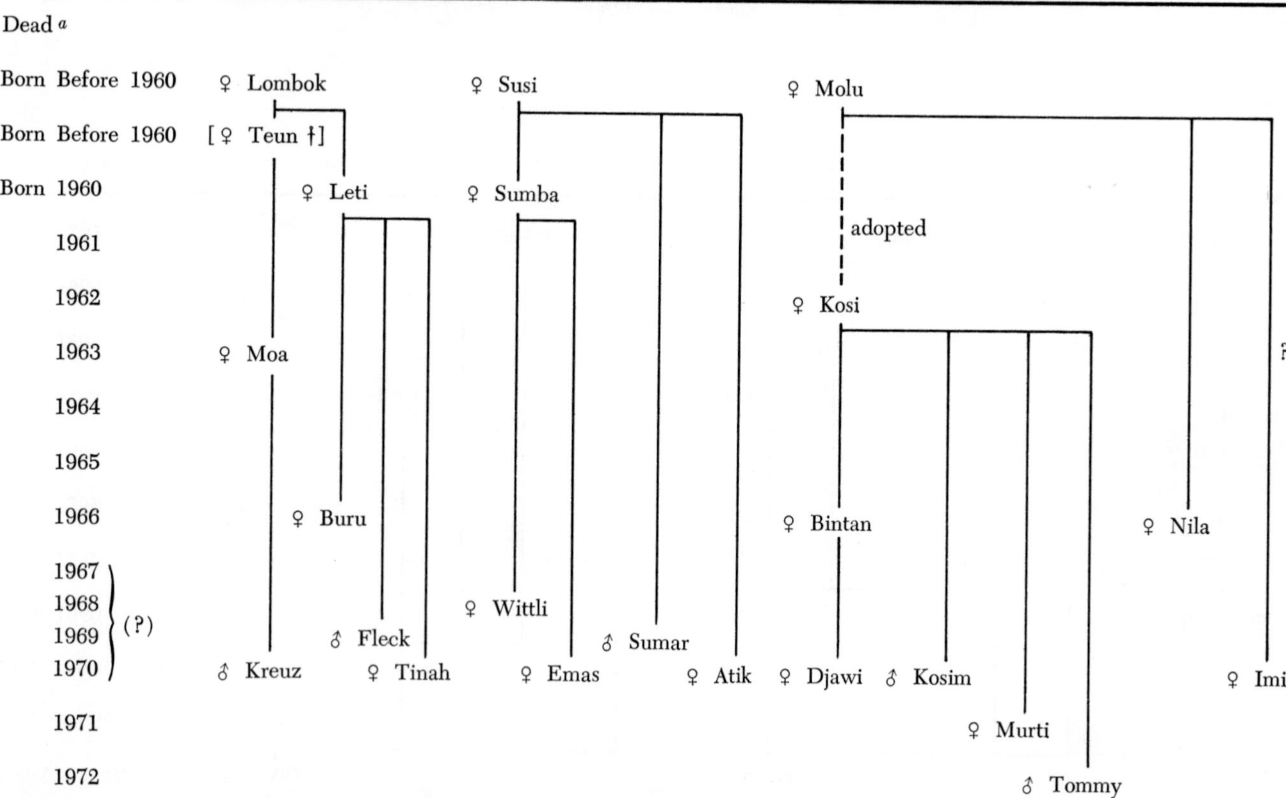

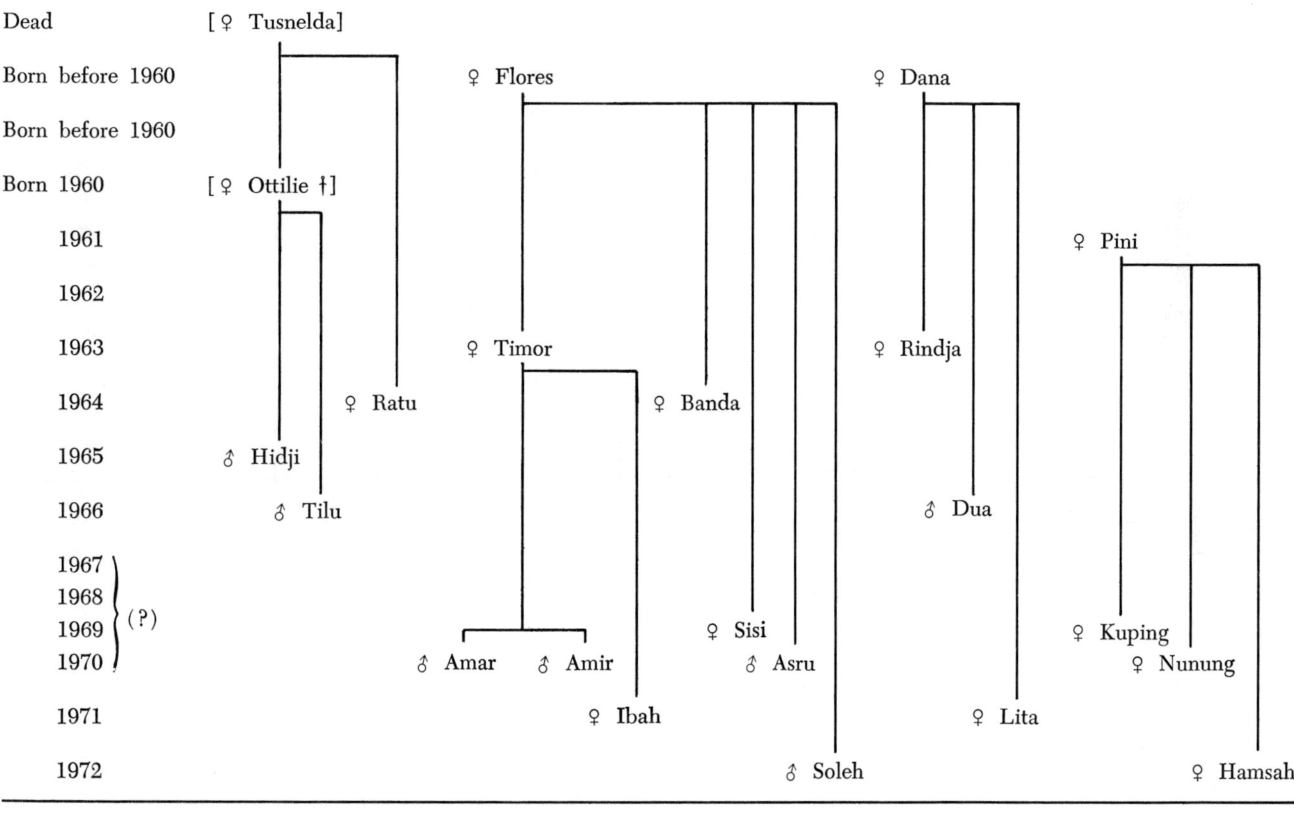
Dead
[♀ Tusnelda]
Born before 1960
♀ Flores
♀ Dana
Born before 1960
Born 1960
[♀ Ottilie †]
1961
♀ Pini
1962
1963
♀ Timor
♀ Rindja
1964
♀ Ratu
♀ Banda
1965
♂ Hidji
1966
♂ Tilu
♂ Dua
1967
1968
1969
1970
(?)
♀ Sisi
♀ Kuping
♂ Amar
♂ Amir
♂ Asru
♀ Nunung
1971
♀ Ibah
♀ Lita
1972
♂ Soleh
♀ Hamsah

(Continued)

TABLE XXVI (Continued)

MATRILINEAR GENEALOGIES OF THE 67 INDIVIDUALS LIVING IN THE BASEL ZOO ON JULY 1, 1972

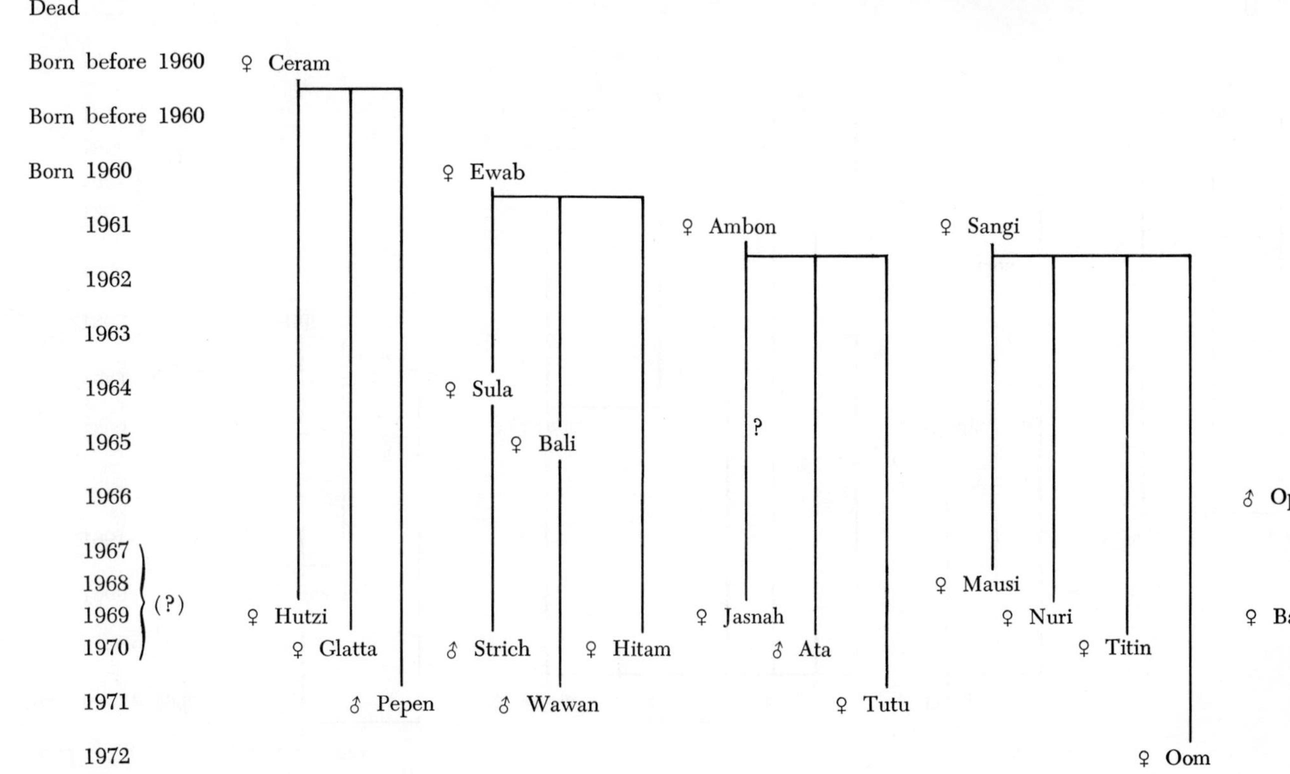

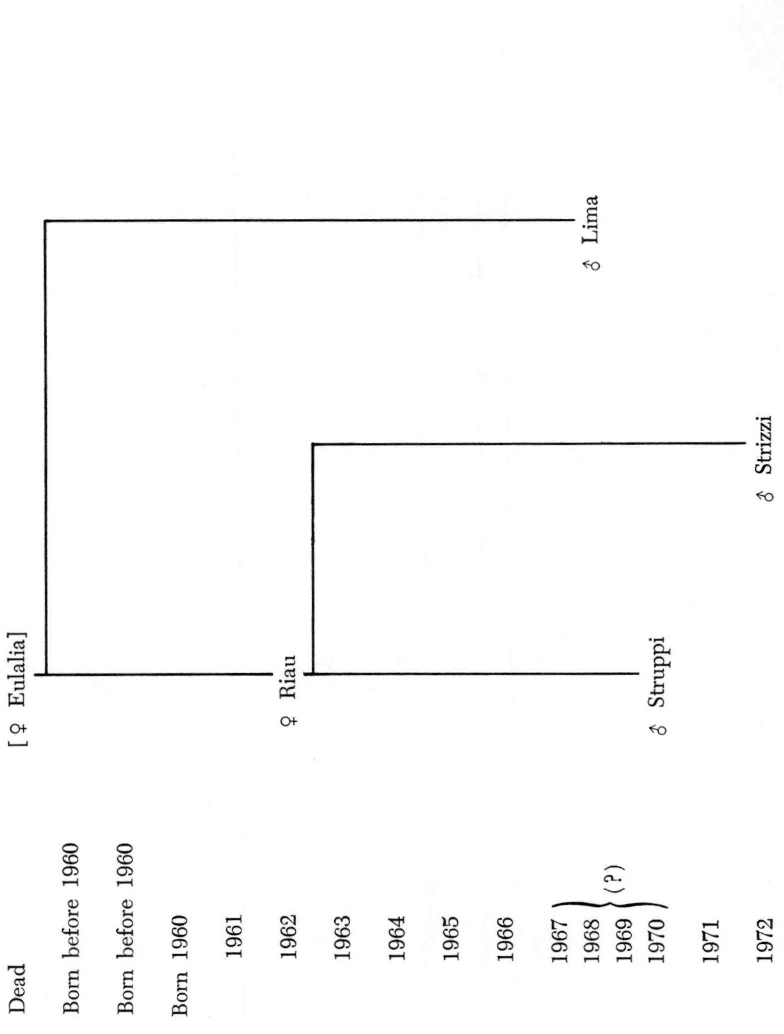

[a] Dead individuals are listed only when necessary for tracing the kinship among living individuals.

TABLE XXVII
FREQUENCY OF AUDIBLE AGGRESSION IN FOREST GROUPS

Location (species)	No. of 0.5-minute units recorded	No. of 0.5-minute units with aggression heard	Proportion aggression: nonaggression	No. of aggressive bouts	Mean interval between aggressive bouts (minutes)
Peutjang island (*Macaca fascicularis*)	4276	74	1 : 58	40	53
Asarori forest (India) (*Macaca mulatta;* Lindburg, 1971)	—	—	—	—	61

iods, sitting quietly in one place. Lindburg (1971) apparently recorded aggression in rhesus monkeys in about the same way. The results show a surprising correspondence (Table XXVII).

VI. SUMMARY

The social organization of *Macaca fascicularis* is described, as observed in Java and Bali, on the one hand, and in the Basel Zoo, on the other hand.

1. The species lives in stable groups, occupying overlapping home ranges.

2. The mean size of Javanese forest groups (48.5 individuals) was the same as in temple-dwelling Balinese groups (47.3).

3. With the exception of two Balinese temple groups, all wild groups observed comprised 2–9 adult males.

4. The description of social organization is based on the concept of "personal relationships," i.e., the dyadic response dispositions, determined by genetics, on the one hand, and by social experience, on the other hand. Social organization is viewed as the whole network of personal relationships, including all interdependencies.

5. The degree of familiarity is assumed to be the most general determinant of the personal relationships.

6. Dominance–inferiority, which is considered as one of several components of a personal relationship, is found by observing unidirectional expressions, such as grimace and screaming.

7. The dominance hierarchy in males before adulthood is mainly based on age. But individuality, kinship, and all kinds of coalitions become increasingly more important and remain efficient throughout adulthood. Senile males drop in rank. It is speculated that males of low descent (mother) are more pushed to the periphery than males of high descent, but, nevertheless, although less easily, are able to become dominant (in dyadic relations) over central males. Some semisolitary males were observed that dominated all central males.

8. The dominance hierarchy of females is mainly based on kinship. Only during youth does age affect status, and senile females drop in rank. The adult females of the same genealogy are more or less equal in rank—their hierarchy seems to be rather random.

9. The "social role" is defined as a functional part of a social system, which is fulfilled by specific behavior of particular individuals. The group-protection role and the alarm role are described.

10. Captive conditions are compared with those in the wild. The impossibility of developing the natural adult male organization in the usual type of enclosures is stressed.

11. Intergroup relations are described and are considered as extensions of the intragroup relationships, ultimately referring to the relationships between the respective central alpha males. But they are also correlated with group size. Especially in competing temple-dwelling groups, a clear dominance–inferiority relation is observed.

Acknowledgments

I express my deep gratitude to Professor Rudolf Schenkel, who encouraged me to work in Udjung Kulon, and introduced me there, for his information and comments. Many thanks also are due Professor Ernst M. Lang, Director of the Basel Zoo, for providing the necessary facilities for my work at the zoo. Furthermore, I wish to acknowledge the valuable cooperation I received in the zoo from Dr. Hans Wackernagel, Vice-Director, and from the two monkey-keepers, Mr. Maximilian Giuliani and Mr. Günther Ruby. I am very grateful to Mr. Jörg Hess for his criticism and for the expert processing of my photographs. In addition, I thank Dr. Frank Du Mond for his hospitality during my visit to Monkey Jungle as well as for his helpful information. Also, I am grateful to Dr. Irwin S. Bernstein, Dr. Jan A. R. A. M. van Hooff, Mrs. Tina de Benedictis, Professor Hans Kummer, Dr. Walter Götz, Mr. Ulrich Nagel, and Dieter Thommen for data and criticisms. Moreover, I express my appreciation to Mrs. Sabine Bousani, who finished my drawings. My gratitude goes also to the two successive Swiss ambassadors in Indonesia, Dr. Réviliod and Dr. Meier, and to Mr. Robert Spinnler, also from the Swiss Embassy, who helped me in every possible way. I wish to thank the Indonesian authorities, who made available the facilities for my work in Indonesia, especially Mr. Hasan Basjarudin, Professor Oto Sumarwoto, General Walman Sinaga, Mr. Widodo Sukohadi Ramono, Mr. Djuhari, Ing. Subroto, M. Ngaru Antara, and Mr. Sutarsono. I am very grateful to numerous guards of the Udjung Kulon and Baluran reserves and villagers of Bali for their indispensable assistance in the field. Finally, I thank Mr. Ulrich Halder for his cooperative and stimulating company during a part of my work in Indonesia.

The work in Indonesia was supported by a grant from the Swiss National Science Foundation.

References

Altmann, S. A. (1962). A field study on the sociobiology of rhesus monkeys, Macaca mulatta. *Ann. N.Y. Acad. Sci.* **102**, 338–435.

Angst, W. (1973). Pilot experiments to test group tolerance to a stranger in wild Macaca fascicularis. *Amer. J. Phys. Anthropol.* **38**, 625–630.

Angst, W. (1974). Das Ausdrucksverhalten des Javaneraffen, Macaca fascicularis Raffles. *Adv. Ethol.* **15**.

Bernstein, I. S. (1964a). Role of the dominant male rhesus monkey in response to external challenges to the group. *J. Comp. Physiol. Psychol.* **57** (3), 404–406.

Bernstein, I. S. (1964b). The integration of rhesus monkeys introduced to a group. *Folia Primatol.* **2**, 50–63.

Bernstein, I. S. (1966). Analysis of a key role in a capuchin (Cebus albifrons) group. *Tulane Stud. Zool.* **13** (2), 49–54.

Bernstein, I. S. (1967). Intertaxa interactions in a Malayan primate community. *Folia Primatol.* **7**, 198–207.

Bernstein, I. S. (1968). Social status of two hybrids in a wild troop of Macaca irus. *Folia Primatol.* **8,** 121–131.

Bernstein, I. S. (1969). Introductory techniques in the formation of pigtail monkey troops. *Folia Primatol.* **10,** 1–19.

Bernstein, I. S. (1970). Primate status hierarchies. *In* "Primate Behavior: Developments in Field and Laboratory Research" (L. A. Rosenblum, ed.), Vol. 1, pp. 71–109. Academic Press, New York.

Bernstein, I. S., and Mason, W. A. (1963). Group formation by rhesus monkeys. *Anim. Behav.* **11,** 28–31.

Bernstein, I. S., and Sharpe, L. G. (1966). Social roles in a rhesus monkey group. *Behaviour* **26,** 91–104.

Boelkins, R. C., and Wilson, A. P. (1972). Intergroup social dynamics of the Cayo Santiago rhesus (*Macaca mulatta*) with special reference to changes in group membership by males. *Primates* **13,** 125–140.

Chance, M. R. A. (1967). Attention structure as the basis of primate rank orders. *Man* **2,** 503–518.

Crook, J. H. (1970). The socio-ecology of primates. *In* "Social Behavior in Birds and Mammals" (J. H. Crook, ed.), pp. 103–166. Academic Press, New York.

Fady, J.–C. (1969). Les jeux sociaux: Le compagnon de jeux chex les jeunes. Observations chez Macaca irus. *Folia Primatol.* **11,** 134–143.

Fooden, J. (1971). Report on primates collected in western Thailand January-April, 1967. *Fieldiana Zool.* **59,** 1–62.

Furuya, Y. (1965). Social organization of the crab-eating monkey. *Primates* **6,** 285–336.

Gartlan, J. S. (1968). Structure and function in primate society. *Folia Primatol.* **8,** 89–120.

Goustard, M. (1961). La structure sociale d'une colonie de Macaca irus. *Ann. Sci. Natur., Zool.* **12,** 297–322.

Hansen, E. W., Harlow, H. F., and Dodsworth, R. O. (1966). Reactions of rhesus monkeys to familiar and unfamiliar peers. *J. Comp. Physiol. Psychol.* **61,** 274–279.

Hausfater, G. (1972). Intergroup behavior of free-ranging rhesus monkeys (Macaca mulatta). *Folia Primatol.* **18,** 78–107.

van Hooff, J. A. R. A. M. (1967). The facial displays of the catarrhine monkeys and apes. *In* "Primate Ethology" (D. Morris, ed.), pp. 7–65. Weidenfeld & Nicolson, London.

Imanishi, K. (1957). Social behavior in Japanese monkeys, Macaca fuscata. *Psychologia* **1,** 47–54. [Reprinted *in* "Primate Social Behavior" (C. H. Southwick, ed.), pp. 68–81. Van Nostrand-Reinhold, Princeton, New Jersey, 1963.]

Imanishi, K. (1965). Identification: A process of socialization in the subhuman society of Macaca fuscata. *In* "Japanese Monkeys" (S. A. Altmann, ed.), pp. 30–51. Published by S. A. Altmann.

Kaufmann, J. H. (1967). Social relations of adult males in a freeranging band of rhesus monkeys. *In* "Social Communication among Primates" (S. A. Altmann, ed.), pp. 73–98. Univ. of Chicago Press, Chicago, Illinois.

Kawai, M. (1960). A field experiment on the process of group formation in the Japanese monkey (Macaca fuscata), and the releasing of the group at Ohirayama. *Primates* **2,** 181–253.

Kawai, M. (1965). On the system of social ranks in a natural troop of Japanese monkeys: 1. Basic rank and dependent rank. *In* "Japanese Monkeys" (S. A. Altmann, ed.), pp. 66–86. Published by S. A. Altmann.

Kawamura, S. (1965). Matriarchal social ranks in the Minoo-B troop: A study of the rank system of Japanese monkeys. *In* "Japanese Monkeys" (S. A. Altmann, ed.), pp. 105–112. Published by S. A. Altmann.

Koyama, N. (1967). On dominance rank and kinship of a wild Japanese monkey troop in Arashiyama. *Primates* **8**, 189–216.

Kummer, H. (1971). "Primate Societies." Aldine, Atherton, Chicago, Illinois, and New York.

Kurland, J. A. (1973). A natural history of kra macaques (Macaca fascicularis Raffles, 1821) at the Kutei Reserve, Kalimantan Timur, Indonesia.

Kurt, F., and Sinaga, W. (1970). Survey ke suaka margasatwa gn. Loeser. Direktorat Pembinaan Hutan, Bogor.

Lindburg, D. G. (1971). The rhesus monkey in north India. An ecological and behavioral study. *In* "Primate Behavior: Developments in Field and Laboratory Research" (L. A. Rosenblum, ed.), Vol. 2, pp. 1–106. Academic Press, New York.

Medway, Lord (1969). "The Wild Mammals of Malaya, and Offshore Islands including Singapore." Oxford Univ. Press, London and New York.

Missakian, E. A. (1972). Genealogical and cross-genealogical dominance relations in a group of free-ranging rhesus monkeys (Macaca mulatta) on Cayo Santiago. *Primates* **13**, 169–180.

Reynolds, V. (1970). Roles and role change in monkey society: the consort relationship of rhesus monkeys. *Man* **5**, 449–465.

Rosenblum, L. A., and Lowe, A. (1971). The influence of familiarity during rearing on subsequent partner preferences in squirrel monkeys. *Psychon. Sci.* **23**, 35–37.

Rowell, T. E. (1966). Hierarchy in the organization of a captive baboon group. *Anim. Behav.* **14**, 430–443.

Rowell, T. E. (1967). Variability in the social organization of primates. *In* "Primate Ethology" (D. Morris, ed.), pp. 219–235. Weidenfeld & Nicolson, London.

Rowell, T. E. (1972). "Social Behavior of Monkeys." Penguin Books, London.

Sade, D. S. (1967). Determinants of dominance in a group of free-ranging rhesus monkeys. *In* "Social Communication among Primates" (S. A. Altmann, ed.), pp. 99–114. Univ. of Chicago Press, Chicago, Illinois.

Sade, D. S. (1969). An algorythm for dominance relations among rhesus monkeys: Rules for adult females and sisters. Annu Meet. Amer. Asso. Phys. Anthropologists, Mexico City.

Shirek-Ellefson, J. (1967). Visual Communication in Macaca irus. Doctoral dissertation, University of California, Berkeley. Unpublished.

Southwick, C. H., and Cadigan, F. C., Jr. (1972). Population studies of Malaysian primates. *Primates* **13**, 1–18.

Southwick, C. H., Beg, M. A., and Siddiqui, M. R. (1965). Rhesus monkeys in north India. *In* "Primate Behavior" (I. DeVore, ed.), pp. 111–159. Holt, New York.

Vandenbergh, J. G. (1967). The development of social structure in free-ranging rhesus monkeys. Behaviour, **29**, 179–194.

Vessey, S. H. (1968). Interactions between free-ranging groups of rhesus monkeys. *Folia Primatol.* **8**, 228–239.

Vessey, S. H. (1973). Night observations of free-ranging rhesus monkeys. *Amer. J. Phys. Anthropol.*

Wynne-Edwards, V. C. (1972). "Animal Dispersion in Relation to Social Behaviour." Oliver & Boyd, Edinburgh.

Zajonc, R. B. (1971). Attraction, affiliation, and attachment. *In* "Man and Beast: Comparative Social Behavior" (J. F. Eisenberg, ed.), pp. 141–179. Smithson. Inst. Press, Washington, D.C.

AUTHOR INDEX

Numbers in italics refer to the pages on which the complete references are listed.

A

B

F

G

H

M

N

O

P

R

S

T

U

V

SUBJECT INDEX

M

N

O

P

R

T

V

3 3208 10041 7492
Friendship Libr/Fairleigh Dickinson Univ
Madison, NJ 07940